Lecture Notes in Statistics

Edited by D. Brillinger, S. Fienberg, J. Gani, J. Hartigan, and K. Krickeberg

13

J. Pfanzagl
With the Assistance of
W. Wefelmeyer

Contributions to a General Asymptotic Statistical Theory

Springer-Verlag
New York Heidelberg Berlin

J. Pfanzagl
Mathematisches Institut
der Universität Zu Köln
Weyertal 86-90
5000 Köln 41
West Germany

AMS Classification: 62A99

Library of Congress Cataloging in Publication Data

Pfanzagl, J. (Johann)
Contributions to a general asymptotic statistical theory.

(Lecture notes in statistics ; 13)
Bibliography: p.
Includes indexes.
1. Mathematical statistics--Asymptotic theory.
I. Wefelmeyer, W. II. Title. III. Series: Lecture notes in statistics (Springer-Verlag) ; v. 13.
QA276.P473 1982 519.5 82-19252

With 2 Illustrations

Printed and bound by R. R. Donnelley & Sons, Harrisonburg, VA.
Printed in the United States of America.

9 8 7 6 5 4 3 2 1

ISBN 0-387-90776-9 Springer-Verlag New York Heidelberg Berlin
ISBN 3-540-90776-9 Springer-Verlag Berlin Heidelberg New York

CONTENTS

0. Introduction 1

0.1. Why asymptotic theory? 1
0.2. The object of a unified asymptotic theory, 2
0.3. Models, 3
0.4. Functionals, 6
0.5. What are the purposes of this book? 8
0.6. A guide to the contents, 10
0.7. Adaptiveness, 14
0.8. Robustness, 16
0.9. Notations, 18

1. The local structure of families of probability measures 22

1.1. The tangent cone $T(P,\mathfrak{P})$, 22
1.2. Properties of $T(P,\mathfrak{P})$ - properties of $\mathfrak{P}$, 27
1.3. Convexity of $T(P,\mathfrak{P})$, 28
1.4. Symmetry of $T(P,\mathfrak{P})$, 30
1.5. Tangent spaces of induced measures, 31

2. Examples of tangent spaces 33

2.1. 'Full' tangent spaces, 33
2.2. Parametric families, 35
2.3. Families of symmetric distributions, 42
2.4. Measures on product spaces, 47
2.5. Random nuisance parameters, 50
2.6. A general model, 54

3. Tangent cones 57

3.1. Introduction, 57
3.2. Order with respect to location, 58
3.3. Order with respect to concentration, 59
3.4. Order with respect to asymmetry, 60
3.5. Monotone failure rates, 61
3.6. Positive dependence, 62

4. Differentiable functionals 65

4.1. The gradient of a functional, 65
4.2. Projections into convex sets, 69
4.3. The canonical gradient, 71
4.4. Multidimensional functionals, 73
4.5. Tangent spaces and gradients under side conditions, 75
4.6. Historical remarks, 76

5. Examples of differentiable functionals 78

5.1. Von Mises functionals, 78
5.2. Minimum contrast functionals, 80
5.3. Parameters, 83
5.4. Quantiles, 85
5.5. A location functional, 86

6. Distance functions for probability measures 90

6.1. Some distance functions, 90
6.2. Asymptotic relations between distance functions, 91
6.3. Distances in parametric families, 96
6.4. Distances for product measures, 97

7. Projections of probability measures 99

7.1. Motivation, 99
7.2. The projection, 100
7.3. Projections defined by distances, 101
7.4. Projections of measures - projections of densities, 104

7.5. Iterated projections, 107
7.6. Projections into a parametric family, 109
7.7. Projections into a family of product measures, 112
7.8. Projections into a family of symmetric distributions, 113

8. Asymptotic bounds for the power of tests 115

8.1. Hypotheses and co-spaces, 115
8.2. The dimension of the co-space, 119
8.3. The concept of asymptotic power functions, 122
8.4. The asymptotic envelope power function, 125
8.5. The power function of asymptotically efficient tests, 131
8.6. Restrictions of the basic family, 135
8.7. Asymptotic envelope power functions using the Hellinger distance, 141

9. Asymptotic bounds for the concentration of estimators 150

9.1. Comparison of concentrations, 151
9.2. Bounds for asymptotically median unbiased estimators, 154
9.3. Multidimensional functionals, 157
9.4. Locally uniform convergence, 163
9.5. Restrictions of the basic family, 168
9.6. Functionals of induced measures, 173

10. Existence of asymptotically efficient estimators for probability measures 177

10.1. Asymptotic efficiency, 177
10.2. Density estimators, 179
10.3. Parametric families, 182
10.4. Projections of estimators, 184
10.5. Projections into a parametric family, 188
10.6. Projections into a family of product measures, 191

11. Existence of asymptotically efficient estimators for functionals 196

11.1. Introduction, 196
11.2. Asymptotically efficient estimators for functionals from asymptotically efficient estimators for probability measures, 197
11.3. Functions of asymptotically efficient estimators are asymptotically efficient, 199
11.4. Improvement of asymptotically inefficient estimators, 200
11.5. A heuristic justification of the improvement procedure, 205
11.6. Estimators with stochastic expansion, 209

12. Existence of asymptotically efficient tests 211

12.1. Introduction, 211
12.2. An asymptotically efficient critical region, 211
12.3. Hypotheses on functionals, 212

13. Inference for parametric families 215

13.1. Estimating a functional, 215
13.2. Variance bounds for parametric subfamilies, 218
13.3. Asymptotically efficient estimators for parametric subfamilies, 220

14. Random nuisance parameters 226

14.1. Introduction, 226
14.2. Estimating a structural parameter in the presence of a known random nuisance parameter, 227
14.3. Estimating a structural parameter in the presence of an unknown random nuisance parameter, 229

15. Inference for symmetric probability measures 237

15.1. Asymptotic variance bounds for functionals of symmetric distributions, 237

15.2. Asymptotically efficient estimators for functionals of symmetric distributions, 241
15.3. Symmetry in two-dimensional distributions, 246

16. Inference for measures on product spaces 249

16.1. Introduction, 249
16.2. Variance bounds, 250
16.3. Asymptotically efficient estimators for product measures, 252
16.4. Estimators for von Mises functionals, 253
16.5. A special example, 255

17. Dependence - independence 258

17.1. Measures of dependence, 258
17.2. Estimating measures of dependence, 260
17.3. Tests for independence, 261

18. Two-sample problems 265

18.1. Introduction, 265
18.2. Inherent relationships between x and y, 266
18.3. The tangent spaces, 268
18.4. Testing for equality, 271
18.5. Estimation of a transformation parameter, 275
18.6. Estimation in the proportional failure rate model, 277
18.7. Dependent samples, 281

19. Appendix 289

19.1. Miscellaneous lemmas, 289
19.2. Asymptotic normality of log-likelihood ratios, 292

References 300
Notation index 311
Author index 312
Subject index 314

0. INTRODUCTION

This book intends to provide a basis for a unified asymptotic statistical theory, comprising parametric as well as non-parametric models.

0.1. Why asymptotic theory?

The purpose of any statistical theory is to analyze the performance of statistical procedures, and to provide methods for the construction of optimal procedures. Non-asymptotic theory meets these requirements in certain special cases, but its success is erratic rather than systematic. (For a collection of illustrative examples see Pfanzagl, 1980a, pp. 1-4.) There is no hope of a non-asymptotic theory meeting such requirements in general. Therefore, we have to be content with approximate solutions (i.e. an approximate evaluation of the performance, and methods for the construction of approximately optimal procedures). The main tool for obtaining such approximate solutions is asymptotic theory, based on approximations by limit distributions, or Edgeworth expansions. Experiences with parametric families suggest that the accuracy obtainable from approximations by limit distributions may be unsatisfactory for moderate sample sizes, so that Edgeworth expansions seem indispensable for obtaining an accuracy sufficient for practical purposes. Our present endeavors to obtain a general foundation of asymptotic theory based on normal approximations are, therefore, not more than a first step.

0.2. The object of a unified asymptotic theory

So far, statistical theory is either 'parametric' or 'nonparametric', i.e. the basic family is either parametrized by a finite number of parameters, or it contains all probability measures fulfilling certain regularity conditions.

There are, of course, certain 'intermediate' models treated in literature, but a general theory applicable to an arbitrary intermediate model is still missing. As examples of such intermediate models think of the family of all symmetric probability measures, or the family of all mixtures with a random nuisance parameter. Up to now, the number of intermediate models available in literature is not too large. Since applied statisticians are confined to represent reality by models which can be handled mathematically, this is not surprising. The availability of a general theory will certainly encourage the use of general models.

To illustrate the kind of problems to which a general statistical theory may contribute, consider the estimation of a quantile. If the basic family is parametric, then, of course, we express the quantile as a function of the parameters and obtain an estimator of the quantile by replacing the parameters in this function by estimators. On the other hand, with nothing known about the basic family, one may be confined to using the quantile of the sample as an estimator for the quantile of the distribution. But situations where absolutely nothing is known are rare. If the distribution is known to have a continuous

Lebesgue density - can this be used to obtain a better estimate of the quantile, for instance the quantile of a c o n t i n u o u s density estimator? And what if the true probability measure is known to be symmetric? If the quantile in question is the median, then asymptotically much better estimators than the sample median exist, for instance medians of symmetrized density estimators. But how is the outlook in case of an arbitrary quantile? Are there estimators which are asymptotically optimal? And what is their asymptotic distribution?

A general theory cannot be considered as successful unless it passes a simple test: Applied to any parametric family, it has to reproduce the well known results. (To illustrate this idea: The minimum distance method, using distances based on distribution functions, is certainly n o t a useful general method, because applied to parametric families it leads to inefficient estimators.)

Moreover, there should be enough interesting problems of a more general type to which this theory applies. To meet these requirements, we pay due attention to parametric families as an important special case, and we include a number of chapters (13 - 18) illustrating the application of the general results (obtained in Chapters 8 - 12).

0.3. Models

Any statistical inference starts from a basic family of probability measures, expressing our prior knowledge about the nature of the probability measure from where the observations originate. Recognizing that these models are to a certain extent arbitrary, asymptotic results should not be substantially influenced by accidental attributes of the model.

a) 'Sensitivity'

Our prior knowledge is necessarily vague. We may be sure that certain probability measures belong to the basic family, but we will be uncertain about others. Even if we are sure that the 'true' probability measure has a rather smooth density, we will usually be unable to specify this smoothness more precisely, for instance by giving bounds for the derivative of the density. As another example, consider the case of a contaminated normal distribution. From general experience we may be sure that the amount of contamination is small, but we shall hardly be able to give a realistic bound.

Being aware of this inherent vagueness of all models, we feel uncomfortable about optimality results which depend in a decisive way on certain aspects of a model which have been chosen ad libitum. (As an example we mention a location estimator suggested by Huber (1964, Section 9, or 1981, Sections 4.5 and 4.6) which minimizes the maximal variance for an ε-contamination model, and which depends heavily on ε.)

b) 'Invariance'

Two models may describe reality in identical, but formally different ways. In such a case meaningful results have to be identical.

The obvious example is that of a parametric family which may be parametrized in different ways. The asymptotic bound for the concentration of estimators of a given functional is, of course, independent of how the family is parametrized, and a corresponding assertion holds for the asymptotic envelope power function of tests (see Remark 5.3.3 in connection with Theorem 9.2.2 resp. Remark 8.4.6).

In the general case of an arbitrary family of probability measures, one can always consider a sample consisting of $n = 2m$ random variables, governed by P, as a sample consisting of m independent (two-dimensional) random variables each governed by $P\times P$. If a result is meaningful, then it must necessarily be the same under both models.

Consider, as an example, the asymptotic bound for the concentration of an estimator κ^n for a functional $\kappa: \mathfrak{P} \to \mathbb{R}$. According to Theorem 9.2.2, we obtain that $P^n * n^{1/2}(\kappa^n - \kappa(P))$ is at most concentrated like $N(0,P(\kappa^{\cdot}(\cdot,P)^2))$, where $\kappa^{\cdot}(\cdot,P)$ is the canonical gradient of κ at P. For $n = 2m$, we may consider the sample x_ν, $\nu = 1,\ldots,2m$, governed by P^{2m}, as a sample $(x_{2\nu-1},x_{2\nu})$, $\nu = 1,\ldots,m$, governed by $(P\times P)^m$, and the functional $\kappa: \mathfrak{P} \to \mathbb{R}$ as a functional $\bar{\kappa}: \{P\times P: P \in \mathfrak{P}\} \to \mathbb{R}$, defined by $\bar{\kappa}(P\times P) := \kappa(P)$. Then the canonical gradient becomes $\bar{\kappa}^{\cdot}(x,y,P\times P) = \frac{1}{2}(\kappa^{\cdot}(x,P) + \kappa^{\cdot}(y,P))$, so that $P\times P(\bar{\kappa}^{\cdot}(\cdot,P\times P)^2) = P(\kappa^{\cdot}(\cdot,P)^2)/2$. Applying Theorem 9.2.2 to the functional $\bar{\kappa}$ and the family $\{P\times P: P \in \mathfrak{P}\}$ we obtain that $(P\times P)^m * m^{1/2}(\bar{\kappa}^m - \bar{\kappa}(P\times P))$ is at most concentrated like $N(0,P(\kappa^{\cdot}(\cdot,P)^2)/2)$. Hence $(P\times P)^m * (2m)^{1/2}(\bar{\kappa}^m - \bar{\kappa}(P\times P))$ is maximally concentrated like $N(0,P(\kappa^{\cdot}(\cdot,P)^2))$. Therefore, the two formally different models lead to the same asymptotic bound for the concentration of estimators.

The same holds true for the asymptotic envelope power function given in Corollary 8.4.4 as a function of $n^{1/2}\Delta(Q;\mathfrak{P}_0)$ (where $\Delta(Q;\mathfrak{P}_0)$ measures the distance of the alternative Q from the hypothesis $\mathfrak{P}_0$). Since $\Delta(Q\times Q;P\times P) = \sqrt{2}\,\Delta(Q;P)(1 + o(\Delta(Q;P))$ (see (6.3.4)), the condition $n^{1/2}\Delta(Q;\mathfrak{P}_0) \leq c$ implies that $m^{1/2}\Delta(Q\times Q;\{P\times P: P \in \mathfrak{P}_0\}) = (2m)^{1/2}\Delta(Q;\mathfrak{P}_0) + o(m^{-1/2})$, so that this theorem, applied to the family $\{Q\times Q: Q \in \mathfrak{P}\}$ and the hypothesis $\{P\times P: P \in \mathfrak{P}_0\}$ leads to the same asymptotic envelope power function.

Another invariance requirement can be described as follows: If in addition to the independent random variables x_ν, $\nu = 1,\ldots,n$, governed by $P \in \mathfrak{P}$, we observe random variables y_ν, $\nu = 1,\ldots,n$, which are governed by a fixed p-measure Q, and which are mutually independent and independent of $x_1,\ldots,x_n$, then our basic family $\mathfrak{P}$ is replaced by $\mathfrak{P}\times Q = \{P\times Q: P\in\mathfrak{P}\}$. Since the observations y_ν, $\nu = 1,\ldots,n$, have no relationship whatsoever to the original family $\mathfrak{P}$, this formally different model should lead to exactly the same result. It is easy to check that our results (like Theorem 9.2.2 and Theorem 8.4.1) pass this test.

Some readers may think that such invariance requirements are too obvious to be mentioned at all. Yet consistency theorems for maximum likelihood estimators use conditions like P_θ-integrability of $x \to \sup\{\log p(x,\tau): \tau \in B\}$ (where B is the complement of some compact set containing θ). This is not a natural condition for an asymptotic theorem, since one may as well consider the family $\{P_\theta \times P_\theta : \theta \in \Theta\}$. The same condition, posed upon the latter family, namely $P_\theta \times P_\theta$-integrability of $(x_1, x_2) \to \sup\{\log p(x_1, \tau) + \log p(x_2, \tau) : \tau \in B\}$ is much weaker in certain cases. (See Kiefer and Wolfowitz, 1956, p. 904.)

c) Dependence on the sample size

Useful conclusions from a sample of size 10 can be obtained only if the prior knowledge is rather precise, i.e. if the basic family of probability measures is comparatively small and the functional to be estimated comparatively simple. From a sample of size 500, useful conclusions can be obtained even if the prior knowledge is rather vague, and the problem under investigation more delicate. Hence a practically useful asymptotic theory should, perhaps, allow for the complexity of the model to increase with increasing sample size.

This idea has been present, at least implicitly, in many investigations, so for instance in nonparametric density estimation. Under the suggestive name of a 'sieve', it has recently met with increasing interest (see Grenander, 1981, Geman, 1981, and Geman and Hwang, 1982). With our attempts at a general asymptotic theory being still at an exploratory stage, we have abstained from including this aspect in our treatise.

0.4. Functionals

An important part of this treatise deals with the estimation of a functional defined on a basic family of probability measures, and with tests of a hypothesis on the value of such a functional. Our considerations are based on the assumption that this functional is given.

For many practical problems this may be an unrealistic idealization. The real problem may be concerned with a certain characteristic of the true probability measure (such as location or spread) which is available only in an intuitive form. In such a situation there are usually several mathematical constructs catching hold of this intuitive notion equally well.

Confronted with a vaguely defined characteristic, one may be tempted to choose the functional representing this characteristic with regard to technical aspects, such as whether it is easy to estimate or not. This seems to be the attitude of Bickel and Lehmann (1975a,b, 1976, 1979) in their remarkable sequence on 'Descriptive Statistics'. But is it really meaningful to use refined techniques for estimating a functional which is chosen more or less arbitrarily?

Some authors take an even more generous attitude. They start from an estimator which, in some intuitive sense, estimates the characteristic in question (for instance location), and think of the functional as being defined by the estimator itself. This attitude naturally leads to the question: What does a given estimator estimate? Is it the median of its distribution, or perhaps the mean? With this approach we run the risk of discovering that the estimator estimates a different functional for each sample size. Huber (1981, p. 6f.) suggests to define the parameter to be estimated in terms of the limit of the estimator as the sample size tends to infinity. We suspect that - whenever this leads to a meaningful construct - this functional could be defined directly, i.e. without intermingling the problem of defining the functional with the problem of how to obtain a good estimator for it. From the abstract point of view there is no relationship whatsoever between the limiting value to which an estimator-sequence converges stochastically, and the concentration of its distribution about this value, say for the sample size 20.

0.5. What are the purposes of this book?

This is a book on methodology, not on mathematics. Our goal is to convey ideas rather than mathematical theorems.

If a theory is applicable to a great variety of special cases, we would like to have the general theorems of this theory under regularity conditions which are 'optimal' in the sense that they are not too far from necessary if specialized to any particular case. It seems doubtful whether this ideal can be achieved here. Therefore, we decided to keep the results and the basic ideas of the proofs transparent at the cost of sometimes unnecessarily restrictive regularity conditions. It would be a bad deal to sacrifice transparency to a slight increase in generality, without coming visibly closer to the ideal of 'optimal' general results. Hence our theorems - together with their proofs - are to be considered as *models* which may be improved in each instance by taking advantage of the particular circumstances.

Even the task of gaining experience with different versions of the basic concepts is still lying before us. As an example we mention the concept of the tangent space. Its definition (see 1.1.1) requires a remainder term to converge to zero. There are different possibilities of making this intuitive notion precise, and it depends on the particular problem which of these options suits best. Moreover, the question is not yet settled whether this definition should be based on the relative densities, as done here, or on their square root (the differentiability in quadratic mean introduced by LeCam, 1966, and

Hájek, 1962, and used by many authors since, in particular also by Levit).

This uncertainty about which are the most appropriate versions of the basic concepts is another reason for being not too particular about regularity conditions.

One could even question whether a complete specification of the regularity conditions is really meaningful from the applied point of view. For example, consider a theorem specifying the limit distribution of an estimator-sequence, assuming among the regularity conditions that the second derivative of the density fulfills a local Lipschitz condition. For the purpose of a numerical approximation such a theorem is useless. What we need is a theorem which furnishes a numerical bound for the difference between the true distribution, and the limit distribution used as an approximation. If such a bound were available, it would depend somehow on the regularity conditions (in particular: on the constant occuring in the local Lipschitz condition). Since the regularity conditions fulfilled by an unknown density are only vaguely known to us, this bound could hardly be used for estimating the approximation error. For practical purposes, numerical trial computations are much more informative than the whole business of regularity conditions.

Yet there is something which can be said in favor of regularity conditions: If they are not only sufficient, but close to necessary, they contain informations about the general structure of the result (e.g. whether the convergence to a limiting distribution takes place u n i f o r m l y over a certain class of probability measures).

0.6. A guide to the contents

The first five chapters are used to introduce the basic concepts of a *tangent cone* $T(P,\mathfrak{P})$ at P of a family of probability measures $\mathfrak{P}$. It appears that by this concept one gets hold of those local properties of the family of probability measures which determine the asymptotic performance of statistical procedures - as long as one confines oneself to approximations of first order, i.e. approximations by limiting distributions.

Another basic concept is the *canonical gradient* $\kappa^{\bullet}(\cdot,P) \in T(P,\mathfrak{P})$ of a functional $\kappa: \mathfrak{P} \to \mathbb{R}$ which enables us to approximate the change of the value of the functional by $\int \kappa^{\bullet}(\xi,P) g(\xi) P(d\xi)$ if the probability measure moves away from P in a certain direction $g \in T(P,\mathfrak{P})$. These concepts are illustrated by a number of examples (see Chapters 2, 3 and 5).

The concepts 'tangent space' and 'gradient' have been used implicitly or explicitly in connection with special problems by a great number of authors, too many to be listed here. But it seems appropriate to mention the name of Levit who was the first to take steps in the direction of a general theory. (See Levit, 1974, 1975, and Koshevnik and Levit, 1976.)

In Chapter 6 certain measures for the distance between probability measures are introduced, and it is shown that several of these are asymptotically equivalent (in the sense that their ratio converges to one if the distances converge to zero).

Chapter 7 introduces the projection of probability measures into certain subfamilies of probability measures and investigates how such projections can be described locally.

The theoretical investigations in Chapters 8 - 12 are restricted to regular cases in which the tangent cones are linear spaces, and the functionals admit gradients. Roughly speaking this excludes all situations in which the best estimators converge at a rate different from $n^{-1/2}$.

Because of the exploratory nature of this treatise, the investigations are restricted to the most simple case, that of a sample of independent, identically distributed random variables. The generalization to two (or more) samples of independent, identically distributed random variables is straightforward. The possibility of other generalizations remains to be explored.

As far as the kind of problems is concerned, we restrict ourselves to estimation of finite-dimensional functionals and testing of hypotheses. Corresponding results can be obtained for confidence procedures.

Chapter 8 deals with a general hypothesis $\mathfrak{P}_o$ (which is not necessarily described in terms of a functional). It is shown that the asymptotic envelope power function of level-α-tests for the hypothesis $\mathfrak{P}_o$ against alternatives Q at a distance of order $n^{-1/2}$ from this hypothesis can be expressed by $\Phi(N_\alpha + n^{1/2}\delta(Q,\mathfrak{P}_o))$ (where δ is the Hellinger distance or any other asymptotically equivalent distance). It is certainly more convenient to formulate results on the asymptotic envelope power function in terms of sequences of alternatives converging to the hypothesis. Our somewhat unusual formulation, based on the distance of the alternative from the hypothesis, is a natural consequence of our endeavors to consider asymptotic theory as a tool for obtaining approximations (rather than as a collection of interesting limit theorems). The reader who prefers sequences of alternatives should not be disturbed by this departure from tradition; it bears no inherent relationship to the main object of this treatise.

Moreover, Chapter 8 contains a theorem (8.5.3) specifying the asymptotic power function of a test which is asymptotically most powerful for alternatives in a certain direction. Such tests are bound to have asymptotic power zero for alternatives deviating from the hypothesis in certain other directions (unless there is only one direction into which alternatives can deviate from the hypothesis). Section 8.6 discusses how the asymptotic envelope power function is influenced by restrictions of the basic family.

Chapter 9 contains corresponding results for estimators. In the introductory Section 9.1 it is suggested to base the comparison of estimators on the comparison of the distributions of their losses rather than on the risk (≡ expected loss), provided these distributions *are* comparable in the sense that one is more concentrated than the other. Section 9.2 contains an asymptotic bound for the concentration of asymptotically median unbiased estimators, Section 9.3 a version of Hájek's convolution theorem for estimators of multidimensional functionals. Section 9.4 demonstrates that for 'large' families of probability measures the convergence of the distribution of estimators to their limiting distribution can be *uniform* only over certain subfamilies. Section 9.5 investigates how a restriction of the basic subfamily influences the asymptotic bound for the concentration of estimators.

Results giving bounds for the asymptotic efficiency of statistical procedures require as a counterpart methods for the construction of statistical procedures which are asymptotically optimal in the sense of attaining these bounds. In this respect we have to offer no more than heuristic principles, together with some basic ideas which may be turned into a proof in each particular instance.

In Chapter 10 the asymptotic efficiency of estimators $P_n(\underline{x},\cdot)$ for probability measures is introduced (by the requirement that $\int f(\xi)P_n(\underline{x},d\xi)$ be an asymptotically efficient estimator for $\int f(\xi)P(d\xi)$, for every $f \in T(P,\mathfrak{P})$). Some suggestions are given as to how such estimator-sequences can be obtained, and it is shown that projections of asymptotically efficient estimators into a s u b f a m i l y are asymptotically efficient for this subfamily.

In Chapter 11 it is shown that asymptotically efficient estimators for functionals can be obtained by applying this functional to asymptotically efficient estimators of the probability measure. If the estimator for the probability measure fails to be asymptotically efficient, an asymptotically efficient estimator for the functional can be obtained by an improvement procedure based on the canonical gradient (see Section 11.4).

In Chapter 12 a heuristic procedure for the construction of asymptotically efficient tests, based on the canonical gradient, is suggested. In Section 12.3 it is shown that for hypotheses on the value of a functional asymptotically efficient tests can be obtained from asymptotically efficient estimators.

Chapters 13 - 18 are devoted to examples illustrating the general results obtained in Chapters 8 - 12.

All results are restricted to normal approximations. An asymptotic theory of higher order, based on Edgeworth expansions, is, so far, available for parametric families. (See Pfanzagl, 1980a, for a survey.) There can be no doubt that a result like 'first order efficiency implies second order efficiency' also holds true in the more general framework adopted here. The proof of this result for minimum contrast functionals, given in Pfanzagl (1981), generalizes immediately to arbitrary differentiable functionals. The characteristics of parametric families occuring in third-order Edgeworth expansions have recently become the subject of geometric interpretations. (See Efron, 1975, Skovgaard, 1981, Amari, 1981, 1982a,b, Kumon and Amari, 1981, Amari and Kumon, 1982.) It is to be hoped that the use of higher order geometric concepts like 'curvature' (together with the first order geometric concept of a 'tangent space') will eventually prove useful for generalizing results from parametric to arbitrary families of probability measures. As a first attempt at a general asymptotic theory, this treatise is restricted to normal approximations.

Many readers will be surprised that a treatise on asymptotic statistics contains nothing about 'adaptiveness' and 'robustness'. In the following Sections 0.7 and 0.8 we try to justify these omissions.

0.7. Adaptiveness

The word 'adaptive' seems to occur in literature with two different meanings. We are of the opinion that neither of these has a proper place in statistical theory.

To illustrate our point, consider a family of probability measures $P_{\theta,\eta}$, where θ is a real-valued parameter, and η an arbitrary (nuisance) parameter (say a vector of real-valued parameters, or a general 'shape' parameter). Assume we have for each family $\{P_{\theta,\eta}: \theta \in \mathbb{R}\}$ an estimator-sequence $\underline{x} \to \theta^n(\underline{x},\eta)$.

Some authors (such as Hogg, 1974) use the word 'adaptive' to denote an estimator-sequence $\underline{x} \to \theta^n(\underline{x},\eta^n(\underline{x}))$ which is obtained if η is replaced by an estimate $\eta^n(\underline{x})$. In certain situations, such a terminology may be natural from the psychological point of view (for instance if $\underline{x} \to \theta^n(\underline{x},\eta)$ is an estimator of location presuming the shape η to be known, which is 'adapted' to the unknown shape by means of an estimator for η). Since this is only a vague idea about how estimators for θ on the family $\mathfrak{P} = \{P_{\theta,\eta}: \theta \in \Theta, \eta \in H\}$ can be obtained, and not a clearly defined method, there is no reason to introduce a special name for such estimators.

Some more theoretically minded authors (like Bickel, 1982, or Fabian and Hannan, 1982) call the estimator $\underline{x} \to \theta^n(\underline{x},\eta^n(\underline{x}))$ 'adaptive' if it is asymptotically efficient as an estimator for θ in each of the families $\{P_{\theta,\eta}: \theta \in \mathbb{R}\}$, with η known. In our opinion this more restrictive use of 'adaptive' should also be abandoned, because it mingles properties of an estimator with properties of the family of probability measures. Assume we are given an estimator for θ which is asymptotically efficient in the basic family $\mathfrak{P} = \{P_{\theta,\eta}: \theta \in \Theta, \eta \in H\}$. Whether this estimator is also asymptotically efficient for each subfamily $\{P_{\theta,\eta}: \theta \in \mathbb{R}\}$ depends on how the family $\mathfrak{P}$ is made up of these subfamilies, and has nothing to do with the estimator: If $\mathfrak{P}$, as a combination of the subfamilies $\{P_{\theta,\eta}: \theta \in \mathbb{R}\}$, has a certain - exceptional - structure, then a n y estimator-sequence which is asymptotically efficient for $\mathfrak{P}$ will be asymptotically efficient for each of the subfamilies $\{P_{\theta,\eta}: \theta \in \mathbb{R}\}$. (See Example 9.5.3 for details.)

To summarize: 'Asymptotic efficiency' is a property of an estimator-sequence, 'adaptiveness' is a consequence of asymptotic efficiency under certain - exceptional - circumstances . There is no method for constructing 'adaptive' estimators, only one for constructing asymptotically efficient estimators.

0.8. Robustness

The assumption that the true probability measure belongs to a certain family $\mathfrak{P}_0$ is often unrealistic. To illustrate the difficulties arising in such a situation, consider the problem of estimating a functional κ, defined on a larger family $\mathfrak{P} \supset \mathfrak{P}_0$. It may turn out that estimators which are asymptotically optimal for estimating κ on $\mathfrak{P}_0$ are useless if the true probability measure is in $\mathfrak{P}-\mathfrak{P}_0$, because they are biased, or have a much larger asymptotic variance than asymptotically optimal estimators for κ on $\mathfrak{P}$.

What to do in such a case if one is quite sure that the true probability measure is close to $\mathfrak{P}_0$, but not absolutely sure that it belongs, in fact, to $\mathfrak{P}_0$? The basic difficulty is how to define what 'close to $\mathfrak{P}_0$' means. If we think of all probability measures in $\mathfrak{P}$ contained in a fixed neighborhood of $\mathfrak{P}_0$, then this is for the probability measures in $\mathfrak{P}_0$ - asymptotically - the same as if we consider neighborhoods consisting of *all* probability measures in $\mathfrak{P}$, because any *asymptotic* condition on the estimator (such as asymptotic median unbiasedness) is effective only if applied to the probability measures in a *shrinking* neighborhood, and such one belongs eventually to any fixed neighborhood. To obtain a nontrivial asymptotic problem one has, therefore, to define 'close to $\mathfrak{P}_0$' in terms of a neighborhood depending on the sample size. This is certainly meaningful from the operational point of view, because it expresses the fact that the prior information about the location of the true probability measure within $\mathfrak{P}$ (namely its being close to $\mathfrak{P}_0$) is not negligible compared to the information about the true probability measure, contained in the sample. Even if this idea is adequate, it is difficult to formalize. Technically speaking we have to consider neighborhoods shrinking with the sample size n like $n^{-1/2}$, and this is hardly possible without fixing essential ingredients of the neighborhood entirely arbitrarily, thus exerting a decisive influence on the 'optimal' estimator.

Although we fully recognize the practical relevance of this problem, we doubt whether mathematics can contribute much to its solution.

Statisticians working on 'robustness' are obviously less pessimistic. A widely accepted definition of robustness is that of Hampel (1971, p. 1890) requiring $P \to P^n * T^n$ to be equicontinuous on $\mathfrak{P}_o$. This definition refers to the whole sequence of estimators. In our opinion, the question whether the estimator for the sample size n = 20 is robust or not has nothing to do with the performance of this estimator for $n \neq 20$. Hampel's definition is weakened by Huber (1981, p. 10) to the requirement that for every $\varepsilon > 0$ there exist $\eta_\varepsilon > 0$ and $n_\varepsilon \in \mathbb{N}$ such that $\delta(P^n * T^n, P_o^n * T^n) < \varepsilon$ whenever $\delta(P, P_o) < \eta_\varepsilon$ and $n \geq n_\varepsilon$, where δ is some distance function metrizing weak convergence. This definition places no restriction whatsoever upon $P \to P^n * T^n$ for n fixed. If we assume that T^n is a consistent estimator of a certain functional (i.e. that $P^n * T^n$ converges stochastically to $\kappa(P)$), then Huber's definition is equivalent to the continuity of $P \to \kappa(P)$ with respect to the topology of weak convergence (see Huber, 1981, p. 41, Theorem 6.2), i.e. it describes a property of the functional, not one of the estimator.

I confess my difficulties in seeing any relationship of these definitions to the original idea of robustness. I do not see why 'closeness' of probability measures is always adequately expressed by the topology of weak convergence. (If $\kappa(P) := \int x P(dx)$, then probability measures with widely differing values of κ are not close together.) And I am completely lost vis a vis the problem of evaluating the performance of an estimator of a certain parameter if the true probability measure is not in the parametric family.

In view of these deficiencies of the author, considerations about robustness have been omitted in this treatise.

0.9. Notations

(1) Probability measures are denoted by letters like P, Q, their densities (with respect to a given dominating measure) by the corresponding small letters. If P is a probability measure on $\mathbb{B}$, its right-continuous distribution function is denoted by F_P.

(2) For the convolution of two probability measures P and Q we write $P \circledast Q$.

(3) $P_n \Rightarrow P$ denotes weak convergence.

(4) To denote the expectation of a function f under a probability measure P, we write $P(f)$ or $\int f(\xi)P(d\xi)$.

(5) $P*f$ denotes the distribution of f under P, defined by $P*f(A) := P(f^{-1}A)$, $A \in \mathscr{A}$.

(6) If $\{P_\theta : \theta \in \Theta\}$, $\Theta \subset \mathbb{R}^k$, is a parametric family dominated by a σ-finite measure μ, we denote (deviating from (1)) the μ-density of P_θ by $p(\cdot,\theta)$, and define

$$\ell(x,\theta) := \log p(\cdot,\theta),$$

$$\ell^{(i)}(x,\theta) := (\partial/\partial\theta_i)\ell(x,\theta),$$

$$L_{i,j}(\theta) := P_\theta(\ell^{(i)}(\cdot,\theta)\ell^{(j)}(\cdot,\theta)),$$

$$L(\theta) := (L_{i,j}(\theta))_{i,j=1,\ldots,k},$$

$$\Lambda(\theta) := L(\theta)^{-1}.$$

(Watch an exception from this rule: For probability measures over $\mathbb{R}^k$, $\ell^{(i)}(x,P)$ is also used to denote $(\partial/\partial x_i)\log p(x)$.)

(7) $N(\mu,\Sigma)$ denotes the multivariate normal distribution with mean vector μ and covariance matrix Σ.

φ,Φ denote the Lebesgue density and distribution function of the (univariate) standard normal distribution, N_α its α-quantile (defined by $\Phi(N_\alpha) = \alpha$).

(8) In a linear space we denote by $[a_i: i = 1,\dots,k]$ the linear span of the vectors $a_1,\dots,a_k$ (i.e. $[a_i: i=1,\dots,k] = \{\sum_{i=1}^{k} \alpha_i a_i: \alpha_i \in \mathbb{R},\ i = 1,\dots,k\}$).

$A+B$ denotes the direct sum of the subspace A and B (i.e. $A+B = \{a+b: a \in A,\ b \in B\}$). We write $A \oplus B$ if A and B are orthogonal.

(9) $\tilde{f}(\underline{x}) := n^{-1/2} \sum_{\nu=1}^{n} (f(x_\nu) - P(f))$.

(10) For sequences $a_n, b_n \in \mathbb{R}$, $n \in \mathbb{N}$, we write $a_n = o(b_n)$ if $a_n/b_n \to 0$ as $n \to \infty$, and $a_n = O(b_n)$ if a_n/b_n, $n \in \mathbb{N}$, is bounded.

(11) For a sequence of measurable functions $f_n: X^n \to \mathbb{R}$ we write $f_n \to 0 \quad (P)$ to denote *convergence in probability*, i.e. $P^n\{|f_n| > \varepsilon\} = o(n^0)$ for all $\varepsilon > 0$. We say that f_n, $n \in \mathbb{N}$, is *bounded in probability* if for every $\varepsilon > 0$ there exists $c > 0$ such that $P\{|f_n| > c\} < \varepsilon$.

(12) For sequences $f_n, g_n: X^n \to \mathbb{R}$ we write $f_n = o_P(g_n)$ if $f_n/g_n \to 0 \quad (P)$, and $f_n = O_P(g_n)$ if f_n/g_n, $n \in \mathbb{N}$, is bounded in probability. If P_θ belongs to a parametric family, we write o_θ for o_{P_θ}, and O_θ for O_{P_θ}.

(13) We use the following *convention*: If in an additive term an index occurs twice, this means summation over all values of the index set.

The author seeks remission of the reader for a few peculiarities in these notations. They result from his endeavors to make things more transparent and have nothing to do with the subject matter.

(i) We write $P(f)$ instead of $E(f)$, since $E(f)$ is ambiguous if more than one probability measure is involved, and $E_P(f)$ or $E(f|P)$ are unnecessarily ponderous.

(ii) We write $P * f$ for the induced measure since this is simpler than $\mathcal{L}(f|P)$.

(iii) We distinguish with great pains between a function f, and f(x), its value at x. This entails that a function without a special symbol like the quadratic has to be written as $x \to x^2$ (because x^2 denotes the value of this function at x).

(iv) We distinguish between the concept of an estimator, κ^n, which is a f u n c t i o n of the sample, and the estimate $\kappa^n(\underline{x})$, which is the value which this function attains for the sample $\underline{x} = (x_1,\dots,x_n)$.

(v) In the chapters presenting the theoretical results we speak of asymptotic properties of estimator - s e q u e n c e s resp. test - s e q u e n c e s (and avoid saying that an estimator is consistent or asymptotically normal.)

(vi) We use the term 'random variable' in an informal way to express the intuitive idea of the outcome of a random experiment, governed by a certain probability measure. In doing so we deviate deliberately from the terminology of probability theory, because sometimes it is persuadingly convenient to describe stochastic models in an intuitive language, and nothing suits this purpose better than the notion of a 'random variable'.

(vii) We write $O(n^o)$ rather than the usual $O(1)$, because $a_{m,n} = O(n^o)$ is easier to interpret than $a_{m,n} = O(1)$.

To the reader

This book is a preliminary publication. Remarks contributing to an improved final version are welcome.

Acknowledgment

The author is indebted to W.Wefelmeyer for his cooperation in the preparation of the manuscript. He worked through several versions of the manuscript, elaborated sketches of proofs, and completed the regularity conditions. In this connection he contributed Propositions 6.2.18 and 7.3.2, Lemma 7.3.1 and Corollaries 19.2.25 and 19.2.26.

The discussions with him had a decisive influence on the final shape of the manuscript and contributed in particular to the final form of the concepts introduced in Chapters 1, 4 and 6. In many instances he was able to bring the results into a form which is definitely superior to my original version. This holds in particular for Propositions 6.2.2 and 7.3.5, for Theorems 7.5.1, 8.4.1, 8.5.3, 19.2.7, and for Corollary 8.4.4. Without his competent assistance I would have been unable to finish the manuscript within a reasonable time.

Thanks are due to W.Droste and K.Bender for their help with some of the computations, and - last, but not least - to E.Lorenz for typing the manuscript (including countless revisions and amendments) with admirable skill and patience. The result speaks for itself.

1. THE LOCAL STRUCTURE OF FAMILIES OF PROBABILITY MEASURES

1.1. The tangent cone $T(P,\mathfrak{P})$

In this section we develop the concept of a tangent cone which seems appropriate for describing the *local structure* of a family of p-measures. Our purpose is to seize upon those local properties which are essential for the asymptotic performance of statistical procedures.

Let $\mathfrak{P}$ be the family of mutually absolutely continuous p-measures on a measurable space $(X,\mathscr{A})$. Let μ denote a σ-finite measure dominating $\mathfrak{P}$, and denote the density of $Q \in \mathfrak{P}$ with respect to μ with the corresponding small letter q. Fix $P \in \mathfrak{P}$, and define

$$\|f\| := \|f\|_p = (P(f^2))^{1/2} ,$$

$$\Delta(Q;P) := \|\frac{q}{p} - 1\| \qquad \text{(see also Section 6.1).}$$

By a *path* P_t, $t \downarrow 0$, in $\mathfrak{P}$ we mean a map $t \to P_t$ from an interval $(0,\varepsilon)$ into $\mathfrak{P}$.

1.1.1. Definition. A path P_t, $t \downarrow 0$, in $\mathfrak{P}$ is *differentiable* at P with *derivative* g if the P-density of P_t can be represented as

(1.1.2) $$1 + t(g + r_t) ,$$

where $\|g\| < \infty$ and r_t converges to 0 as $t \downarrow 0$ in some appropriate sense. We use the symbol $P_{t,g}$, $t \downarrow 0$, to denote a path with derivative g.

If $P(|r_t|) = o(t^0)$, then $P(g) = 0$ and $P(r_t) = 0$ because of

$$P(g + r_t) = t^{-1} P(\frac{p_t}{p} - 1) = 0 .$$

Let $\mathscr{L}_*(P)$ denote the class of all functions $g \in \mathscr{L}_2(P)$ with $P(g) = 0$.

The *tangent cone* $T(P, \mathfrak{P})$ is the class of all derivatives $g \in \mathscr{L}_*(P)$ of paths $P_{t,g}$, $t \downarrow 0$, in $\mathfrak{P}$ which are differentiable at P.

Observe that $T(P, \mathfrak{P})$ is, in fact, a cone: If $g \in T(P, \mathfrak{P})$ and $s > 0$, then $sg \in T(P, \mathfrak{P})$, since the representation (1.1.2) can be rewritten with $\bar{P}_{t,sg} := P_{st,g}$ and $\bar{r}_{t,sg} := s r_{t,g}$.

As a trivial tangent cone, we mention $T(P, \{P\}) = \{0\}$.

Up to now it remained open in which sense the error term r_t converges to zero. In fact, it seems difficult to specify conditions which suit all technical purposes equally well. For different families $\mathfrak{P}$, technically different conditions may be fulfilled which can be used for the same purpose. Moreover, applications to testing theory require conditions on the error terms different from the conditions required in estimation theory.

In the following we suggest two different sets of conditions on the error term which turned out to be useful.

The path P_t, $t \downarrow 0$, is *differentiable in the weak sense* if

(1.1.3) $$P(|r_t| 1_{\{t|r_t| > 1\}}) = o(t) ,$$

(1.1.4) $$P(r_t^2 \, 1_{\{t|r_t| \leq 1\}}) = o(t^0) .$$

The path P_t, $t \downarrow 0$, is *differentiable in the strong sense* if

(1.1.5) $$\| r_t \| = o(t^0) .$$

It is easy to see that differentiability in the strong sense implies differentiability in the weak sense.

In the following, $T_w(P,\mathfrak{P})$ and $T_s(P,\mathfrak{P})$ denote the tangent cones consisting of all weak resp. strong derivatives.

<u>1.1.6. Remark</u>. It is tempting to replace the definition of the strong derivative by one that avoids the use of paths.

With cl denoting the $\| \ \|$-closure, let

$$S_*(P,\mathfrak{P}) := \bigcap_{t>0} \mathrm{cl}\{\Delta(Q;P)^{-1}(q/p-1) : Q \in \mathfrak{P},\ \Delta(Q;P) \le t\},$$

and let $T_*(P,\mathfrak{P})$ denote the cone generated by $S_*(P,\mathfrak{P})$.

(Notice that $\|g\| = 1$ for $g \in S_*(P,\mathfrak{P})$. This can be seen as follows. $g \in S_*(P,\mathfrak{P})$ implies for every $t > 0$ the existence of $Q_t \in \mathfrak{P}$ with

$$\|\Delta(Q_t;P)^{-1}(q_t/p-1) - g\| \le t .$$

Since $\|\Delta(Q_t;P)^{-1}(q_t/p-1)\| = 1$, this implies $\|g\| = 1$.)

As an intersection of closed sets, $S_*(P,\mathfrak{P})$ is closed and so is $T_*(P,\mathfrak{P})$. It is straightforward to show that $T_s(P,\mathfrak{P}) \subset T_*(P,\mathfrak{P})$: For $g \in T_s(P,\mathfrak{P})$ there exists a path $P_{t,g}$, $t \downarrow 0$, with P-density $1 + t(g + r_t)$ and $\|r_t\| = o(t^0)$. Hence $\Delta(P_{t,g};P) = t(\|g\| + o(t^0))$, and

$$\big\|g/\|g\| - \Delta(P_{t,g};P)^{-1}(p_{t,g}/p-1)\big\| = o(t^0) ,$$

so that $g/\|g\| \in S_*(P,\mathfrak{P})$.

In the definition of a path we presumed that P_t is defined for all $t \downarrow 0$ in a neighborhood of 0. Such a property holds true in all natural examples. It is, however, not necessary for the definition of the derivative. For this purpose, it suffices that P_t is defined for t in a null-sequence. With this weaker concept of a path, we obtain immediately that $S_*(P,\mathfrak{P}) \subset T_s(P,\mathfrak{P})$, and therefore $T_*(P,\mathfrak{P}) = T_s(P,\mathfrak{P})$.

(For $g \in S_*(P,\mathfrak{P})$ there exists a sequence Q_n, $n \in \mathbb{N}$, such that $\Delta(Q_n;P) \le n^{-1}$ and $\|r_n\| \le n^{-1}$, where $r_n := \Delta(Q_n;P)^{-1}(q_n/p-1) - g$.

Hence Q_n, $n \in \mathbb{N}$, may be considered as a strongly differentiable path, defined for $t_n = \Delta(Q_n;P)$.)

This implies in particular that $T_s(P,\mathfrak{P})$ - defined with this weaker concept of a path - is $\| \ \|$-closed.

<u>1.1.7. Remark</u>. The specification of the tangent cone $T(P,\mathfrak{P})$ has two aspects. First, that $T(P,\mathfrak{P})$ contains *all* functions of a certain type, second, that it contains *no other* function. In many examples the first part is easier to prove, and for certain applications it is enough to know that certain functions belong to $T(P,\mathfrak{P})$. Perhaps a statement of this kind is also more adequate as a description of our prior knowledge about $\mathfrak{P}$: we know for sure that all p-measures of a certain kind belong to $\mathfrak{P}$ (and therefore all functions of a certain kind to $T(P,\mathfrak{P})$), but it may be asking too much to describe $\mathfrak{P}$ more precisely. In such situations it may be welcome to us that such a description is not needed for certain applications.

The elements of a tangent cone $T(P,\mathfrak{P})$ will be used to approximate densities of p-measures in the neighborhood of P, with an approximation error becoming arbitrarily small if the measures are arbitrarily close to P. This can be made precise as follows.

<u>1.1.8. Definition</u>. Let $T(P) \subset \mathscr{L}_*(P)$ be a $\| \ \|$-closed cone. For each $Q \in \mathfrak{P}$ let g_Q denote the projection of $q/p - 1$ into $T(P)$.

$\mathfrak{P}$ is *at* P *approximable by* $T(P)$ if uniformly for $Q \in \mathfrak{P}$

$$\|q/p - 1 - g_Q\| = o(\Delta(Q;P)).$$

We say that $\mathfrak{P}$ is *approximable by* T if this condition holds uniformly for $Q \in \mathfrak{P}$ and $P \in \mathfrak{P}$.

The following proposition shows that the $\| \ \|$-closure of the strong tangent cone $T_s(P,\mathfrak{P})$ (defined by (1.1.5)) is the smallest cone suitable for approximating $\mathfrak{P}$ at P in the sense of Definition 1.1.8.

1.1.9. Proposition. $T_s(P,\mathfrak{P}) \subset T(P)$, *if* $\mathfrak{P}$ *is at* P *approximable by* $T(P)$.

Proof. For $g \in T_s(P,\mathfrak{P})$, there exists a path $P_t \in \mathfrak{P}$, $t \downarrow 0$, with P-density $1 + t(g + r_t)$ such that $\|r_t\| = o(t^0)$. Let g_t denote the projection of $p_t/p - 1$ into $T(P)$. By assumption, $\|p_t/p - 1 - g_t\| = o(\|g_t\|)$. Since $\|g_t\| \leq \|p_t/p - 1\| = \|t(g+r_t)\| = O(t)$, we have $\|p_t/p - 1 - g_t\| = o(t)$. Hence $\|tg-g_t\| \leq t\|r_t\| + o(t) = o(t)$, and therefore $\|g-t^{-1}g_t\| = o(t^0)$. Since $t^{-1}g_t \in T(P)$, and since $T(P)$ is $\| \ \|$-closed, this implies $g \in T(P)$.

1.1.10. Remark. A successful as. theory requires to exclude local irregularities of $\mathfrak{P}$. Approximability of $\mathfrak{P}$ at P by $T_s(P,\mathfrak{P})$ goes in one direction only: It guarantees that - locally - all densities are approximable by elements of $T_s(P,\mathfrak{P})$. There are other potential irregularities which go into the opposite direction. For instance, $\mathfrak{P}$ may contain 'gaps' in any neighborhood of P. To exclude this, one may require that uniformly for $g \in T_s(P,\mathfrak{P})$,

(1.1.11) $\qquad \inf\{\|q/p-1-g\| : Q \in \mathfrak{P}\} = o(\|g\|).$

Such a condition is needed in the Addendum to Proposition 6.2.8.

For some results we need the condition that $T(P,\mathfrak{P})$ changes continuously with P, i.e., that the cone $T(Q,\mathfrak{P})$ resembles $T(P,\mathfrak{P})$ if Q is close to P in the sense that for every $g \in T(P,\mathfrak{P})$ there exists $g^* \in T(Q,\mathfrak{P})$ which is close to g.

<u>1.1.12. Definition</u>. $Q \to T(Q) \in \mathscr{L}_*(Q)$ is *continuous at* P if uniformly for $g \in T(P)$ and $Q \in \mathfrak{P}$

$$\inf\{\|g - Q(g) - h\|: h \in T(Q)\} = \|g\| o(\Delta(Q;P)^{\circ}) .$$

Notice that this definition is asymmetric. It permits that $T(Q)$ blows up as Q moves away from P.

<u>1.1.13. Convention</u>. Unless otherwise stated, the tangent spaces are assumed $\| \ \|$-closed in the following.

1.2. Properties of $T(P,\mathfrak{P})$ - properties of $\mathfrak{P}$

In Section 1.1 it was shown that $T(P,\mathfrak{P})$ is, by definition, a cone. For a successful theory we occasionally need additional properties, so, for instance, that $T(P,\mathfrak{P})$ is a linear space. Sometimes the functions in $T(P,\mathfrak{P})$ are also required to fulfill certain regularity conditions.

In this section we indicate how such properties of the tangent cone are related to corresponding properties of the family $\mathfrak{P}$. Thus we pay our tribute to tradition which requires to express our prior knowledge about the experiment as properties of the family $\mathfrak{P}$. This traditional way is, perhaps, not the most appropriate one. Whereas our prior knowledge may be reasonably accurate as far as the shape of the distributions is concerned (e.g., their symmetry, unimodality, etc.), it will hardly ever contain precise information about more technical properties such as the existence of moments, differentiability of densities and the like. Though our prior knowledge may include that the densities are 'smooth', more specific assumptions (like a

Lipschitz condition, bounds for the derivative, etc.) will be dictated by the type of mathematical analysis applied to the model.

In our approach, such *technical* conditions refer to the tangent cone rather than to the family $\mathfrak{P}$. Since technical conditions have no firm base in our prior knowledge anyway, we may forgo to derive them from - unfounded - technical assumptions on $\mathfrak{P}$, and impose them immediately on the tangent space itself.

In the following sections we shall discuss how convexity and symmetry of $T(P,\mathfrak{P})$ follow from certain properties of $\mathfrak{P}$. This discussion serves mainly the purpose of a better understanding of the intuitive meaning of these properties. Given a particular case, it will usually be easier to check directly whether $T(P,\mathfrak{P})$ has these properties, than to derive them from corresponding properties of $\mathfrak{P}$.

1.3. Convexity of $T(P,\mathfrak{P})$

A family of p-measures $\mathfrak{P}$ is *convex* if $P_o, P_1 \in \mathfrak{P}$ implies $(1-a)P_o + aP_1 \in \mathfrak{P}$ for $a \in [0,1]$. Only few families $\mathfrak{P}$ are convex in this general sense, but many have a somewhat weaker property, which we call *local asymptotic convexity*, namely: For arbitrary p-measures P_o, P_1 in a neighborhood of P, the convex combination $(1-a)P_o + aP_1$ can be *approximated* by a measure in $\mathfrak{P}$, say P_a, in the sense that the distance between $(1-a)P_o + aP_1$ and P_a is small compared to the distance between P_o and P_1. The following definition makes this precise.

<u>1.3.1. Definition</u>. $\mathfrak{P}$ is *loc. as. convex* at P if uniformly for $a \in (0,1)$ and $P_o, P_1 \in \mathfrak{P}$

$$\inf\{\Delta(Q,(1-a)P_o + aP_1;P) : Q \in \mathfrak{P}\} = o(\Delta(P_o;P) + \Delta(P_1;P)) .$$

<u>1.3.2. Proposition</u>. *If* $\mathfrak{P}$ *is loc. as. convex at* P, *then* $T_s(P,\mathfrak{P})$ *is convex.*

<u>Proof</u>. Let $g_o, g_1 \in T_s(P,\mathfrak{P})$ and $a \in (0,1)$. By definition of $T_s(P,\mathfrak{P})$, for i = 0,1 there exist paths $P_{t,i} \in \mathfrak{P}$, $t \downarrow 0$, with P-densities $1+t(g_i+r_{t,i})$ and $\|r_{t,i}\| = o(t^o)$. Observe that

$$\Delta(P_{t,i};P) = \|\frac{p_{t,i}}{p} - 1\| = t\|g_i + r_{t,i}\| = O(t) .$$

Since $\mathfrak{P}$ is loc. as. convex at P, there exist $P_t \in \mathfrak{P}$ with

$$\Delta(P_t,(1-a)P_{t,o} + aP_{t,1};P) = o(\Delta(P_{t,o};P) + \Delta(P_{t,1};P)) = o(t).$$

Hence the path P_t, $t \downarrow 0$, has P-density $1+t((1-a)g_o + ag_1 + r_t)$ with

$$\begin{aligned}
\|r_t\| &= \|t^{-1}(\frac{p_t}{p} - 1) - (1-a)g_o - ag_1\| \\
&\le t^{-1}\|\frac{p_t}{p} - (1-a)\frac{p_{t,o}}{p} - a\frac{p_{t,1}}{p}\| \\
&\quad + (1-a)\|t^{-1}(\frac{p_{t,o}}{p} - 1) - g_o\| \\
&\quad + a\|t^{-1}(\frac{p_{t,1}}{p} - 1) - g_1)\| \\
&= t^{-1}\Delta(P_t,(1-a)P_{t,o} + aP_{t,1};P) \\
&\quad + (1-a)\|r_o\| + a\|r_1\| \\
&= o(t^o).
\end{aligned}$$

This implies $(1-a)g_o + ag_1 \in T_s(P,\mathfrak{P})$.

1.4. Symmetry of $T(P,\mathfrak{P})$

The tangent cone $T(P,\mathfrak{P})$ is *symmetric* if $g \in T(P,\mathfrak{P})$ implies $-g \in T(P,\mathfrak{P})$. A property of $\mathfrak{P}$ which implies symmetry of $T(P,\mathfrak{P})$ is local as. symmetry.

<u>1.4.1. Definition</u>. $\mathfrak{P}$ is *loc. as. symmetric* at P if uniformly for $R \in \mathfrak{P}$

$$\inf\{\Delta(P,\tfrac{1}{2}Q + \tfrac{1}{2}R;P) : Q \in \mathfrak{P}\} = o(\Delta(R;P)) .$$

<u>1.4.2. Proposition</u>. *If $\mathfrak{P}$ is loc. as. symmetric at P, then $T_s(P,\mathfrak{P})$ is symmetric.*

<u>Proof</u>. Let $g \in T_s(P,\mathfrak{P})$. By definition, there exists a path P_t, $t \downarrow 0$, $\mathfrak{P}$ with P-density $1 + t(g + r_t)$ and $\|r_t\| = o(t^0)$. We have $\Delta(P_t;P) = O(t)$. Since $\mathfrak{P}$ is loc. as. symmetric at P, there exist $P_t^- \in \mathfrak{P}$ with

$$\Delta(P,\tfrac{1}{2}P_t^- + \tfrac{1}{2}P_t;P) = o(\Delta(P_t;P)) .$$

Hence the path P_t^-, $t \downarrow 0$, has P-density $1 + t(-g + r_t^-)$ with

$$\begin{aligned}
\|r_t^-\| &= \|t^{-1}(\frac{p_t^-}{p} - 1) + g\| \\
&\leq t^{-1}\|\frac{p_t^-}{p} + \frac{p_t}{p}\| + \|t^{-1}(\frac{p_t}{p} - 1) - g\| \\
&= 2t^{-1}\Delta(P,\tfrac{1}{2}P_t^- + \tfrac{1}{2}P_t;P) + \|r_t\| \\
&= o(t^0) .
\end{aligned}$$

This implies $-g \in T_s(P,\mathfrak{P})$.

Whereas loc. as. convexity will usually be fulfilled, loc. as. symmetry will fail if P is on the 'boundary' of $\mathfrak{P}$. Such a situation may occur in hypothesis testing if P belongs to the hypothesis, and only one-sided alternatives are of interest.

Hence $T(P,\mathfrak{P})$ will usually be a convex cone. If it is, in addition, symmetric, then it is a linear space. In this case we speak of a *tangent space*. Our general considerations are mostly concerned with tangent spaces. An extension of the whole theory to general tangent cones is desirable.

1.5. Tangent spaces of induced measures

Let $\mathfrak{P}|\mathscr{A}$ be a family of p-measures, $T: (X,\mathscr{A}) \to (Y,\mathscr{B})$ a measurable map, and $\mathfrak{P}*T = \{P*T: P \in \mathfrak{P}\}$ the family of induced p-measures. Let P^T denote the conditional expectation operator, given T, i.e.: For any $g \in \mathscr{L}_1(P)$, $(P^T g)\circ T$ is a conditional expectation of g, given T, with respect to P. Then

(1.5.1) $\qquad P^T T_s(P,\mathfrak{P}) \subset T_s(P*T,\mathfrak{P}*T) \qquad$ for every $P \in \mathfrak{P}$.

(If $g \in T_s(P,\mathfrak{P})$, then there exists a path $P_t \in \mathfrak{P}$, $t \downarrow 0$, with P-density $1+t(g+r_t)$ such that $\|r_t\| = o(t^0)$. Since $P^T(p_t/p) \in dP_t*T/dP*T$, we have $1+t(P^T g + P^T r) \in dP_t*T/dP*T$. Since $P*T((P^T r_t)^2) \leq P(r_t^2) = o(t^0)$, this implies $P^T g \in T_s(P*T,\mathfrak{P}*T)$.)

The converse relation,

(1.5.2) $\qquad T_s(P*T,\mathfrak{P}*T) \subset P^T T_s(P,\mathfrak{P})$

holds under additional conditions only. Since this relation will not be needed in the following, we confine ourselves to sketching the proof.

Given $k \in T_s(P*T, \mathfrak{P}*T)$, there exists a path P_t*T, $t \downarrow 0$, such that

$$dP_t*T/dP*T = 1 + t(k + s_t),$$

with $P((s_t \circ T)^2) = o(t^o)$. The essential point is that P_t, $t \downarrow 0$, can be chosen such that $t^{-1}(p_t/p-1)$ admits a $\| \ \|$-accumulation point $g \in \mathscr{L}_*(P)$. Then there exist a null-sequence $t_n \downarrow 0$ and $r_n \in \mathscr{L}_*(P)$ with $P(r_n^2) = o(n^o)$ such that

$$t_n^{-1}(p_{t_n}/p-1) = g + r_n .$$

We have $P^T(p_{t_n}/p) \in dP_{t_n}*T/dP*T$; hence $P^T(g + r_n) = k + s_{t_n}$ $P*T$-a.e. Since $P*T((P^T r_n)^2) \leq P(r_n^2) = o(n^o)$ and $P*T(s_{t_n}^2) = o(n^o)$, this implies $k = P^T g$ $P*T$-a.e.

It remains to show that g is the derivative of a path in $\mathfrak{P}$. This is true for the weaker concept of a path using only a s e q u e n c e $t_n \downarrow 0$. For interpolating between P_{t_n}, $n \in \mathbb{N}$, we need additional local connectedness conditions on $\mathfrak{P}$.

<u>1.5.3. Remark</u>. If T is sufficient for $\mathfrak{P}$, then there exist relative densities which are contractions of T. Hence for any $P \in \mathfrak{P}$, the tangent space $T_s(P,\mathfrak{P})$ consists of contractions of T, and we have $T_s(P*T,\mathfrak{P}*T) \circ T = T_s(P,\mathfrak{P})$ (in the sense that any function $g \in T_s(P,\mathfrak{P})$ can be written as $g = g_o \circ T$ with $g_o \in T_s(P*T,\mathfrak{P}*T)$, and, conversely, for every $g_o \in T_s(P*T, \mathfrak{P}*T)$, we have $g_o \circ T \in T_s(P,\mathfrak{P})$).

2. EXAMPLES OF TANGENT SPACES

For many families of p-measures occuring in the literature the tangent cones are, in fact, linear spaces. This chapter contains some examples of this kind. Several families do, however, not share all of the additional properties discussed in Chapter 1, such as loc. as. symmetry or loc. as. convexity.

2.1. 'Full' tangent spaces

In this section we present examples in which the tangent space $T_s(P,\mathfrak{P})$ equals $\mathscr{L}_*(P)$.

<u>2.1.1. Example</u>. The obvious example for which $T_s(P,\mathfrak{P}) = \mathscr{L}_*(P)$ is the family which contains with P all p-measures Q equivalent to P with $\Delta(Q;P)$ sufficiently small.

<u>Proof</u>. For arbitrary $g \in \mathscr{L}_*(P)$ and $t > 0$ sufficiently small, let

$$r_t := -g1_{\{tg < -1/2\}} + P(g1_{\{tg < -1/2\}}) ,$$

and define $P_{t,g}$ by its P-density

(2.1.2) $\quad 1 + t(g + r_t)$.

We have $\mu(p_{t,g}) = 1$ and $p_{t,g} \geq 0$ for all sufficiently small $t > 0$. Moreover,

$$\Delta(P_{t,g};P)^2 \leq t^2 P(g^2) ,$$

so that $P_{t,g} \in \mathfrak{P}$ if t is sufficiently small.

Finally,

$$\|r_t\|^2 = P(r_t^2) \leq P(g^2 1_{\{t|g| > 1/2\}}) = o(t^0) .$$

If $\mathfrak{P}$ contains with P all p-measures Q equivalent to P with $\Delta(Q;P) < \varepsilon$, then the ε-neighborhood of P in $\mathfrak{P}$ is (exactly) convex. In particular, $\mathfrak{P}$ is loc. as. convex at P, and $T_s(P,\mathfrak{P}) = T_*(P,\mathfrak{P})$. Furthermore, $\mathfrak{P}$ is at P approximable in the strongest possible sense: For sufficiently small $\Delta(Q;P)$,

$$q/p - 1 \in T_s(P,\mathfrak{P}) .$$

<u>2.1.3. Example</u>. In papers on robustness, it is common to consider for given P only neighborhoods consisting of p-measures $(1-\alpha)P + \alpha Q$, with arbitrary Q, and $0 \leq \alpha \leq \varepsilon$. Even in this case the tangent space $T_s(P,\mathfrak{P})$ equals $\mathscr{L}_*(P)$.

The proof is a slight modification of the proof given above. Let $P_{t,g}$ be the p-measure defined in (2.1.2). Then $Q_{t,g} := (1-\varepsilon)P + P_{t/\varepsilon,g} \in \mathfrak{P}$ has the P-density $1 + t(g + r_{t/\varepsilon})$, and $\|r_{t/\varepsilon}\| = o(t^0)$.

<u>2.1.4. Example</u>. If $\mathfrak{P}|\mathbb{B}^k$ consists of all p-measures with positive and continuous Lebesgue density, then for every $P \in \mathfrak{P}$ the tangent space $T_s(P,\mathfrak{P})$ equals $\mathscr{L}_*(P)$.

<u>Proof</u>. For $g \in \mathscr{L}_*(P)$ and $t > 0$ there exists a continuous function $f_t: \mathbb{R}^k \to \mathbb{R}$ with bounded support and $t|f_t| \leq 1/2$ such that

$$\| f_t - g1_{\{t|g| \le 1/2\}} \| \le t .$$

(See, e.g. Hewitt and Stromberg, 1965, p. 197, Theorem (13.21).)

Since $P(g) = 0$, we obtain

$$|P(f_t)| \le P(|f_t - g|) \le t + P(|g| 1_{\{t|g| > 1/2\}}) = o(t) .$$

Define

$$r_t := f_t - g - P(f_t)$$

and a path $P_{t,g}$, $t \downarrow 0$, with Lebesgue density

$$p_{t,g} := p(1 + t(f_t - P(f_t))) = p(1 + t(g + r_t)) .$$

Then $p_{t,g}$ is continuous and positive for $t > 0$ sufficiently small, and $P_{t,g}(\mathbb{R}^k) = 1$. Furthermore,

$$\| r_t \| \le \| f_t - g1_{\{t|g| \le 1/2\}} \| + \| g1_{\{t|g| > 1/2\}} \| + |P(f_t)|$$

$$= o(t^0) .$$

2.2. Parametric families

<u>2.2.1. Proposition</u>. *Let $\mathfrak{P} = \{P_\theta : \theta \in \Theta\}$, $\Theta \subset \mathbb{R}^k$, be a parametrized family of mutually absolutely continuous p-measures. Let $p(\cdot,\theta)$ denote the density of P_θ with respect to some dominating measure μ.*

Assume that $p^{(i)}(\cdot,\theta) := (\partial/\partial\theta_i) p(\cdot,\theta)$ fulfill for $i = 1,\dots,k$ at an inner point θ of Θ a local Lipschitz condition

$$|p^{(i)}(x,\tau) - p^{(i)}(x,\theta)| \le |\tau - \theta| p(x,\theta) M(x,\theta)$$

for $|\tau-\theta| < \varepsilon$, where $M(\cdot,\theta)$ and $\ell^{(i)}(\cdot,\theta) := p^{(i)}(\cdot,\theta)/p(\cdot,\theta)$ are P_θ-square integrable.

Assume that P_τ does not converge to P_θ if τ tends to the boundary of Θ, and that the matrix

$$L(\theta) := (L_{i,j}(\theta))_{i,j=1,\ldots,k} \qquad \textit{with}$$

$$L_{i,j}(\theta) := P_\theta(\ell^{(i)}(\cdot,\theta)\ell^{(j)}(\cdot,\theta)) \qquad \textit{is nonsingular.}$$

Then the tangent space $T_s(P_\theta,\mathfrak{P})$ *is the linear space spanned by* $\ell^{(i)}(\cdot,\theta)$, $i = 1,\ldots,k$.

<u>Proof</u>. (i) By a Taylor expansion we obtain for $a \in \mathbb{R}^k$

$$\frac{p(x,\theta+ta)}{p(x,\theta)} = 1 + t(a'\ell^{(\cdot)}(x,\theta) + r_{t,a}(x,\theta))$$

with $\|r_{t,a}(\cdot,\theta)\| = O(t)$. Hence $a'\ell^{(\cdot)}(\cdot,\theta) \in T_s(P_\theta,\mathfrak{P})$.

(ii) It remains to prove that $T_s(P_\theta,\mathfrak{P})$ contains no other functions. Let $g \in T_s(P_\theta,\mathfrak{P})$. For $t > 0$ there exists $\theta(t) \neq \theta$ such that

$$p(x,\theta(t)) = p(x,\theta)(1 + t(g(x) + r_t(x)))$$

with $\|r_t\| = o(t^0)$. We have

$$\Delta(P_{\theta(t)};P_\theta) = t\|g+r_t\| = t\|g\| + o(t) ,$$

hence $\theta(t) = \theta + o(t^0)$. Since $\|r_t\| = o(t^0)$, there exists a sequence $t_n \to 0$ such that

$$t_n^{-1}(p(\cdot,\theta(t_n)) - p(\cdot,\theta)) \to p(\cdot,\theta)g \qquad \mu\text{-a.e.} \tag{2.2.2}$$

We show that $t_n^{-1}|\theta(t_n) - \theta|$, $n \in \mathbb{N}$, remains bounded.

A Taylor expansion yields

$$|\theta(t) - \theta|^{-1}(p(\cdot,\theta(t)) - p(\cdot,\theta) - (\theta(t) - \theta)'p^{(\cdot)}(\cdot,\theta)) \to 0 . \tag{2.2.3}$$

Assume that $t_n^{-1}|\theta(t_n) - \theta| \to \infty$ for some subsequence. Then from (2.2.2) we obtain for this subsequence

$$|\theta(t_n) - \theta|(p(\cdot,\theta(t_n)) - p(\cdot,\theta)) \to 0 \qquad \mu\text{-a.e.}$$

Hence from (2.2.3), with $a_n := |\theta(t_n) - \theta|^{-1}(\theta(t_n) - \theta)$,

$$a_n'p^{(\cdot)}(\cdot,\theta) \to 0 \qquad \mu\text{-a.e.}$$

Since $|a_n| = 1$, there exists a subsequence converging to some $a \in \mathbb{R}^k$ with $|a| = 1$, so that $a'p^{(\cdot)}(\cdot,\theta) = 0$ μ-a.e., a contradiction to the assumption that $L(\theta)$ is nonsingular and hence positive definite.

Since $t_n^{-1}|\theta(t_n) - \theta|$, $n \in \mathbb{N}$, is bounded, we obtain from (2.2.3)

$$t_n^{-1}(p(\cdot,\theta(t_n)) - p(\cdot,\theta) - (\theta(t_n) - \theta)'p^{(\cdot)}(\cdot,\theta)) \to 0,$$

hence from (2.2.2)

$$t_n^{-1}(\theta(t_n) - \theta)'p^{(\cdot)}(\cdot,\theta) \to p(\cdot,\theta)g \qquad \mu\text{-a.e.}$$

Since $t_n^{-1}|\theta(t_n) - \theta|$, $n \in \mathbb{N}$, is bounded, there exists a subsequence converging to $a \in \mathbb{R}^k$, say. Therefore,

$$a'p^{(\cdot)}(\cdot,\theta) = p(\cdot,\theta)g \qquad \mu\text{-a.e.},$$

which is the assertion.

Observe that $\mathfrak{P}$ is loc. as. convex and symmetric at θ:

$$\inf\{\Delta(P_\tau,(1-a)P_\rho + aP_\sigma;P_\theta): \tau \in \Theta\}$$
$$\leq \Delta(P_{(1-a)\rho+a\sigma},(1-a)P_\rho + aP_\sigma;P_\theta)$$
$$= o(|\rho-\theta|^2 + |\sigma-\theta|^2)$$
$$= o(\Delta(P_\rho;P_\theta)^2 + \Delta(P_\sigma;P_\theta)^2),$$

$$\inf\{\Delta(\tfrac{1}{2}P_\tau + \tfrac{1}{2}P_\sigma;P_\theta): \tau \in \Theta\}$$
$$\leq \Delta(\tfrac{1}{2}P_{2\theta-\sigma} + \tfrac{1}{2}P_\sigma;P_\theta)$$
$$= o(|\sigma-\theta|^2)$$
$$= o(\Delta(P_\sigma;P_\theta)^2).$$

Under appropriate regularity conditions, $T_s(\cdot,\mathfrak{P})$ is continuous at P_θ:

$$\|\ell^{(i)}(\cdot,\theta) - P_\tau(\ell^{(i)}(\cdot,\theta)) - \ell^{(i)}(\cdot,\tau)\| = O(|\tau-\theta|).$$

Furthermore, $\mathfrak{P}$ is at P approximable by $T_s(P_\theta,\mathfrak{P})$: By Proposition 2.2.1, $T_s(P_\theta,\mathfrak{P})$ consists of the functions

$$g_a := (a'L(\theta)a)^{-1/2}a'\ell^{(\cdot)}(\cdot,\theta), \quad a \in \mathbb{R}^k.$$

Hence

$$\inf\{ \|\frac{p(\cdot,\tau)}{p(\cdot,\theta)} - 1 - \Delta(P_\tau;P_\theta)g\| : \ g \in T_s(P_\theta,\mathfrak{P})\}$$

$$\leq \|\frac{p(\cdot,\tau)}{p(\cdot,\theta)} - 1 - \Delta(P_\tau;P_\theta)g_{\tau-\theta}\|$$

$$= \|\frac{p(\cdot,\tau)}{p(\cdot,\theta)} - 1 - (\tau-\theta)'\ell^{(\cdot)}(\cdot,\theta)\| + o(|\tau-\theta|^2)$$

$$= o(|\tau-\theta|^2)$$

$$= o(\Delta(P_\tau;P_\theta)^2) \ .$$

Since $L(\theta)$ is nonsingular, the dimension of $T(P_\theta,\mathfrak{P})$ equals the number of parameters.

That $T(P_\theta,\mathfrak{P})$ is a linear space depends fundamentally on the fact that θ is an interior point of Θ. If θ is a boundary point, $T(P_\theta,\mathfrak{P})$ will, in general, be a convex cone only. As an illustration, consider the case that Θ, as a subset of $\mathbb{R}^k$, admits a tangent hyperplane at θ, say $c'\tau \geq c'\theta$ for all $\tau \in \Theta$ in a neighborhood of θ. Then

$$T(P_\theta,\mathfrak{P}) = \{a'\ell^{(\cdot)}(\cdot,\theta): \ a \in \mathbb{R}^k, \ c'a \geq 0\} \ .$$

<u>2.2.4. Example</u>. Let $P|\mathbb{B}$ be a p-measure with positive Lebesgue density p, the derivative p' of which fulfills a Lipschitz condition

$$|p'(y) - p'(x)| \leq |y - x| p(x) M(x)$$

for $x \in \mathbb{R}$ and $|y-x| < \varepsilon$, with $\int (x^4+1) M(x)^2 P(dx)$ finite.

For $a \in \mathbb{R}$, $b > 0$ let $P_{a,b} := P*(x \to a+bx)$. Let $p_{a,b}^{(1)}$ resp. $p_{a,b}^{(2)}$ denote the derivative with respect to a resp. b of the Lebesgue density $p_{a,b}(x) = b^{-1}p(b^{-1}(x-a))$ of $P_{a,b}$. We show below that for all (a',b') in a neighborhood of (a,b):

$$|p_{a',b'}^{(i)}(x) - p_{a,b}^{(i)}(x)| \leq |(a',b') - (a,b)| p_{a,b}(x) M_{a,b}(x)$$

with $P_{a,b}(M_{a,b}^2)$ finite.

Furthermore, $p^{(1)}_{a,b}$ and $p^{(2)}_{a,b}$ are linearly independent, hence $L(a,b)$ is nonsingular. We have $P_{a,b} \neq P_{a',b'}$ for $(a,b) \neq (a',b')$, and $P_{a,b}$ degenerates as (a,b) tends to the boundary of $\mathbb{R} \times (0,\infty)$.

Hence the assumptions of Proposition 2.2.1 are fulfilled for $\mathfrak{P} = \{P_{a,b}: a \in \mathbb{R},\ b > 0\}$.

<u>Proof</u>. In order to simplify our notations, we assume w.l.g. that $a = 0$, $b = 1$. We have

$$p^{(1)}_{a,b}(x) = -b^{-2}ap'(b^{-1}(x-a)) ,$$

$$p^{(2)}_{a,b}(x) = -b^{-2}p_{a,b} - b^{-3}(x-a)p'(b^{-1}(x-a)) .$$

We use

$$b^{-n}(x-a) - x = (b^{-n}-1)x - b^{-n}a$$

and the inequality

$$|p'(x)| = p(x)|\ell'(x)| = p(x) \lim_{y\to x} |y-x|^{-1}\left|\frac{p(y)}{p(x)} - 1\right|$$

$$\leq p(x)M(x) .$$

Then

$$|p^{(1)}_{a,b}(x) - p^{(1)}_{o,1}(x)|$$

$$\leq b^{-2}|a||p'(b^{-1}(x-a)) - p'(x)| + b^{-2}|ap'(x)|$$

and

$$|p^{(2)}_{a,b}(x) - p^{(2)}_{o,1}(x)|$$

$$\leq |b^{-2}p_{a,b}(x) - p_{o,1}(x)| + |b^{-3}(x-a)p'(b^{-1}(x-a) - p'(x)|$$

$$\leq b^{-2}|p_{a,b}(x) - p_{o,1}(x)| + |b^{-2} - 1|p(x)$$

$$+ b^{-3}|x-a||p'(b^{-1}(x-a)) - p'(x)|$$

$$+ |b^{-3}(x-a) - x||p'(x)|.$$

It remains to consider the first term in the last sum:

$$|p_{a,b}(x) - p_{o,1}(x)| \leq b^{-1}|p(b^{-1}(x-a)) - p(x)| + |b^{-1}-1|p(x).$$

By a Taylor expansion, there exists ξ between $b^{-1}(x-a)$ and x such that

$$|p(b^{-1}(x-a)) - p(x)|$$

$$\leq |b^{-1}(x-a) - x|(|p'(x)| + |p'(\xi) - p'(x)|) .$$

2.2.5. Example. Starting from a one-dimensional parametric family $\mathfrak{P} = \{P_\theta : \theta \in \Theta\}$, $\Theta \subset \mathbb{R}$, we define another parametric family $\mathfrak{Q}$ consisting of all p-measures

$$Q_{a,\theta,\tau} := (1-a)P_\theta + aP_\tau , \qquad a \in [0,1], \quad \theta,\tau \in \Theta.$$

In general, $T(Q_{a,\theta,\tau},\mathfrak{Q})$ is the linear space spanned by the three functions

$$\frac{p'(\cdot,\theta)}{(1-a)p(\cdot,\theta)+ap(\cdot,\tau)} , \qquad \frac{p'(\cdot,\tau)}{(1-a)p(\cdot,\theta)+ap(\cdot,\tau)} ,$$

and

$$\frac{p(\cdot,\tau)-p(\cdot,\theta)}{(1-a)p(\cdot,\theta)+ap(\cdot,\tau)} ,$$

and is, therefore, three-dimensional. (p' denotes the derivative with respect to the parameter.)

In the particular case $a = 0$, we have $Q_{0,\theta,\tau} = P_\theta$. Since $t \downarrow 0$ implies $Q_{t,\theta,\tau} \to P_\theta$ for any $\tau \in \Theta$, the tangent cone $T(P_\theta,\mathfrak{Q})$ is generated by $\ell'(\cdot,\theta)$ and $\frac{p(\cdot,\tau)}{p(\cdot,\theta)} - 1$, $\tau \in \Theta$, and is, therefore, infinite-dimensional. This cone fails in general to be convex. This defect is explained as follows (see Section 1.3): If $t \downarrow 0$, we have $\Delta(Q_{t,\theta,\tau'},Q_{t,\theta,\tau''};P_\theta) = t\Delta(P_{\tau'},P_{\tau''};P_\theta)$. For $(1-\alpha)Q_{t,\theta,\tau'} + \alpha Q_{t,\theta,\tau''} = (1-t)P_\theta + t(\alpha P_{\tau'} + (1-\alpha)P_{\tau''})$ there exists no element of $\mathfrak{Q}$ for which the distance to $(1-t)P_\theta + t(\alpha P_{\tau'} + (1-\alpha)P_{\tau''})$ tends to zero at a rate higher than t (excepting the case that $\mathfrak{P}$ is convex).

Consider now the subfamily $\mathfrak{Q}_0 = \{Q_{a_0,\theta,\tau} : \theta,\tau \in \Theta\}$ with $a_0 \in (0,1)$ fixed. In general, $T(Q_{a_0,\theta,\tau},\mathfrak{Q}_0)$ is two-dimensional, namely the linear space spanned by $\ell'(\cdot,\theta)$ and $\ell'(\cdot,\tau)$. For $\theta = \tau$, however, we have $Q_{a_0,\theta,\theta} = P_\theta$, and $T(P_\theta,\mathfrak{Q}_0)$ is one-dimensional, namely the space spanned by $\ell'(\cdot,\theta)$, so that $T(P_\theta,\mathfrak{Q}_0) = T(P_\theta,\mathfrak{P})$. This example shows that the extension of a family (from $\mathfrak{P}$ to $\mathfrak{Q}_o$) may leave the tangent space unchanged.

2.2.6. Remark. If we parametrize the same family in a different way, the tangent space should remain unchanged. This is, in fact, the case, since the derivatives with respect to the new parameters are linear combinations of the derivatives with respect to the original parameters.

2.2.7. Remark. Occasionally, we have to consider subfamilies $\mathfrak{P}_o$ of parametric families $\mathfrak{P}$. Usually, $T(P,\mathfrak{P}_o)$ is a genuine subspace of $T(P,\mathfrak{P})$ (for $P \in \mathfrak{P}_o$). There are two natural ways of selecting a subfamily:

(i) *Curved subfamilies*: Let $\Theta \subset \mathbb{R}^k$ and $T \subset \mathbb{R}^m$ with $m < k$. Given $c: T \to \Theta$, let $\mathfrak{P}_o := \{P_{c(\tau)}: \tau \in T\}$. Under appropriate regularity conditions on the map c, the tangent space $T(P_{c(\tau)},\mathfrak{P}_o)$ is the linear space spanned by $c_{ij}(\tau)\ell^{(j)}(\cdot,c(\tau))$, $i = 1,\dots,m$, where $c_{ij}(\tau) = \frac{\partial}{\partial\tau_i} c_j(\tau)$. This m-dimensional space is, in general, a genuine subspace of the k-dimensional space spanned by $\ell^{(i)}(\cdot,c(\tau))$, $i = 1,\dots,k$.

(ii) *Subfamilies specified by side conditions*. Given a function $F: \Theta \to \mathbb{R}^q$, let

$$\Theta_o := \{\theta \in \Theta,\ F(\theta) = 0\}, \quad \text{and} \quad \mathfrak{P}_o := \{P_\theta: \theta \in \Theta_o\}.$$

Under appropriate regularity conditions on the map F, we have $T(P_\theta,\mathfrak{P}_o) = \{a_i\ell^{(i)}(\cdot,\theta): a \in \mathbb{R}^k,\ d_{ij}(\theta)a_j = 0 \text{ for } i = 1,\dots,q\}$, where $d_{ij}(\theta) := \frac{\partial}{\partial\theta_j} F_i(\theta)$. In general, $T(P_\theta,\mathfrak{P}_o)$ is of dimension k-q.

Neglecting technicalities, the two modes of specifying subfamilies are equivalent: An m-dimensional curved subfamily corresponds to a subfamily selected by $q = k-m$ side conditions.

2.3. Families of symmetric distributions

Let $\mathfrak{P}$ be the family of all symmetric distributions over $\mathbb{R}$ which are equivalent to the Lebesgue measure λ. Let p denote the Lebesgue density of P, $\ell(\cdot,P) := \log p$, and $\ell'(x,P) := (d/dx)\ell(x,P)$. If m(P) denotes the median of P, then p is symmetric about m(P) and $\ell'(\cdot,P)$ skew-symmetric about m(P).

Let $\Psi(P)$ denote the class of all functions in $\mathscr{L}_*(P)$ which are symmetric about m(P). Notice that $\psi \in \Psi(P)$ implies

$$\int_{-\infty}^{m(P)} \psi(x)p(x)dx = 0 ,$$

since symmetry of ψ and p about m(P) implies

$$\int_{-\infty}^{m(P)} \psi(x)p(x)dx = \frac{1}{2} \int_{-\infty}^{+\infty} \psi(x)p(x)dx = 0 .$$

In the following we indicate conditions under which the tangent space consists essentially of functions $c\ell'(\cdot,P) + \psi$, with $c \in \mathbb{R}$ and $\psi \in \Psi(P)$. We remark that the two components, $c\ell'(\cdot,P)$ and ψ, are orthogonal. They admit straightforward interpretations: $\ell'(\cdot,P)$ corresponds to a *shift* of the given density, ψ corresponds to a change of the shape, preserving symmetry about m(P).

The following proposition uses a rather stringent regularity condition on ψ, because we try to get along with a single version of this proposition. It is certainly possible to obtain the result under weaker regularity conditions by restriction to a certain subclass of Ψ (for instance of smooth functions, or of bounded functions with a finite number of discontinuities).

2.3.1. Proposition. *Let* $P \in \mathfrak{P}$ *and* $\psi \in \Psi(P)$ *be such that for all* $x, y \in \mathbb{R}$ *with* $|y-x| \leq \varepsilon$

$$|p'(y) - p'(x)| \leq |y-x| p(x) M(x) ,$$

$$|\psi(y)| \leq K(x) ,$$

where $M, K, \ell'(\cdot,P) \in \mathscr{L}_4(P)$.

Then $c\ell'(\cdot,P) + \psi$, $c \in \mathbb{R}$, *are contained in* $T_s(P,\mathfrak{P})$.

Proof. To simplify our notations, we assume that $m(P) = 0$ and write $\ell' := \ell'(\cdot,P)$.

(i) For $t > 0$ define

$$q_t := p(1 + t(\psi + R_t))$$

with

$$R_t := -\psi 1_{\{t|\psi|>\frac{1}{2}\}} + P(\psi 1_{\{t|\psi|>\frac{1}{2}\}}) .$$

Then q_t is positive for t sufficiently small. Moreover, since ψ is symmetric about 0, so is R_t and therefore q_t. For the path P_t with density $q_t(x + ct)$ we have

$$\frac{q_t(x+ct)}{p(x)} - 1 = t(c\ell'(x) + \psi(x) + r_t(x))$$

with

$$(2.3.2) \qquad r_t(x) := \left(\frac{p(x+ct)-p(x)}{tp(x)} - c\ell'(x)\right) + \frac{p(x+ct)}{p(x)}(\psi(x+ct) - \psi(x))$$

$$+ \left(1 - \frac{p(x+ct)}{p(x)}\right)\psi(x) + \frac{p(x+ct)}{p(x)}\left(-\psi(x+ct)1_{\{t|\psi|>\frac{1}{2}\}}(x+ct)\right.$$

$$\left. + P(\psi 1_{\{t|\psi|>\frac{1}{2}\}})\right) .$$

(ii) It remains to show that $\|r_t\| = o(t^0)$. For the first summand on the right side of (2.3.2), observe that the Lipschitz condition on p' yields

$$\left|\frac{p(x+ct)-p(x)}{tp(x)} - c\ell'(x)\right| \leq ctM(x) .$$

For the second summand, use Hölder's inequality, the Lipschitz condition on p', and Lemma 19.1.4 on ψ. For the third summand, observe that

$\psi \in \mathscr{L}_4(P)$ because of $g \in \mathscr{L}_4(P)$. For the last summand use

$$|\psi(x+ct)1_{\{t|\psi|>\frac{1}{2}\}}(x+ct)| \leq g(x)1_{\{t|g|>\frac{1}{2}\}}(x)$$

for t sufficiently small.

2.3.3. Proposition. *Let* $T(P,\mathfrak{P})$ *be generated by the paths with remainder terms* r_t *fulfilling* $P(|r_t|) = o(t^o)$.

Then every $g \in T(P,\mathfrak{P})$ *is of the form* $c\ell'(\cdot,P) + \psi$ *with* $\psi \in \Psi(P)$.

Proof. To simplify our notations, we assume that $m(P) = 0$ and write $\ell' = \ell'(\cdot,P)$.

(i) Let $g \in T(P,\mathfrak{P})$. There exists a path P_t in $\mathfrak{P}$ with λ-density $p_t = p(1 + t(g+r_t))$ such that $P(|r_t|) = o(t^o)$, i.e.,

$$P(|t^{-1}(\frac{p_t}{p} - 1) - g|) = o(t^o) .$$

Hence there exists a sequence $t_n \downarrow 0$ such that

$$t_n^{-1}(\frac{p_{t_n}}{p} - 1) \to g \qquad P\text{-a.e.},$$

so that

$$(2.3.4) \qquad t_n^{-1}(p_{t_n} - p) \to pg \qquad \lambda\text{-a.e.}$$

For notational convenience we write p_n for p_{t_n} and m_n for $m(P_{t_n})$.

(ii) We start with the remark that $m_n \to 0$. Since $t_n^{-1}(p_n - p) \to gp$ λ-a.e., we have $p_n \to p$ λ-a.e. Assume first that $m_n \to \infty$ for some subsequence. Let $a > 0$ be so large that $P(-\infty,a) > \frac{1}{2}$. Since $m_n > a$ for sufficiently large n, symmetry of P_n about m_n implies $P_n(-\infty,a) \leq P_n(-\infty,m_n) \leq \frac{1}{2}$. Since $p_n 1_{(-\infty,a)} \to p1_{(-\infty,a)}$ λ-a.e., Fatou's lemma implies $P(-\infty,a) \leq \underline{\lim}\, P_n(-\infty,a)$, which is a contradiction. Hence m_n, $n \in \mathbb{N}$, is bounded. Let m_n, $n \in \mathbb{N}_o$, be a subsequence converging to c, say. Since p is continuous and $p_n \to p$ λ-a.e., Lemma 19.1.3 implies the existence of a subsequence such that $p_n(x+m_n) \to p(x+c)$ for λ-a.a. $x \in \mathbb{R}$. Since $x \to p_n(x+m_n)$ is symmetric about 0 and $x \to p(x+c)$ is symmetric about $-c$, this implies $c = 0$. Since any convergent subsequence of m_n contains a subsequence converging to 0, m_n itself converges to 0.

(iii) Next we show that $t_n^{-1} m_n$ remains bounded. Assume that $t_n^{-1} m_n \uparrow \infty$ for some subsequence. Multiplying both sides of (2.3.4) by $t_n m_n^{-1}$ we obtain $m_n^{-1}(p_n - p) \to 0$ λ-a.e. Since $m_n^{-1}(p(x-m_n)-p(x)) \to -p'(x)$, we have

$$m_n^{-1}(p_n(x)-p(x-m_n))-p'(x) \to 0 \qquad \text{for } \lambda\text{-a.a. } x \in \mathbb{R}$$

By Lemma 19.1.3, there exists a subsequence such that

$$m_n^{-1}(p_n(x+m_n)-p(x))-p'(x+m_n) \to 0 \qquad \text{for } \lambda\text{-a.a. } x \in \mathbb{R}.$$

Since p' is continuous,

$$m_n^{-1}(p_n(x+m_n)-p(x)) \to p'(x) \qquad \text{for } \lambda\text{-a.a. } x \in \mathbb{R}.$$

Since the left side is symmetric about 0 and the right side skew-symmetric about 0, this is contradictory. Hence $t_n^{-1} m_n$ remains bounded.

(iv) Finally, we choose a subsequence of $t_n^{-1} m_n$ converging to c, say. Since

$$t_n^{-1}(p(x-m_n)-p(x)) = t_n^{-1} m_n m_n^{-1}(p(x-m_n)-p(x)) \to -cp'(x),$$

(2.3.4) implies

$$t_n^{-1}(p_n(x)-p(x-m_n)) \to p(x)(g(x)+c\ell'(x)) \qquad \text{for } \lambda\text{-a.a. } x \in \mathbb{R}.$$

Since $m_n \to 0$, Lemma 19.1.3 implies the existence of a subsequence such that

$$t_n^{-1}(p_n(x+m_n)-p(x)) \to p(x)(g(x)+c\ell'(x)) \qquad \text{for } \lambda\text{-a.a. } x \in \mathbb{R}.$$

Since the left side is symmetric about 0, the right side is symmetric about 0. Since p is symmetric about 0, this implies that $g + c\ell'$ is symmetric about 0, q.e.d.

We remark that $\Psi(P)$, and therefore the linear space spanned by $\ell'(\cdot,P)$ and $\Psi(P)$, is $\|\ \|_P$-closed.

2.3.5. Remark. Our result is n e g a t i v e as far as the approximability of p-measures close to P by elements of $T(P,\mathfrak{P})$ is concerned. As an example, consider the path with density

$$p_t(x) = p(x-t) + t\sin(\pi x/2t)1_{(-1,1)}(x-t),\quad t \downarrow 0.$$

Let p be symmetric about 0. Then p_t is symmetric about t. We have

$$t^{-1}\left(\frac{p_t(x)}{p(x)} - 1\right) = t^{-1}\left(\frac{p(x-t)}{p(x)} - 1\right) + \frac{\sin(\pi x/2t)}{p(x)} 1_{(-1,1)}(x-t)$$

$$= \ell'(x,P) + \frac{\sin(\pi x/2t)}{p(x)} 1_{(-1,1)}(x-t) + r_t(x).$$

The remainder term

$$r_t(x) = t^{-1}\left(\frac{p(x-t)}{p(x)} - 1\right) - \ell'(x,P)$$

will be sufficiently small if $\ell'(\cdot,P)$ is sufficiently regular. However, the function $x \to \sin(\pi x/2t)/p(x)$ is s k e w - s y m m e t r i c about 0. Hence the approximation of $t^{-1}((p_t/p) - 1)$ by elements $c\ell'(\cdot,P) + \psi$, with ψ symmetric about 0, leaves us with an error term

$$\frac{\sin(\pi x/2t)}{p(x)} 1_{(-1,1)}(x-t) + r_t(x)$$

which fails to converge to zero in a technically useful sense.

The way out of this dilemma is either to restrict $\mathfrak{P}$ to p-measures with a sufficiently smooth density, or to introduce a more stringent distance function with respect to which the path P_t, $t \downarrow 0$, of our example fails to converge to P.

2.4. Measures on product spaces

For $i \in \{1,\dots,m\}$ let $\mathfrak{Q}_i$ be a family of p-measures on a measurable space $(X_i,\mathcal{A}_i)$. In the following, sums Σ and products $\times,\Pi$ always run over i from 1 to m.

<u>2.4.1. Proposition</u>. *(i) The family* $\times\mathfrak{Q}_i := \{\times Q_i : Q_i \in \mathfrak{Q}_i,\ i = 1,\dots,m\}$ *on* $(\times X_i,\times\mathcal{A}_i)$ *has the following tangent space:*

$$T_s(\times Q_i,\times\mathfrak{Q}_i) = \{(x_1,\dots,x_m) \to \Sigma g_i(x_i) : g_i \in T_s(Q_i,\mathfrak{Q}_i),\ i = 1,\dots,m\}.$$

(ii) If $T_s(Q_i,\mathfrak{Q}_i)$ *is linear and* $\|\ \|_{Q_i}$*-closed for* $i = 1,\dots,m$, *then* $T_s(\times Q_i,\times\mathfrak{Q}_i)$ *is* $\|\ \|_{\times Q_i}$*-closed.*

<u>Addendum</u>. *If* $(X_i,\mathcal{A}_i) = (X,\mathcal{A})$ *and* $\mathfrak{Q}_i = \mathfrak{Q}$ *for* $i = 1,\dots,m$, *and* $\mathfrak{Q}^m := \{Q^m : Q \in \mathfrak{Q}\}$, *then*

$$T_s(Q^m,\mathfrak{Q}^m) = \{(x_1,\dots,x_m) \to \Sigma g(x_i) : g \in T_s(Q,\mathfrak{Q})\}.$$

<u>Proof</u>. For notational convenience, we give the proof for $m = 2$. The extension to arbitrary m is straightforward.

(i) For $g_i \in T_s(Q_i,\mathfrak{Q}_i)$ let $Q_{i,t}$ denote a path with density

$$q_{i,t} = q_i(1 + t(g_i + r_{i,t})).$$

We show that the path $Q_{1,t}\times Q_{2,t}$ has the asserted properties. For the following, it is convenient to consider $q_{i,t}$ etc. as functions on $X_1\times X_2$. For this purpose we define

$$\bar{g}_i(x_1,x_2) := g_i(x_i), \qquad \text{etc.}$$

Notice that $Q_1\times Q_2(\bar{g}_i) = 0$, $Q_1\times Q_2(\bar{g}_i^2) = Q_i(g_i^2)$, $Q_1\times Q_2(\bar{g}_1^2\bar{g}_2^2) = Q_1(g_1^2)Q_2(g_2^2)$, etc. We have

$$\bar{q}_{1,t}\bar{q}_{2,t} = q_1q_2(1 + t(g_1 + g_2 + r_t))$$

with $r_t = \bar{r}_{1,t} + \bar{r}_{2,t} + t(\bar{g}_1 + \bar{r}_{1,t})(\bar{g}_2 + \bar{r}_{2,t})$. It follows easily that $\|r_t\|_{Q_1\times Q_2} = o(t^0)$. This proves that $\bar{g}_1 + \bar{g}_2 \in T_s(Q_1\times Q_2,\mathfrak{Q}_1\times\mathfrak{Q}_2)$.

Now we shall show that for every path P_t of product measures with $Q_1 \times Q_2$-density $1 + t(g+r_t)$ we have $g = \bar{g}_1 + \bar{g}_2$ $\quad Q_1 \times Q_2$-a.e., where

$$\bar{g}_1(x_1) = \int g(x_1, \xi_2) Q_2(d\xi_2) ,$$
$$\bar{g}_2(x_2) = \int g(\xi_1, x_2) Q_1(d\xi_1) .$$

Let

$$\bar{r}_{1,t}(x_1) = \int r_t(x_1, \xi_2) Q_2(d\xi_2) ,$$
$$\bar{r}_{2,t}(x_2) = \int r_t(\xi_1, x_2) Q_1(d\xi_1) .$$

Then the i-th marginal distribution of P_t has Q_i-density $1 + t(\bar{g}_i + \bar{r}_{i,t})$. Since P_t is a product measure, we have the following identity in t:

$$(1 + t(\bar{g}_1+\bar{r}_{1,t}))(1 + t(\bar{g}_2+\bar{r}_{2,t})) = 1 + t(g+r_t) \qquad Q_1 \times Q_2\text{-a.e.},$$

whence

$$\bar{g}_1 + \bar{g}_2 + \bar{r}_{1,t} + \bar{r}_{2,t} + t(\bar{g}_1 + \bar{r}_{1,t})(\bar{g}_2+\bar{r}_{2,t}) = g + r_t \qquad Q_1 \times Q_2\text{-a.e.}$$

Since r_t, $\bar{r}_{1,t}$, $\bar{r}_{2,t}$ converge to 0 in $\| \; \|_{Q_1 \times Q_2}$-norm, there exists a sequence $t_n \downarrow 0$ such that $r_{t_n} \to 0$, $\bar{r}_{1,t_n} \to 0$, $\bar{r}_{2,t_n} \to 0$ $\quad Q_1 \times Q_2$-a.e. This implies $g = \bar{g}_1 + \bar{g}_2$ $\quad Q_1 \times Q_2$-a.e., the assertion.

(ii) Follows from general results on orthogonal products of linear spaces.

For the purpose of future applications we now discuss certain families of p-measures on product spaces which contain all product measures (but not only these). Such families may serve as models for alternatives to the hypothesis of independence.

For $i = 1,\dots,m$ let $\mathfrak{Q}_i$ be a family of p-measures $Q_i | \mathbb{B}$ with Lebesgue density q_i, and let $\mathfrak{Q}$ denote the family of all p-measures with Lebesgue density

$$(x_1,\dots,x_m) \to \int \Pi q_i(x_i - \xi_i) N(0,\Sigma)(d(\xi_1,\dots,\xi_m)) ,$$

where $Q_i \in \mathfrak{Q}_i$, $i = 1,\dots,m$, and $N(0,\Sigma)$ is a normal distribution on $\mathbb{B}^m$ with covariance matrix Σ, possibly degenerate. In other words, the observations are m-tuples $(x_1+\xi_1,\dots,x_m+\xi_m)$ with stochastically independent variables $x_1,\dots,x_m$ and correlated normally distributed errors $\xi_1,\dots,\xi_m$. $\mathfrak{P}$ includes, in particular, all product measures $\times Q_i$ with $Q_i \in \mathfrak{Q}_i$, $i = 1,\dots,m$. Related models have been considered by Bhuchongkul (1964, p. 141), Hájek and Šidák (1967, Section II.4.11) and others.

What is the tangent space of a product measure in this larger family? It will be larger than the tangent space in the family of product measures, of course, since the measures of the new family can deviate from $\times Q_i$ in directions which are incompatible with the product structure. The extension of the tangent space is, however, rather modest. We have

(2.4.2) $$T(\times Q_i,\mathfrak{P}) = \{(x_1,\dots,x_m) \to \sum_i g_i(x_i) + \frac{1}{2}\sum_i \sigma_{ii}(\ell'(x_i,Q_i)^2 + \ell''(x_i,Q_i)) + \sum_{i\neq j}\sigma_{ij}\ell'(x_i,Q_i)\ell'(x_j,Q_j):$$
$$g_i \in T(Q_i,\mathfrak{Q}_i),\ i=1,\dots,m;\ (\sigma_{ij})_{i,j=1,\dots,m} \text{ nonnegative definite}\},$$

where $\ell'(x,Q) = (d/dx)\log q(x)$ and $\ell''(x,Q) = (d/dx)\ell'(x,Q)$.

This can be seen as follows. The family $\mathfrak{P}$ is of the form considered in Section 2.6. Hence the tangent space at $\times Q_i$ is the direct sum of $T(\times Q_i,\times\mathfrak{Q}_i)$ and the derivatives of paths with densities

$$(x_1,\dots,x_m) \to \int \Pi q_i(x_i-\xi_i)N(0,t\Sigma)(d(\xi_1,\dots,\xi_m)) .$$

A Taylor expansion yields

$$\int \Pi q_i(x_i-\xi_i)N(0,t\Sigma)(d(\xi_1,\dots,\xi_m))$$
$$= \Pi q_i(x_i)(1 + t(\frac{1}{2}\sum_i \sigma_{ii}(\ell'(x_i,Q_i) + \ell''(x_i,Q_i))$$
$$+ \sum_{i\neq j}\sigma_{ij}\ell'(x_i,Q_i)\ell'(x_j,Q_j) + r_t(x_1,\dots,x_m))) .$$

(2.4.2) now follows with Proposition 2.4.1.

Notice that $(x_1,\dots,x_m) \to \sum_{i\neq j}\sigma_{ij}\ell'(x_i,Q_i)\ell'(x_j,Q_j)$ is orthogonal to any function $(x_1,\dots,x_m) \to f_i(x_i)$ for $i = 1,\dots,m$, if $x_1,\dots,x_m$ are independent, hence in particular to $f_i = g_i + \frac{1}{2}\sigma_{ii}(\ell'(\cdot,Q_i)^2 + \ell''(\cdot,Q_i))$. There are some cases in which $\ell'(\cdot,Q_i)^2 + \ell''(\cdot,Q_i) \in T(Q_i,\mathfrak{Q}_i)$, so that

$T(\times Q_i, \mathfrak{P})$ is the orthogonal sum of $T(\times Q_i, \times \mathfrak{Q}_i)$ and the space generated by

$$(x_1,\dots,x_m) \to \sum_{i \neq j} \sigma_{ij} \ell'(x_i,Q_i)\ell'(x_j,Q_j) .$$

This is, for instance, the case if $\mathfrak{Q}_i$ is a 'full' family of p-measures, or the family of all one-dimensional normal distributions.

It is the case of full families $\mathfrak{Q}_i$ where the model becomes useful, for instance as a model for alternatives close to the hypothesis that the true p-measure is a product measure (with sufficiently regular marginal measures). A particularly useful feature of this model is that the transition from the family of all product measures to the more general family allowing some kind of dependence adds only a finite number of dimensions to the (infinite) dimension of the tangent space of the family of all product measures. For the special case $m = 2$, the dimension increases only by one: The tangent space is the orthogonal sum of $T(Q_1 \times Q_2, \mathfrak{Q}_1 \times \mathfrak{Q}_2)$ and the linear space spanned by the function $(x_1,x_2) \to \ell'(x_1,Q_1)\ell'(x_2,Q_2)$.

2.5. Random nuisance parameters

Consider a parametric family $\mathfrak{P} = \{P_{\theta,\eta} : (\theta,\eta) \in \Theta \times H\}$ on $(X,\mathscr{A})$ with $\Theta \subset \mathbb{R}^p$ and H arbitrary. We are interested in the (structural) parameter θ. The value of the nuisance parameter η changes from observation to observation, being a random variable distributed according to some unknown p-measure Γ on $(H,\mathscr{B})$ which belongs to a family $\mathscr{G}$. In other words: The observation x_ν is a realization governed by P_{θ,η_ν}, where η_ν is a realization governed by Γ. By the product measure theorem (see, e.g., Ash, 1972, p. 92), the p-measure $Q_{\theta,\Gamma}|\mathscr{A}\times\mathscr{B}$ governing (x_ν,η_ν) is uniquely defined by

(2.5.1) $$Q_{\theta,\Gamma}(A\times B) = \int_B P_{\theta,\eta}(A)\Gamma(d\eta), \qquad A\in\mathcal{A},\ B\in\mathcal{B}.$$

Hence our basic family of p-measures, say

$$\mathfrak{Q} := \{Q_{\theta,\eta}:\ \theta\in\Theta,\ \Gamma\in\mathcal{G}\},$$

contains two parameters, θ and Γ.

As a particular example, think of $P_{\theta,\eta}$ as a p-measure over $\{0,1\}$, $P_{\theta,\eta}\{1\}$ being the probability that a subject with ability at level η solves a task of difficulty θ. Then the same task, posed to different subjects, produces realizations $x_\nu \in \{0,1\}$ governed by P_{θ,η_ν}, $\nu = 1,\dots,n$. A familiar model of this kind, due to Rasch (1961, p. 323), presumes $P_{\theta,\eta}\{1\} = \theta\eta/(1+\theta\eta)$.

To determine the tangent space $T(Q_{\theta,\Gamma},\mathfrak{Q})$, we proceed as follows. For $\Gamma\in\mathcal{G}$ resp. $\theta\in\Theta$ we define the subfamilies

$$\mathfrak{P}_\Gamma := \{Q_{\theta,\Gamma}:\ \theta\in\Theta\},$$

$$\mathfrak{Q}_\theta := \{Q_{\theta,\Gamma}:\ \Gamma\in\mathcal{G}\}.$$

<u>2.5.2. Proposition</u>. *Assume that* $p^{(i)}(\cdot,\theta,\eta) := (\partial/\partial\theta_i)p(\cdot,\theta,\eta)$ *fulfill for* $i = 1,\dots,p$ *at an inner point* θ *of* Θ *a local Lipschitz condition*

$$|p^{(i)}(x,\tau,\eta) - p^{(i)}(x,\theta,\eta)| \le |\tau-\theta|\, p(x,\theta,\eta)M(x,\theta,\eta)$$

for $|\tau-\theta| < \varepsilon$ *and* $x\in X$, *and that* $P_{\theta,\eta}(\ell^{(i)}(\cdot,\theta,\eta)^2)$ *and* $P_{\theta,\eta}(M(\cdot,\theta,\eta)^2)$ *are bounded in* η *(where* $\ell^{(i)}(\cdot,\theta,\eta) := p^{(i)}(\cdot,\theta,\eta)/p(\cdot,\theta,\eta)$*). Then for every* $\Gamma\in\mathcal{G}$,

(i) $[(x,\eta)\to \ell^{(i)}(x,\theta,\eta):\ i = 1,\dots,p] \subset T_s(Q_{\theta,\Gamma},\mathfrak{P}_\Gamma)$

(ii) $\{(x,\eta)\to k(\eta):\ k\in T_s(\Gamma,\mathcal{G})\} \subset T_s(Q_{\theta,\Gamma},\mathfrak{Q}_\theta)$

(iii) the orthogonal sum

$$T_o(Q_{\theta,\Gamma},\mathfrak{Q}) := [(x,\eta)\to \ell^{(i)}(x,\theta,\eta):\ i = 1,\dots,p] \oplus \{(x,\eta)\to k(\eta):\ k\in T_s(\Gamma,\mathcal{G})\}$$

is contained in $T_s(Q_{\theta,\Gamma},\mathfrak{Q})$.

Proof. (i) Let $a \in \mathbb{R}^p$. The path $Q_{\theta+ta,\Gamma}$, $t \downarrow 0$, in $\mathfrak{P}_\Gamma$, has $\mu\times\Gamma$-density $(x,\eta) \to p(x,\theta+ta,\eta)$. A Taylor expansion yields

$$p(x,\theta+ta,\eta) = p(x,\theta,\eta)(1 + t(a_i \ell^{(i)}(x,\theta,\eta) + r_t(x,\theta,\eta))),$$

where

$$r_t(x,\theta,\eta) = a_i \int_0^1 (p^{(i)}(x,\theta+uta,\eta) - p^{(i)}(x,\theta,\eta))du/p(x,\theta,\eta).$$

By assumption,

$$Q_{\theta,\Gamma}(r_t(\cdot,\theta,\cdot)^2) \leq t|a|^2 \int P_{\theta,\eta}(M(\cdot,\theta,\eta)^2)\Gamma(d\eta) = O(t).$$

Hence $a_i \ell^{(i)}(\cdot,\theta,\cdot) \in T_s(Q_{\theta,\Gamma},\mathfrak{P}_\Gamma)$.

(ii) Let $k \in T_s(\Gamma,\mathcal{G})$. There exists a path Γ_t, $t \downarrow 0$, with Γ-density $1 + t(k+s_t)$ and $\Gamma(s_t^2) = o(t^0)$. The path Q_{θ,Γ_t}, $t \downarrow 0$, in $\mathfrak{Q}_\theta$ has $\mu\times\Gamma$-density

$$(x,\eta) \to p(x,\theta,\eta)(1 + t(k(\eta) + s_t(\eta)))$$

with

$$\int s_t(\eta)^2 Q_{\theta,\Gamma}(d(x,\eta)) = \int s_t(\eta)^2 p(x,\theta,\eta)\mu(dx)\Gamma(d\eta)$$
$$= \Gamma(s_t^2) = o(t^0).$$

Hence $(x,\eta) \to k(\eta) \in T_s(Q_{\theta,\Gamma},\mathfrak{Q}_\theta)$.

(iii) By (i) and (ii), the path $Q_{\theta+ta,\Gamma_t}$, $t \downarrow 0$, has $Q_{\theta,\Gamma}$-density

$$(x,\eta) \to (1 + t(a_i \ell^{(i)}(x,\theta,\eta) + r_t(x,\theta,\eta)))(1 + t(k(\eta) + s_t(\eta))).$$

It is easy to see from the independence of x and η that the path $Q_{\theta+ta,\Gamma_t}$, $t \downarrow 0$, has derivative

$$(x,\eta) \to a_i \ell^{(i)}(x,\theta,\eta) + k(\eta)$$

with a remainder term $o(t^0)$ in $\mathscr{L}_*(Q_{\theta,\Gamma})$. Furthermore, $P_{\theta,\eta}(\ell^{(i)}(\cdot,\theta,\eta)) = 0$ for all $\eta \in H$ implies

$$\int \ell^{(i)}(x,\theta,\eta)k(\eta)Q_{\theta,\Gamma}(d(x,\eta)) = \int P_{\theta,\eta}(\ell^{(i)}(\cdot,\theta,\eta))\Gamma(d\eta) = 0.$$

If the nuisance parameter is not observable, the p-measure governing the observations is the first marginal of $Q_{\theta,\Gamma}$, say $Q'_{\theta,\Gamma} := Q_{\theta,\Gamma} * \pi_1$, where $\pi_1(x,\eta) := x$. From (2.5.1),

$$Q'_{\theta,\Gamma}(A) = \int P_{\theta,\eta}(A)\Gamma(d\eta)\,, \qquad A \in \mathscr{A}. \tag{2.5.3}$$

If $P_{\theta,\eta}$ has μ-density $p(\cdot,\theta,\eta)$, then $Q'_{\theta,\Gamma}$ has μ-density $x \to \int p(x,\theta,\eta)\Gamma(d\eta)$. Let

$$\mathfrak{Q}' := \mathfrak{Q} * \pi_1 = \{Q'_{\theta,\Gamma} : \theta \in \Theta,\ \Gamma \in \mathscr{G}\}\,.$$

For any p-measure $Q|\mathscr{A}\times\mathscr{B}$ let Q^{π_1} denote the conditional expectation operator, given π_1, i.e.: For any Q-integrable function $f: X \times H \to \mathbb{R}$, $(Q^{\pi_1} f)\circ\pi_1$ is a conditional expectation of f, given π_1, with respect to Q. It is easy to see that

$$(Q_{\theta,\eta}^{\pi_1} f)(x) = \int f(x,\eta)p(x,\theta,\eta)\Gamma(d\eta) / \int p(x,\theta,\eta)\Gamma(d\eta)\,. \tag{2.5.4}$$

From Proposition 2.5.2 and (1.6.1) we obtain that

$$Q_{\theta,\Gamma}^{\pi_1}(a_i \ell^{(i)}(\cdot,\theta,\cdot) + k) \in T_s(Q'_{\theta,\Gamma},\mathfrak{Q}') \tag{2.5.5}$$

for $a \in \mathbb{R}^p$ and $k \in T_s(\Gamma,\mathscr{G})$.

The family $\mathfrak{Q}$ is of the form considered in Section 2.6. Hence it is plausible that equality holds in Proposition 2.5.2. Moreover, it is suggested in Section 1.6 that $P^T T_s(P,\mathfrak{P})$ equals $T_s(P*T,\mathfrak{P}*T)$. Hence we expect that the inclusions given in Proposition 2.5.2 and (2.5.5) are, in fact, equalities. Since only the inclusions are needed (see Sections 14.2 and 14.3), we refrain from specifying regularity conditions under which equality takes place.

2.6. A general model

In the examples of the preceding Sections 2.2 - 2.5, the tangent space decomposes into a direct sum of certain subspaces. In this section we suggest a general model which entails such a decomposition.

Let Θ and H be two arbitrary sets, and $\mathfrak{P} := \{P_{\theta,\eta} : (\theta,\eta) \in \Theta \times H\}$ a family of mutually absolutely continuous p-measures. In this model, each p-measure $P_{\theta,\eta}$ can be considered as an element of two subfamilies

$$\mathfrak{Q}_\theta := \{P_{\theta,\eta} : \eta \in H\}, \qquad \mathfrak{P}_\eta := \{P_{\theta,\eta} : \theta \in \Theta\} .$$

The models discussed in the preceding Sections 2.2 - 2.5 can be subsumed under this general model by appropriate 'parametrization'. This is obvious for the parametric family discussed in Section 2.2. The family of symmetric distributions in Section 2.3 can be parametrized as $(P,\theta) \to P*(x \to x+\theta)$, with P running through all distributions symmetric about zero. For the product measures considered in Section 2.4 the parametrization (for m = 2) is $(Q_1,Q_2) \to Q_1 \times Q_2$. The families introduced in Section 2.5 are already written in the form of the general model.

The following considerations suggest that the tangent space in $\mathfrak{P}$ will be the direct sum of the tangent spaces in these two subfamilies.

If $T(P_{\theta,\eta},\mathfrak{P})$ is linear, the inclusion

(2.6.1) $\quad T(P_{\theta,\eta},\mathfrak{Q}_\theta) + T(P_{\theta,\eta},\mathfrak{P}_\eta) \subset T(P_{\theta,\eta},\mathfrak{P})$

is immediate: $\mathfrak{Q}_\theta \subset \mathfrak{P}$ implies $T(P_{\theta,\eta},\mathfrak{Q}_\theta) \subset T(P_{\theta,\eta},\mathfrak{P})$, and $\mathfrak{P}_\eta \subset \mathfrak{P}$ implies $T(P_{\theta,\eta},\mathfrak{P}_\eta) \subset T(P_{\theta,\eta},\mathfrak{P})$. For conditions on $\mathfrak{P}$ which entail linearity of $T(P_{\theta,\eta},\mathfrak{P})$ see Sections 1.3 and 1.4.

The converse inclusion,

(2.6.2) $\quad T(P_{\theta,\eta},\mathfrak{P}) \subset T(P_{\theta,\eta},\mathfrak{Q}_\theta) + T(P_{\theta,\eta},\mathfrak{P}_\eta)$,

holds under additional regularity conditions only. It is, therefore, preferable to prove it in each case separately, taking advantage of the individual properties. (See Sections 2.2 - 2.5 for such proofs.)

The following considerations are to indicate the general nature of these regularity conditions. For $g \in T(P_{\theta,\eta},\mathfrak{P})$ there exists a path P_{θ_t,η_t}, $t \downarrow 0$, in $\mathfrak{P}$ with $P_{\theta,\eta}$-density $1+t(g+r_t)$, where r_t converges to zero in some appropriate sense. We have

(2.6.3) $$p_{\theta_t,\eta_t}/p_{\theta,\eta} - 1 = (p_{\theta_t,\eta_t}/p_{\theta_t,\eta} - 1) + (p_{\theta_t,\eta}/p_{\theta,\eta} - 1) + (p_{\theta_t,\eta_t}/p_{\theta_t,\eta} - 1)(p_{\theta_t,\eta}/p_{\theta,\eta} - 1) .$$

If the distance of $P_{\theta_t,\eta}$ from $P_{\theta,\eta}$ is of the same order of magnitude as the distance of P_{θ_t,η_t} from $P_{\theta,\eta}$, and if $T(P_{\theta,\eta},\mathfrak{Q}_\theta)$ and $T(P_{\theta,\eta},\mathfrak{P}_\eta)$ are linearly independent, then one can expect that $t^{-1}(p_{\theta_t,\eta}/p_{\theta,\eta} - 1)$, $t \downarrow 0$, converges to an element in $T(P_{\theta,\eta},\mathfrak{P}_\eta)$, say h. Multiplying (2.6.3) by t^{-1} and taking the limits for $t \downarrow 0$, we therefore obtain under appropriate regularity conditions that

(2.6.4) $$g = \lim_{t\downarrow 0} t^{-1}(p_{\theta_t,\eta_t}/p_{\theta_t,\eta} - 1) + h .$$

Hence (2.6.2) follows if $t^{-1}(p_{\theta_t,\eta_t}/p_{\theta_t,\eta} - 1)$, $t \downarrow 0$, converges to an element in $T(P_{\theta,\eta},\mathfrak{Q}_\theta)$. This requires local uniformity in θ of the convergence of $t^{-1}(p_{\theta',\eta_t}/p_{\theta',\eta} - 1)$ to an element of $T(P_{\theta',\eta},\mathfrak{Q}_{\theta'})$, and continuity of $T(P_{\theta',\eta},\mathfrak{Q}_{\theta'})$ for $\theta' \to \theta$.

That (2.6.2) is not true in general can be seen as follows.

<u>2.6.5. Example</u>. For $\theta \in [0,1]$ and $\eta \in \mathbb{R}$, let $P_{\theta,\eta}|\mathbb{B}$ be the p-measure with Lebesgue-density

$$x \to (1-\theta)p(x) + \theta p(x-\eta) ,$$

where p is a sufficiently regular Lebesgue density of a p-measure.

In this case,

$$\mathfrak{Q}_0 = \{P_{0,0}\} \quad \text{and} \quad \mathfrak{P}_0 = \{P_{0,0}\},$$

so that $T(P_{0,0},\mathfrak{Q}_0) = \{0\}$ and $T(P_{0,0},\mathfrak{P}_0) = \{0\}$. On the other hand, the path $P_{\sqrt{t},\sqrt{t}}$, $t \downarrow 0$, leads to $[p'/p] \subset T(P_{0,0},\mathfrak{P})$.

3. TANGENT CONES

3.1. Introduction

The as. theory developed in Chapters 8 - 12 presumes that the tangent cones are linear spaces. In the present chapter we collect a few natural examples where the tangent cone fails to be a linear space. These examples are to remind the reader that an extension of the theory to convex tangent cones is wanted. Since the results are not needed in the rest of the book, we are more generous about regularity conditions.

The common feature of the examples is the following: Given a pre-order (i.e., a reflexive and transitive order relation) on a family of p-measures, and a subfamily $\overline{\mathfrak{P}}$ of order equivalent p-measures, the family $\mathfrak{P}$ consists of p-measures comparable with the elements of $\overline{\mathfrak{P}}$. This usually leads to a (convex) tangent cone if only p-measures larger (or smaller) than those in $\overline{\mathfrak{P}}$ are considered, or to a tangent cone consisting of a convex cone and its reflexion about 0 if both smaller and larger p-measures are allowed. For partial orders (i.e., antisymmetric pre-orders), $\overline{\mathfrak{P}}$ reduces to a single p-measure.

We do not assume the p-measures in $\mathfrak{P}$ to be pairwise comparable. This would lead to a linearly ordered family, and hence to one-dimensional tangent cones.

In the following sections we discuss some examples of order relations. For further order relations see van Zwet (1964), Barlow and Proschan (1966), Doksum (1969), Barlow et al. (1972), and Lawrence (1975).

3.2. Order with respect to location

The following definitions try to compare p-measures $P_i|\mathbb{B}$ with respect to their location. F_i denotes the distribution function of P_i.

We offer three alternative definitions for $P_1 \leq P_2$:

(3.2.1) $\quad F_2(x) \leq F_1(x) \quad$ for all $x \in \mathbb{R}$,

(3.2.2) $\quad F_2(x)/F_1(x)$ and $(1-F_2(x))/(1-F_1(x))$ are nondecreasing in x,

(3.2.3) $\quad (F_2(x'')-F_2(x'))/(F_1(x'')-F_1(x'))$ is nondecreasing in x',x'', whenever $F_1(x') < F_1(x'')$.

These three partial orders are of increasing stringency (see Pfanzagl, 1964, p. 1218).

For $P \in \mathfrak{P}$, let $P_{t,g}$, $t \downarrow 0$, denote a differentiable path converging to P with derivative $g \in \mathscr{L}_*(P)$. If $P \leq P_{t,g}$ for all sufficiently small $t > 0$ in one of these order relations, then this imposes a certain condition on g, namely:

ad (3.2.1) $\quad \int_{-\infty}^{x} g(\xi)P(d\xi) \leq 0 \quad$ for all $x \in \mathbb{R}$,

ad (3.2.2) $\quad \int_{-\infty}^{x} g(\xi)P(d\xi)/\int_{-\infty}^{x} P(d\xi)$ and $\int_{x}^{\infty} g(\xi)P(d\xi)/\int_{x}^{\infty} P(d\xi)$ are nondecreasing in x,

ad (3.2.3) $\quad g$ is nondecreasing.

For $i = 1,2,3$ let $C_i(P)$ denote the class of all $g \in \mathscr{L}_*(P)$ fulfilling condition (3.2.i). It is easy to check that $C_1(P) \supset C_2(P) \supset C_3(P)$, corresponding to the increasing stringency of the order relations (3.2.1), (3.2.2) and (3.2.3). $C_i(P)$ is a convex cone. Even the largest of these cones, $C_1(P)$, is smaller than a half-space, so that the double cone $\{g \in \mathscr{L}_*(P): g \in C_1(P) \text{ or } -g \in C_1(P)\}$ fails to be a linear space. Hence, whatever the family $\mathfrak{P}$, if its elements are comparable with P according to order relation (3.2.i), its tangent space $T(P,\mathfrak{P}_i)$, being a subspace of $\{g \in \mathscr{L}_*(P): g \in C_i(P) \text{ or } -g \in C_i(P)\}$, will be a double cone, but not a linear space (excepting degenerate cases like linearly ordered families).

3.3. Order with respect to concentration

The following definition tries to compare p-measures $P_i|\mathbb{B}$ with respect to their concentration. F_i denotes the distribution function of P_i.

P_2 is *more concentrated* than P_1 if

(3.3.1) $$F_2^{-1}(\beta)-F_2^{-1}(\alpha) \leq F_1^{-1}(\beta)-F_1^{-1}(\alpha) \qquad \text{for } 0 < \alpha < \beta < 1 .$$

This pre-order was introduced by Bickel and Lehmann (1979, p. 34), where the reader can also find an intuitive justification. Recall that $F_2(x) = F_1((x-a)/b)$ implies $F_2^{-1}(\beta)-F_2^{-1}(\alpha) = b(F_1^{-1}(\beta)-F_1^{-1}(\alpha))$, so that p-measures differing in location and scale only are always comparable with respect to this order.

Under suitable regularity conditions, the order relation defined by (3.3.1) is equivalent to either of the following conditions

(3.3.2) $$\frac{d}{d\alpha}F_2^{-1}(\alpha) \leq \frac{d}{d\alpha}F_1^{-1}(\alpha) \qquad \text{for all } \alpha \in (0,1)$$

or

(3.3.3) $$p_1(F_1^{-1}(\alpha)) \leq p_2(F_2^{-1}(\alpha)) \qquad \text{for all } \alpha \in (0,1)$$

(with p_i denoting the Lebesgue density of P_i).

For $P \in \mathfrak{P}$ let $P_{t,g}$, $t \downarrow 0$, denote a differentiable path converging to P with derivative $g \in \mathscr{L}_*(P)$. If $P_{t,g}$ is less concentrated than P for all sufficiently small $t > 0$, then

(3.3.4) $\int_{-\infty}^{x} g(\xi)p(\xi)d\xi/p(x)$ is nonincreasing in x.

It is straightforward that the set of all $g \in \mathscr{L}_*(P)$ fulfilling (3.3.4) is a convex cone, but not a half-space.

3.4. Order with respect to asymmetry

We consider p-measures on $\mathcal{B}$ with Lebesgue density. The hypothesis of symmetry about 0 can be formulated by means of the distribution function as

(3.4.1) $F(x) + F(-x) = 1$ for all $x \in \mathbb{R}$.

Natural alternatives to the hypothesis of symmetry about zero are p-measures P which are stochastically larger than their reflection about zero, $P*(x \to -x)$ (with distribution function $x \to 1-F(-x)$). Using the weakest concept of 'stochastically larger', the one given by (3.2.1), we obtain for the alternatives the condition

(3.4.2) $F(x) + F(-x) \leq 1$ for all $x \in \mathbb{R}$.

Let $\mathfrak{P}$ denote the family of all distributions fulfilling (3.4.2). $\mathfrak{P}$ contains on its boundary the family $\mathfrak{P}_o$ of all distributions fulfilling (3.4.1). If P is symmetric about zero, and $P_{t,g}$, $t \downarrow 0$, a differentiable path converging to P with derivative $g \in \mathscr{L}_*(P)$, of p-measures fulfilling (3.4.2), this imposes on g the condition

(3.4.3) $\int_{-\infty}^{x} (g(\xi)-g(-\xi))p(\xi)d\xi \leq 0$ for all $x \in \mathbb{R}$.

For $P \in \mathfrak{P}_o$, $T(P,\mathfrak{P})$ consists of all functions $g \in \mathscr{L}_*(P)$ fulfilling (3.4.3) and is, therefore, a convex cone, smaller than a half-space (i.e., the union of $T(P,\mathfrak{P})$ and its reflection about 0 is not a linear space).

Since $\int_{-\infty}^{x} \ell'(\xi,P)p(\xi)d\xi = p(x)$, $T(P,\mathfrak{P})$ contains, in particular, all functions $g = c\ell'(\cdot,P)$ with $c < 0$ (which corresponds to the fact that, for P symmetric about 0, $P*(x \to x-c)$ is stochastically larger than its reflection about 0, which equals in this case $P*(x \to x+c)$). Since $T(P,\mathfrak{P}_0) = \{g \in \mathscr{L}_*(P): g(x) = g(-x)\}$, functions $g \in T(P,\mathfrak{P})$ orthogonal to $T(P,\mathfrak{P}_0)$ are skew-symmetric about 0, so that (3.4.3) reduces to

$$(3.4.4) \qquad \int_{-\infty}^{x} g(\xi)p(\xi)d\xi \leq 0 \qquad \text{for all } x \in \mathbb{R}.$$

Because gp is skew-symmetric about 0, we have

$$\int_{-\infty}^{x} g(\xi)p(\xi)d\xi = \int_{-\infty}^{-x} g(\xi)p(\xi)d\xi ,$$

so that it suffices to require condition (3.4.4) for all $x < 0$.

The reader interested in other concepts of positive biasedness (as opposed to symmetry about zero) may consult Yanagimoto and Sibuya (1972b).

3.5. Monotone failure rates

For any p-measure $P|\mathbb{B} \cap (0,\infty)$ with Lebesgue density p, the *failure rate* r at x is defined as

$$(3.5.1) \qquad r(x) := p(x)/\int_{x}^{\infty} p(\xi)d\xi .$$

P has *nondecreasing failure rate* if the function r is nondecreasing. A particular case are exponential distributions P_λ, $\lambda > 0$, with Lebesgue density $p(x,\lambda) = \lambda \exp[-\lambda x]$, $x > 0$, which have constant failure rate λ.

For testing the hypothesis that the true p-measure is exponential, alternatives with nondecreasing failure rate are natural in certain applications. Let $\mathfrak{P}$ denote the class of all p-measures with nondecreasing failure rate, $\mathfrak{P}_o$ the family of exponential distributions.

To simplify our notations, we write P for P_1. (It suffices to consider the particular case $\lambda = 1$, since any condition on $g \in \mathscr{L}_*(P_1)$ corresponds to an equivalent condition on the function $x \to g(\lambda x)$ in $\mathscr{L}_*(P_\lambda)$.) Let $P_{t,g}$, $t \downarrow 0$, denote a differentiable path converging to P with derivative $g \in \mathscr{L}_*(P)$. If $P_{t,g}$ has nondecreasing failure rate for all sufficiently small $t > 0$, then

$$(3.5.2) \qquad g(x) + e^x \int_0^x g(\xi) e^{-\xi} d\xi \qquad \text{is nondecreasing in } x.$$

Hence for $P \in \mathfrak{P}_o$, $T(P,\mathfrak{P})$ consists of all functions $g \in \mathscr{L}_*(P)$ fulfilling (3.5.2). It is straightforward to see that $T(P,\mathfrak{P})$ is a convex cone which is n o t a half-space. On its boundary, this cone contains the functions $g(x) = c(1-x)$, for which (3.5.2) equals c. This is the direction in which the exponential family with constant failure rates extends.

Another widely used condition is that of *nondecreasing average failure rates*, i.e. that

$$\int_0^x r(\xi) d\xi / x \quad \text{is a nondecreasing function of } x.$$

The reader interested in such conditions on the failure rate is referred to Barlow et al. (1972) and the references cited there.

3.6. Positive dependence

Let $P|\mathbb{B}^2$ be the distribution of (x,y). In testing the independence of (x,y), it is often natural to restrict attention to the alternative of positive (or negative) dependence. We shall discuss two definitions of positive dependence. To simplify our notations, we assume that P has a Lebesgue density, say p. Let $p_1(x) := \int p(x,\eta) d\eta$, $p_2(y) := \int p(\xi,y) d\xi$.

Positive regression dependence:

$$(3.6.1)\qquad \int_y^\infty p(x,\eta)d\eta/p_1(x) \quad \text{is nondecreasing in } x \text{ for every } y\in\mathbb{R}.$$

Positive quadrant dependence:

$$(3.6.2)\qquad \int_x^\infty\int_y^\infty p(\xi,\eta)d\xi d\eta \ge \int_x^\infty p_1(\xi)d\xi\int_y^\infty p_2(\eta)d\eta \qquad \text{for all } x,y\in\mathbb{R}.$$

It is well known (Lehmann, 1966, p. 1143) that regression dependence is a property stronger than quadrant dependence.

Let $\mathfrak{P}_o$ denote the family of all product measures, and $\mathfrak{P}_1$ and $\mathfrak{P}_2$ the families of all p-measures with positive regression dependence and quadrant dependence, respectively. We have $\mathfrak{P}_o\subset\mathfrak{P}_1\subset\mathfrak{P}_2$.

According to Proposition 2.4.1,

$$T(P_1\times P_2,\mathfrak{P}_o) = \{(x,y)\to g_1(x)+g_2(y):\ g_i\in\mathscr{L}_*(P_i),\ i=1,2\}.$$

To obtain convenient expressions for the tangent spaces $T(P_1\times P_2,\mathfrak{P}_i)$, we use the representation of $g\in\mathscr{L}_*(P_1\times P_2)$ by

$$(3.6.3)\qquad g(x,y) = g_1(x)+g_2(y)+\bar{g}(x,y)$$

with

$$g_1(x) = \int g(x,\eta)P_2(d\eta)\ ,$$

$$g_2(y) = \int g(\xi,y)P_1(d\xi)\ .$$

Notice that $\bar{g}$, thus defined, is orthogonal to $T(P_1\times P_2,\mathfrak{P}_o)$ and fulfills $\int\bar{g}(x,\eta)P_2(d\eta) = \int\bar{g}(\xi,y)P_1(d\xi) = 0$ for all $x,y\in\mathbb{R}$. With this representation, we obtain the following expressions for the tangent cones:

$$(3.6.4)\qquad T(P_1\times P_2,\mathfrak{P}_1) = \{g\in\mathscr{L}_*(P_1\times P_2):\ x\to\int_y^\infty\bar{g}(x,\eta)P_2(d\eta) \text{ is nondecreasing for all } y\in\mathbb{R}\},$$

$$(3.6.5)\qquad T(P_1\times P_2,\mathfrak{P}_2) = \{g\in\mathscr{L}_*(P_1\times P_2):\ \int_x^\infty\int_y^\infty\bar{g}(\xi,\eta)P_1(d\xi)P_2(d\eta)\ge 0 \text{ for all } x,y\in\mathbb{R}\}.$$

It is straightforward to see that $T(P_1\times P_2,\mathfrak{P}_i)$ is a convex cone. Since positive regression dependence implies positive quadrant dependence, we have $\mathfrak{P}_1\subset\mathfrak{P}_2$ and therefore $T(P_1\times P_2,\mathfrak{P}_1)\subset T(P_1\times P_2,\mathfrak{P}_2)$. This relation can also be seen directly from the representation given by (3.6.4) and (3.6.5).

(Hint: $x \to \int_y^\infty \bar{g}(x,\eta) P_2(d\eta)$ nondecreasing

implies

$$x \to \int_x^\infty \int_y^\infty \bar{g}(\xi,\eta) P_1(d\xi) P_2(d\eta) / \int_x^\infty P_1(d\xi) \quad \text{nondecreasing.}$$

Applied for x and x = $-\infty$ this yields

$$\int_x^\infty \int_y^\infty \bar{g}(\xi,\eta) P_1(d\xi) P_2(d\eta) \geq 0. \;)$$

For the purpose of illustration, consider the family $\mathfrak{Q}$ of p-measures generated from independent variables by random nonnegatively correlated normally distributed disturbances, i.e. the family with λ^2-densities

$$(3.6.6) \qquad (x,y) \to \int p_1(x-\xi) p_2(y-\eta) N(0,\Sigma)(d(\xi,\eta)),$$

where the correlation coefficient of $N(0,\Sigma)$ is nonnegative. We have (see Section 2.4)

$$T(P_1 \times P_2, \mathfrak{Q}) = \{(x,y) \to g_1(x) + g_2(y) + c\ell'(x,P_1)\ell'(y,P_2): \\ c \geq 0,\ g_i \in \mathscr{L}_*(P_i),\ i = 1,2\}.$$

Hence $\bar{g}(x,y) = c\ell'(x,P_1)\ell'(y,P_2)$. According to (3.6.4), p-measures near $P_1 \times P_2$ have approximate positive regression dependence for $c > 0$ if

$$x \to \ell'(x,P_1) \int_y^\infty \ell'(\eta,P_2) p_2(\eta) d\eta = -\ell'(x,P_1) p_2(y) \text{ is nondecreasing.}$$

This requires that $\ell'(\cdot,P_1)$ be nonincreasing. The measures will be approximately quadrant dependent without further assumption, since

$$\int_x^\infty \int_y^\infty \ell'(\xi,P_1)\ell'(\eta,P_2) p_1(\xi) p_2(\eta) d\xi d\eta = p_1(x) p_2(y) > 0.$$

For further concepts of positive dependence see also Yanagimoto (1972).

4. DIFFERENTIABLE FUNCTIONALS

4.1. The gradient of a functional

Let $\kappa: \mathfrak{P} \to \mathbb{R}$ be a functional. For asymptotic theory, the *local properties* of this functional are essential, i.e., its behavior in contiguous neighborhoods. These local properties determine how good optimal tests and estimators can be asymptotically. The mathematical construct suitable for this purpose is the gradient.

4.1.1. Definition. A function $\kappa^{\cdot}(\cdot,P) \in \mathcal{L}_*(P)$ is a *gradient* of κ at P for $\mathfrak{P}$ if for every $g \in T(P,\mathfrak{P})$ and every path $P_{t,g}$, $t \downarrow 0$, in $\mathfrak{P}$ with derivative g,

$$\kappa(P_{t,g}) - \kappa(P) = tP(\kappa^{\cdot}(\cdot,P)g) + o(t) .$$

If κ admits a gradient at P we call κ *differentiable* at P.

The existence of a gradient implies that the function $t \to \kappa(P_{t,g})$ is differentiable in $t = 0$, its derivative being $P(\kappa^{\cdot}(\cdot,P)g)$. For fixed g, the derivative of $t \to \kappa(P_{t,g})$ does not depend on the particular path $P_{t,g}$, $t \downarrow 0$. The remainder term $o(t)$ for the functional depends, however, on the particular path. It need not even be uniform over a class of paths with uniformly vanishing remainder terms.

Considered as a function of g, the derivative is linear and continuous under $\| \ \|$-convergence (i.e., $\|g_n - g\| \to 0$ implies $P(\kappa^{\cdot}(\cdot,P)g_n) \to P(\kappa^{\cdot}(\cdot,P)g)$).

Another possible approach to the gradient is through directional derivatives, say $\hat{\kappa}(g,P)$, defined as derivatives of $t \to \kappa(P_{t,g})$ at $t=0$. If these directional derivatives, considered as a functional on the Hilbert space $T(P,\mathfrak{P})$, are linear and bounded, there exists $\kappa^{\bullet}(\cdot,P) \in T(P,\mathfrak{P})$ such that $\hat{\kappa}(g,P) = P(\kappa^{\bullet}(\cdot,P)g)$.

For certain applications we need that the error term $o(t)$ in Definition 4.1.1 holds *uniformly* over all $g \in T(P,\mathfrak{P})$ with $\|g\|=1$, say, in the sense that for every g there exists a path such that the error term is uniform over this class of paths. This uniform version is automatically fulfilled if we start from the following stronger concept of a gradient, based on a distance function δ.

4.1.2. Definition. A function $\kappa^{\bullet}(\cdot,P) \in \mathscr{L}_*(P)$ is a *strong gradient* of κ at P for $\mathfrak{P}$ if

$$\kappa(Q) - \kappa(P) = Q(\kappa^{\bullet}(\cdot,P)) + o(\delta(Q,P)) .$$

The definition $\kappa(Q) - \kappa(P) = \int \kappa^{\bullet}(\xi,P)(Q-P)(d\xi) + o(\delta(Q,P))$ may look more familiar to some readers, but is the same, since $\kappa^{\bullet}(\cdot,P) \in \mathscr{L}_*(P)$ requires $P(\kappa^{\bullet}(\cdot,P)) = 0$.

Definition 4.1.2 has the following obvious interpretation: If we approximate $\kappa(Q)$ by $\kappa(P) + Q(\kappa^{\bullet}(\cdot,P))$, then the error of this approximation tends to zero faster than the distance $\delta(Q,P)$.

4.1.3. Remark. Let $\kappa^{\bullet}(\cdot,P)$ be a strong gradient with $\delta(Q,P) := \Delta(Q;P)$ in the remainder term. Then $\kappa^{\bullet}(\cdot,P)$ is a gradient for the tangent cone $T_s(P,\mathfrak{P})$.

Proof. Let $P_{t,g}$, $t \downarrow 0$, be a path in $\mathfrak{P}$ with P-density $1 + t(g + r_t)$ and $\|r_t\| = o(t^0)$. Then

$$\Delta(P_{t,g};P) = \left\| \frac{p_{t,g}}{p} - 1 \right\| = t\|g+r_t\| = O(t) .$$

Hence

$$\begin{aligned}\kappa(P_{t,g}) - \kappa(P) &= P_{t,g}(\kappa^{\bullet}(\cdot,P)) + o(\Delta(P_{t,g};P)) \\ &= tP(\kappa^{\bullet}(\cdot,P)(g+r_t)) + o(t) \\ &= tP(\kappa^{\bullet}(\cdot,P)g) + o(t).\end{aligned}$$

<u>4.1.4. Remark</u>. Distance functions δ which approximate Δ in the sense of Definition 6.2.1, fulfill $t^{-1}\delta(P_{t,g},P) \to a_o$ as $t \downarrow 0$, where a_o does not depend on $g \in T(P,\mathfrak{P})$ if $\|g\| = 1$. Hence, for such directions g, the distance $\delta(P_{t,g},P)$ is approximately proportional to t. By Definition 4.1.1, $P(\kappa^{\bullet}(\cdot,P)g)$ measures in this case the rate of change of κ, in relation to the distance, if the p-measure moves away from P in the direction g. If $P(\kappa^{\bullet}(\cdot,P)g)$ is particularly large for a certain g, then a difference between $\kappa(P_{t,g})$ and $\kappa(P)$ will be particularly difficult to detect, since then the corresponding distance $\delta(P_{t,g},P)$, which is essential for the discrimination power of tests, is particularly small. The direction g_o for which $g \to P(\kappa^{\bullet}(\cdot,P)g)$ becomes maximal (subject to the condition $\|g\| = 1$) is, therefore, least favorable. In Sections 8.1 and 9.2 it will be shown that these least favorable directions play a distinguished role in connection with optimality of statistical procedures. The idea of considering a one-dimensional parametric family P_{t,g_o}, $t \in \mathbb{R}$, which passes through the true p-measure P and for which every estimation problem is asymptotically at least as difficult as for any other parametric family passing through P, was already used by Stein (1956).

<u>4.1.5. Remark</u>. It may happen that the derivative of every path vanishes at P. This is the case, e.g., if the functional has an extremum or a saddlepoint at P. Such functionals occur in connection with testing hypotheses consisting of a single p-measure P. Then the gradients at P are orthogonal to the tangent space, and the canonical gradient vanishes. In order to find least favorable directions it is then necessary

to consider higher order derivatives of paths at P. This aspect will be pursued elsewhere.

Numerous examples of differentiable functionals will be discussed in Chapter 5. We conclude this section with two examples of functionals which are not differentiable.

4.1.6. Example. Let $\mathfrak{P}$ be the family of all p-measures $Q|\mathbb{B}$ with positive and continuous Lebesgue density. Then $T(P,\mathfrak{P}) = \mathscr{L}_*(P)$ by Example 2.1.4. Let $\kappa(Q)$ be the value of the density of Q at 0. Let $P_{t,g}$ be any p-measure with P-density $1+t(g+r_t)$. Then $t^{-1}(\kappa(P_{t,g})-\kappa(P)) = p(0)g(0) + p(0)r_t(0)$. Hence the limit of $t^{-1}(\kappa(P_{t,g})-\kappa(P))$ for $t \downarrow 0$ will not exist for arbitrary differentiable paths $P_{t,g}$, $t \downarrow 0$. If we could solve the difficulty with the remainder term $p(0)r_t(0)$ by restricting $\mathfrak{P}$ somehow, then the derivative in direction g would be $p(0)g(0)$. This derivative would be linear but unbounded and therefore not representable by a gradient.

4.1.7. Example. Let $\mathfrak{P}$ be the family of all p-measures $Q|\mathbb{B}$ with twice differentiable and strongly unimodal Lebesgue density. Let $\kappa(Q)$ be the - unique - mode of Q. The mode of a path $P_{t,g}$ with P-density $1+t(g+r_t)$ is the solution in x of

$$\frac{d}{dx}\, p(x)\,(1+t(g(x)+r_t(x))) = 0 .$$

Again, there are difficulties with the remainder r_t. If these can be solved by appropriate restrictions on $\mathfrak{P}$, then we obtain as derivative of κ in direction g the value

$$-g(\kappa(P))\,\frac{p(\kappa(P))}{p''(\kappa(P))} .$$

Considered as a function of g, this derivative is linear, but unbounded, and is, therefore, not representable by a gradient.

4.2. Projections into convex sets

In this section we collect a few well-known auxiliary results concerning projections into closed convex sets.

Let H be a real Hilbert space with inner product (,) and norm $\| \ \|$.

4.2.1. Proposition. *(i) Let* $C \subset H$ *be a closed convex set. Then for any* $a \in H$ *there is a unique* $\bar{a} \in C$ *such that*

$$\|a-\bar{a}\| = \inf\{\|a-c\| : c \in C\}.$$

$\bar{a}$ *is called the projection of* a *into* C. *It is uniquely determined by the relation*

$$(a,\bar{a}-c) \geq (\bar{a},\bar{a}-c) \qquad \text{for all } c \in C.$$

(ii) If C *is a closed convex cone, then the projection* $\bar{a}$ *of* a *is uniquely determined by the relations*

$$(a,\bar{a}) = (\bar{a},\bar{a})$$

and $$(a,c) \leq (\bar{a},c) \qquad \text{for all } c \in C.$$

In particular, $(\bar{a},\bar{a}) \leq (a,a)$.

(iii) If C *is a closed subspace, then the projection* $\bar{a}$ *of* a *is uniquely determined by*

$$(a,c) = (\bar{a},c) \qquad \text{for all } c \in C.$$

Proof. See Barlow et al. (1972, p. 314, Theorem 7.2, and p. 315, Theorem 7.3, for (i); p. 318, Theorem 7.8, for (ii)).

(iii) follows by an application of (ii) for -c.

4.2.2. Proposition. *Assume that* $C \subset B \subset H$*, where* C *is a closed convex set and* B *a linear space. For any* $a \in H$ *let* $\bar{a}$ *denote the projection into* B*, and* a^+ *the projection of* $\bar{a}$ *into* C*. Then* a^+ *is the projection of* a *into* C*.*

Proof. By Proposition 4.2.1(i),

$$(\bar{a}, a^+ - c) \geq (a^+, a^+ - c) \qquad \text{for all } c \in C.$$

Since $a^+ - c \in B$, Proposition 4.2.1(iii) implies $(a, a^+ - c) = (\bar{a}, a^+ - c)$.

Hence $$(a, a^+ - c) \geq (a^+, a^+ - c) \qquad \text{for all } c \in C.$$

4.2.3. Proposition. *For given linearly independent* $b_i \in H$, $i = 1, \ldots, m$*, let* B *be the linear space spanned by* $b_1, \ldots, b_m$*. Then the projection of* $a \in H$ *into* B *is*

$$\bar{a} = D_{ij}(a, b_j) b_i \, ,$$

where the matrix D *is the inverse of* $((b_i, b_j))_{i,j=1,\ldots,m}$.

Proof. Immediate consequence of Proposition 4.2.1(iii).

4.2.4. Proposition. *For given* $b_i \in H$, $i = 1, \ldots, m$*, let* $B^{\perp} = \{a \in H: (a, b_i) = 0,\ i = 1, \ldots, m\}$*. Then the projection* $\bar{a}$ *of* $a \in H$ *into* $B^{\perp}$ *is the regression residual of* a *with respect to* $b_1, \ldots, b_m$*, i.e.,*

$$\bar{a} = a - D_{ij}(a, b_j) b_i \, .$$

Proof. We have $\bar{a} \in B^{\perp}$, and $a - \bar{a}$ is a linear combination of b_i and therefore orthogonal to $B^{\perp}$. Hence the assertion follows from Proposition 4.2.1(iii).

4.2.5. Remark. Let H be written as an orthogonal sum $B \oplus B^{\perp}$. Propositions 4.2.3 and 4.2.4 imply that every $a \in H$ can be represented as the orthogonal sum of its projections into B and $B^{\perp}$.

4.2.6. Proposition. *Let* $B \subset H$ *be a closed subspace. For* $i = 1, \ldots, m$ *let* $a_i \in H$ *and denote by* $\bar{a}_i$ *the projection of* a_i *into* B*.*

Then $((a_i, a_j) - (\bar{a}_i, \bar{a}_j))_{i,j=1,\ldots,m}$ *is nonnegative definite.*

Proof. By Proposition 4.2.1(iii) we have $(a_i, \bar{a}_j) = (\bar{a}_i, \bar{a}_j)$ for $i,j = 1, \dots, m$. Hence

$$0 \leq \|u_i(a_i - \bar{a}_i)\|^2 = u_i u_j (a_i - \bar{a}_i, a_j - \bar{a}_j) = u_i u_j ((a_i, a_j) - (\bar{a}_i, \bar{a}_j)).$$

4.3. The canonical gradient

Let $\mathfrak{P}$ be a family of p-measures. Write $C(P,\mathfrak{P})$ for the tangent cone of $\mathfrak{P}$ at P, and let $\kappa: \mathfrak{P} \to \mathbb{R}$ be a functional which is differentiable at P. Then the gradient $\kappa^\cdot(\cdot,P)$ determines a linear functional $g \to P(\kappa^\cdot(\cdot,P)g)$ on the P-square closure $\bar{C}(P,\mathfrak{P})$ of the tangent cone. This linear functional remains unchanged if we add to $\kappa^\cdot(\cdot,P)$ a function orthogonal to the tangent cone. Hence the gradient is not unique.

4.3.1. Example. Let $\mathfrak{P} = \{N(\mu,1): \mu \in \mathbb{R}\}$ and $\kappa(N(\mu,1)) := \mu$. It is easy to see that

$$\kappa^\cdot(x,\mu) := f(x-\mu)/\int \xi f(\xi) N(0,1)(d\xi)$$

is a gradient if f is an odd function which fulfills appropriate integrability conditions. In fact, such a gradient is even a strong gradient, i.e.,

$$\kappa(N(\mu,1)) - \kappa(N(\mu_o,1)) - \int \kappa^\cdot(\xi,\mu_o) N(\mu,1)(d\xi) = o(\delta(N(\mu,1),N(\mu_o,1))).$$

This equality follows easily from

$$\int \kappa^\cdot(\xi,\mu_o) N(\mu,1)(d\xi) = \int \kappa^\cdot(\xi,\mu_o)(N(\mu,1) - N(\mu_o,1))(d\xi)$$
$$= \mu - \mu_o + o(|\mu-\mu_o|)$$

and

$$\delta(N(\mu,1),N(\mu_o,1)) = O(|\mu-\mu_o|)\ .$$

Let $B(P,\mathfrak{P})$ denote the smallest closed linear space containing $C(P,\mathfrak{P})$. The following is an immediate consequence of the results on projections presented in Section 4.2.

<u>4.3.2. Proposition</u>. *(i) Let* $\kappa^{\cdot}(\cdot,P)$ *be a gradient. Then the class of all gradients for* $\mathfrak{P}$ *is*

$$\{f \in \mathscr{L}_*(P): P(fg) = P(\kappa^{\cdot}(\cdot,P)g) \quad \textit{for all } g \in C(P,\mathfrak{P})\}.$$

(ii) Among the gradients for $\mathfrak{P}$ *there is one and only one gradient which belongs to* $B(P,\mathfrak{P})$, *say* $\kappa^*(\cdot,P)$. *This gradient will be called the* <u>*canonical gradient*</u>. $\kappa^*(\cdot,P)$ *is the projection of any gradient* $\kappa^{\cdot}(\cdot,P)$ *into* $B(P,\mathfrak{P})$. *It is uniquely determined in* $B(P,\mathfrak{P})$ *by the relation*

$$(4.3.3) \qquad P(\kappa^*(\cdot,P)g) = P(\kappa^{\cdot}(\cdot,P)g) \qquad \textit{for all } g \in C(P,\mathfrak{P}).$$

(iii) If $\kappa^{\cdot}(\cdot,P) \in \mathscr{L}_*(P)$ *and* $\kappa^*(\cdot,P) \in B(P,\mathfrak{P})$ *are gradients for* $\mathfrak{P}$, *then*

$$(4.3.4) \qquad P(\kappa^*(\cdot,P)^2) \le P(\kappa^{\cdot}(\cdot,P)^2).$$

<u>Proof</u>. (i) is an immediate consequence of the definition. (ii) and (iii) follow from Proposition 4.2.1. Observe that by linearity and continuity (4.3.3) must hold for all $g \in B(P,\mathfrak{P})$.

<u>4.3.5. Proposition</u>. *All gradients* $\kappa^{\cdot}(\cdot,P)$ *for* $\mathfrak{P}$ *have the same projection* $\kappa^+(\cdot,P)$ *into* $\bar{C}(P,\mathfrak{P})$. $\kappa^+(\cdot,P)$ *is uniquely determined by*

$$P(\kappa^{\cdot}(\cdot,P)\kappa^+(\cdot,P)) = P(\kappa^+(\cdot,P)^2)$$

and

$$P(\kappa^{\cdot}(\cdot,P)g) \le P(\kappa^+(\cdot,P)g) \qquad \textit{for all } g \in C(P,\mathfrak{P}).$$

Furthermore,

$$P(\kappa^+(\cdot,P)^2)^{1/2} = \sup\{P(g^2)^{-1/2}P(\kappa^{\cdot}(\cdot,P)g): g \in C(P,\mathfrak{P})\}.$$

<u>Proof</u>. Proposition 4.2.1(i), (ii).

<u>4.3.6. Remark</u>. If the projection $\kappa^+(\cdot,P)$ of a gradient $\kappa^{\cdot}(\cdot,P)$ into

$B(P,\mathfrak{P})$ falls into $\bar{C}(P,\mathfrak{P})$, it coincides with the projection of $\kappa^{\bullet}(\cdot,P)$ into $\bar{C}(P,\mathfrak{P})$.

4.4. Multidimensional functionals

The concept of a gradient introduced in Section 4.1 for one-dimensional functionals can be applied to the components κ_i, $i=1,\dots,k$, of any k-dimensional functional $\kappa: \mathfrak{P} \to \mathbb{R}^k$ with $\kappa = (\kappa_1,\dots,\kappa_k)$.

The following proposition corresponds to Koshevnik and Levit (1976, p. 744, Theorem 1).

<u>4.4.1. Proposition</u>. *If* $\kappa_i^{\bullet}(\cdot,P)$ *is an arbitrary gradient and* $\kappa_i^{*}(\cdot,P)$ *the canonical gradient of* κ_i *at* P *for* $\mathfrak{P}$, $i=1,\dots,k$, *then the matrix*

$$(P(\kappa_i^{\bullet}(\cdot,P)\kappa_j^{\bullet}(\cdot,P)) - P(\kappa_i^{*}(\cdot,P)\kappa_j^{*}(\cdot,P)))_{i,j=1,\dots,k}$$

is nonnegative definite.

<u>Proof</u>. Proposition 4.2.6.

<u>4.4.2. Proposition</u>. *If* $\kappa_i^{\bullet}(\cdot,P)$ *is a gradient of* κ_i *at* P *for* $\mathfrak{P}$, $i=1,\dots,k$, *and if* $K: \mathbb{R}^k \to \mathbb{R}$ *is differentiable with continuous partial derivatives, then the functional* $K\circ\kappa$ *(defined by* $P \to K(\kappa_1(P),\dots,\kappa_k(P))$*) has at* P *the gradient* $K^{(i)}(\kappa(P))\kappa_i^{\bullet}(\cdot,P)$.

<u>Addendum</u>. *If* $\kappa_i^{*}(\cdot,P)$ *is the* <u>*canonical*</u> *gradient of* κ_i *at* P *for* $\mathfrak{P}$, $i=1,\dots,k$, *then* $K^{(i)}(\kappa(P))\kappa_i^{*}(\cdot,P)$ *is an element of* $T(P,\mathfrak{P})$ *and, therefore, the* <u>*canonical*</u> *gradient of* $K\circ\kappa$ *at* P.

<u>Proof</u>. Let P_t, $t \downarrow 0$, be a path in $\mathfrak{P}$ with derivative g. Since κ_i is differentiable,

$$\kappa_i(P_{t,g}) - \kappa_i(P) = tP(\kappa_i^{\bullet}(\cdot,P)g) + o(t) .$$

Using the continuity of $K^{(i)}$, we obtain

$$K(\kappa(P_{t,g})) - K(\kappa(P))$$

$$= (\kappa_i(P_{t,g}) - \kappa_i(P)) \int_0^1 K^{(i)}(\kappa(P) + u(\kappa(P_{t,g}) - \kappa(P)))du$$

$$= tP(K^{(i)}(\kappa(P))\kappa_i^{\bullet}(\cdot,P)g) + o(t) .$$

<u>4.4.3. Remark</u>. It may happen that $K^{(i)}(\kappa(P))\kappa_i^{\bullet}(\cdot,P)$ is canonical even though the gradients $\kappa_i^{\bullet}(\cdot,P)$, $i = 1,\ldots,k$, are not canonical. To see this, let $\mathfrak{P}$ be a location parameter family, say $\mathfrak{P} = \{P_o*(x \to x+\theta): \theta \in \mathbb{R}\}$, where P_o is symmetric about 0. Let $\kappa_\beta(P)$ be the β-quantile of P. Then

$$(4.4.4) \qquad \kappa^{\bullet}(\cdot,P_o) = -\ell'(\cdot,P_o)/P_o(\ell'(\cdot,P_o)^2)$$
$$+ [2\beta - 1 + 1_{(-\infty,\kappa_{1-\beta}(P_o))} - 1_{(-\infty,\kappa_\beta(P_o))}]/2p_o(\kappa_\beta(P_o))$$

is a gradient of κ_β at P_o for $\mathfrak{P}$, but not a canonical one. $T(P_o,\mathfrak{P})$ is the linear space spanned by $\ell'(\cdot,P_o)$, and $\kappa_\beta^{\bullet}(\cdot,P_o)$ contains an additive term orthogonal to $T(P_o,\mathfrak{P})$. (See (15.1.3).)

From (4.4.4) we obtain for the functional $P \to \frac{1}{2}\kappa_{1-\beta}(P) + \frac{1}{2}\kappa_\beta(P)$ according to Proposition 4.4.2 (applied for $\kappa_1 = \kappa_{1-\beta}$, $\kappa_2 = \kappa_\beta$ and $K(t_1,t_2) = \frac{1}{2}t_1 + \frac{1}{2}t_2$) the gradient

$$\tfrac{1}{2}\kappa_{1-\beta}^{\bullet}(\cdot,P_o) + \tfrac{1}{2}\kappa_\beta^{\bullet}(\cdot,P_o) = -\ell'(\cdot,P_o)/P_o(\ell'(\cdot,P_o)^2) .$$

This gradient belongs to $T(P_o,\mathfrak{P})$ and is, therefore, canonical.

<u>4.4.5. Remark</u>. We restrict ourselves in this monograph to finite-dimensional functionals. For further applications, we draw the attention of the reader to functionals attaining their values in a function space, say $\kappa(P) = (\kappa_t(P))_{t\in\mathbb{R}}$.

Examples of such functionals are the following:

(i) for $P|\mathbb{B}$ the *distribution function*

$$\kappa_t(P) = P(-\infty,t] , \quad t \in \mathbb{R};$$

(ii) for $P|\mathbb{B} \cap [0,\infty)$ the *failure rate*

$$\kappa_t(P) = p(t)/P[t,\infty), \qquad t \geq 0;$$

or the *mean residual life time*,

$$\kappa_t(P) = \int_t^\infty (u-t)P(du)/P[t,\infty), \qquad t \geq 0;$$

(iii) for $P|\mathbb{B} \cap [0,\infty)$ the *Lorenz curve*

$$\kappa_t(P) = \int_0^t F^{-1}(u)du/\int_0^1 F^{-1}(u)du, \qquad 0 \leq t \leq 1,$$

where F is the distribution function of P;

(iv) For $P|\mathbb{B}^2$ the *regression function* of y on x

$$\kappa_x(P) = \int yp(x,y)dy/\int p(x,y)dy, \qquad x \in \mathbb{R}.$$

4.5. Tangent spaces and gradients under side conditions

4.5.1. Proposition. *Let* $\kappa: \mathfrak{P} \to \mathbb{R}$ *be a functional which is differentiable at* P *with gradient* $\kappa^\bullet(\cdot,P)$. *Let* $\mathfrak{P}_o := \{Q \in \mathfrak{P}: \kappa(Q) = \kappa(P)\}$.

Then $T(P,\mathfrak{P}_o) \subset \{g \in T(P,\mathfrak{P}): P(\kappa^\bullet(\cdot,P)g) = 0\}$, *with equality under suitable uniformity conditions on the remainder terms of the paths.*

Proof. We prove only the first part of the assertion. Let $g \in T(P,\mathfrak{P}_o)$. There exists a path $P_{t,g}$, $t \downarrow 0$, in $\mathfrak{P}_o$ with derivative g. Since the path is in $\mathfrak{P}_o$ and κ is differentiable at P, we have

$$0 = \kappa(P_{t,g}) - \kappa(P) = tP(\kappa^\bullet(\cdot,P)g) + o(t) .$$

Hence $P(\kappa^\bullet(\cdot,P)g) = 0$.

4.5.2. Proposition. *For* $i = 0,1$ *let* $\kappa_i: \mathfrak{P} \to \mathbb{R}$ *be functionals which are differentiable at* $P \in \mathfrak{P}_o = \{Q \in \mathfrak{P}: \kappa_o(Q) = c_o\}$ *with canonical gradients* $\kappa_i^*(\cdot,P)$ *for* $\mathfrak{P}$. *Assume that* $T(P,\mathfrak{P}_o) = \{g \in T(P,\mathfrak{P}): P(\kappa_o^*(\cdot,P)g) = 0\}$.

Then the canonical gradient of κ_1 *at* P *for* $\mathfrak{P}_o$ *is*

$$\kappa_1^+(\cdot,P) = \kappa_1^*(\cdot,P) - P(\kappa_o^*(\cdot,P)^2)^{-1/2}P(\kappa_1^*(\cdot,P)\kappa_o^*(\cdot,P))\kappa_o^*(\cdot,P).$$

Proof. Proposition 4.2.4.

We remark that $P(\kappa_1^+(\cdot,P)^2) \leq P(\kappa_1^*(\cdot,P)^2)$, and that this inequality is strict unless $P(\kappa_1(\cdot,P)\kappa_o(\cdot,P)) = 0$. Functionals κ_o,κ_1 with orthogonal canonical gradients will be called *unrelated*. For applications of this concept see Examples 8.6.11, 9.5.1 and 13.2.3.

4.6. Historical remarks

For statistical purposes, derivatives of a functional have been introduced by von Mises (1936, 1938, 1947, 1952) and used by many authors since. Almost all authors (a notable exception being Stein, 1956) confine themselves to the case that $\mathfrak{P}$ is the family of all p-measures, so that $T(P,\mathfrak{P}) = \mathscr{L}_*(P)$. In this case, the gradient is uniquely determined. Our approach deals with the more general case of a functional κ defined on some smaller - perhaps even parametric - family of p-measures and, consequently, with tangent spaces $T(P,\mathfrak{P})$ which are genuine subspaces of $\mathscr{L}_*(P)$. It is this generalization which opens the possibility of a g e n e r a l asymptotic theory. A paper closely related to our approach is Koshevnik and Levit (1976).

In considerations about robustness, the gradient runs under the name 'influence curve', with a different interpretation. Let δ_x denote the p-measure with mass 1 in $\{x\}$ (i.e., $\delta_x(B) = 1_B(x)$). Assuming that the functional κ is defined for all p-measures, hence in particular for the p-measure $(1-t)P + t\delta_x$, the influence curve of κ at P and x is defined as $\lim_{t\to 0} t^{-1}(\kappa((1-t)P+t\delta_x)-\kappa(P))$. If we assume that the relation

$$\kappa(Q) - \kappa(P) = \int\kappa^{\cdot}(x,P)Q(dx) + o(\delta(P,Q))$$

also holds true for $Q = (1-t)P + t\delta_x$, and that the distance δ is such that $\delta(P,(1-t)P + t\delta_x) = O(t)$, then

$$\kappa((1-t)P + t\delta_x) - \kappa(P) = \int \kappa^{\bullet}(\cdot,P)\,d((1-t)P + t\delta_x) + o(t)$$
$$= t\kappa^{\bullet}(x,P) + o(t) ,$$

hence

$$\lim_{t\to 0} t^{-1}(\kappa((1-t)P + t\delta_x) - \kappa(P)) = \kappa^{\bullet}(x,P) .$$

This leads to the following interpretation: $t\kappa^{\bullet}(x,P)$ approximates the change of the functional if we diminish the mass of P by fraction t and concentrate this mass in the point x.

5. EXAMPLES OF DIFFERENTIABLE FUNCTIONALS

5.1. Von Mises functionals

Let $(X,\mathscr{A})$ be a measurable space and $k: X^m \to \mathbb{R}$ a measurable function. For a p-measure $P|\mathscr{A}$ the *von Mises functional* is defined by

$$\kappa(P) := \int k(x_1,\dots,x_m)P(dx_1)\dots P(dx_m).$$

Many widely used functionals are of this type:

5.1.1. *The expectation*: $P|\mathbb{B}$, $m=1$, $k(x)=x$

5.1.2. *Measures of dispersion:* $P|\mathbb{B}$, $m=2$, $k(x_1,x_2)=\frac{1}{2}k_o(x_1-x_2)$ with k_o symmetric about 0. With $k_o(x) = x^2$ we obtain the *variance*, with $k_o(x) = 2|x|$ Gini's *mean difference*.

5.1.3. *Measures of dependence*: $P|\mathbb{B}^2$, $m=2$, $k((x_1,y_1),(x_2,y_2)) = \frac{1}{2}k_o(x_1-x_2,y_1-y_2)$, with $k_o(-u,v) = -k_o(u,v)$, $k_o(u,-v) = -k_o(u,v)$, and $k_o(u,v) > 0$ if $u>0$ and $v>0$. The interpretation of

$$\kappa(P) := \int k((x_1,y_1),(x_2,y_2))P(d(x_1,y_1))P(d(x_2,y_2))$$

as a measure of dependence is supported by $\kappa(P_1 \times P_2) = 0$ (i.e., the functional attains the values 0 for independent variables). This is an immediate consequence of the fact that $P_1^2*((x_1,x_2) \to (x_1-x_2))$ is symmetric about 0.

In most applications, $k_o(u,v) = k_1(u)k_1(v)$ with $k_1(-u) = -k_1(u)$. For $k_1(u) = u$ we obtain the *covariance*, for $k_1(u) = \sqrt{2}$ sign u the *quadrant correlation coefficient* (≡ Kendall's difference sign covariance).

5.1.4. *The expected life time*: Consider an aggregate consisting of r different components. The aggregate fails if one of the components fails. Let $P|\mathbb{B}_+^r$ (with $\mathbb{B}_+ := \mathbb{B} \cap [0,\infty)$) denote the joint distribution of the life times of r components. Then

$$\kappa(P) = \int \min\{x_1,\dots,x_r\} P(d(x_1,\dots,x_r))$$

is the *expected life time* of the aggregate. This is a von Mises functional with $m = 1$ and $k(x_1,\dots,x_r) = \min\{x_1,\dots,x_r\}$.

W.l.g. we may assume that k is symmetric in its arguments (for otherwise we may replace k by the function $(x_1,\dots,x_m) \to \frac{1}{m!} \Sigma\, k(x_{i_1},\dots,x_{i_m})$, where the summation extends over all permutations $(i_1,\dots,i_m)$ of $(1,\dots,m)$).

The following result occurs already in von Mises (1936).

5.1.5. Proposition. *If* $\int k(x_1,\dots,x_m)^2 \times_{i=1}^{m} P(dx_i) < \infty$, *then the von Mises functional is defined in a* Δ*-neighborhood of* P *and has at* P *the strong gradient*

$$\kappa^{\cdot}(x,P) = m\Big(\int k(x,x_2,\dots,x_m) \times_{i=2}^{m} P(dx_i) - \kappa(P)\Big),$$

with a remainder $o(\Delta(Q;P)^2)$.

Proof. We have

$$\int k(x_1,\dots,x_m) \times_{i=1}^{m} Q(dx_i) = \int k(x_1,\dots,x_m) \prod_{j=1}^{m} \frac{q(x_j)}{p(x_j)} \times_{i=1}^{m} P(dx_i)$$

$$= \int k(x_1,\dots,x_m) \prod_{j=1}^{m} \Big(1 + \frac{q(x_j)}{p(x_j)} - 1\Big) \times_{i=1}^{m} P(dx_i)$$

$$= \int k(x_1,\dots,x_m) \times_{i=1}^{m} P(dx_i) + m\int\Big(\int k(x,x_2,\dots,x_m) \times_{i=2}^{m} P(dx_i) - \kappa(P)\Big)Q(dx)$$

$$+ \sum_{r=2}^{m} \binom{m}{r} \int k(x_1,\dots,x_m) \prod_{j=1}^{r} \Big(\frac{q(x_j)}{p(x_j)} - 1\Big) \times_{i=1}^{m} P(dx_i)\,.$$

By Hölder's inequality,

$$\left|\int k(x_1,\ldots,x_m) \prod_{j=1}^{r} \left(\frac{q(x_j)}{p(x_j)} - 1\right) \times_{i=1}^{m} P(dx_i)\right|$$

$$\leq \left(\int k(x_1,\ldots,x_m)^2 \times_{i=1}^{m} P(dx_i)\right)^{1/2} \Delta(Q;P)^r .$$

Hence

$$\int k(x_1,\ldots,x_m) \times_{i=1}^{m} Q(dx_i) = \int k(x_1,\ldots,x_m) \times_{i=1}^{m} P(dx_i) + Q(\kappa^{\cdot}(\cdot,P)) + O(\Delta(Q;P)^2).$$

<u>5.1.6. Example</u>. For $k(x_1,x_2) = \frac{1}{2}(x_1-x_2)$ we obtain the functional

$$\kappa(P) = \int (\xi-\mu(P))^2 P(d\xi)$$

and the gradient

$$\kappa^{\cdot}(x,P) = (x-\mu(P))^2 - \kappa(P), \quad \text{with } \mu(P) := \int \xi P(d\xi).$$

In this case, the gradient depends on P only through $\mu(P)$ and $\kappa(P)$.

5.2. Minimum contrast functionals

Let $(X,\mathscr{A})$ be a measurable space and $f: X \times \mathbb{R}^k \to \mathbb{R}$ a measurable function. For a p-measure $P|\mathscr{A}$ define the functional $\kappa(P) = (\kappa_1(P),\ldots,\kappa_k(P))$ as the solution in t of

(5.2.1) $\quad \int f(x,t)P(dx) = \min .$

The function f will be called *contrast function*, κ a *minimum contrast functional*. To avoid technicalities, we assume that the family of p-measures is such that κ is uniquely defined by (5.2.1).

Under suitable regularity conditions on f, (5.2.1) is equivalent to

(5.2.2) $\quad \int f^{(i)}(x,\kappa(P))P(dx) = 0 \qquad$ for $i = 1,\ldots,k$.

Relation (5.2.2) may be meaningful even in cases where the integral (5.2.1) does not exist. Defining the median by the contrast functional $f(x,t) = |x-t|$ presumes the existence of the first moment, whereas the condition $P(-\infty,\kappa(P)] = P[\kappa(P),\infty)$, corresponding to (5.2.2), does not.

A version of the following proposition is already proved in Koshevnik and Levit (1976, p. 747, Theorem 4). Moreover, the gradient occurs as the influence curve of M-estimators in a number of papers; see, e.g., Huber (1972, p. 1048).

5.2.3. Proposition. *Let* κ *be a minimum contrast functional in the sense of (5.2.2) which fulfills a local Lipschitz condition at* P *with respect to the sup-distance* V. *Assume that the contrast function* f *has second derivatives which fulfill a local Lipschitz condition*

$$|f^{(ij)}(x,\kappa(Q)) - f^{(ij)}(x,\kappa(P))| \leq |\kappa(Q) - \kappa(P)|M(x)$$

with $Q(M)$ *bounded for* Q *in a* V*-neighborhood of* P, *and that the matrix-valued function* $F(Q) := Q(f^{(\cdot\cdot)}(\cdot,Q))$ *is nonsingular and* V*-continuous at* P. *Then* κ *has at* P *the strong gradient*

$$\kappa^{\cdot}(\cdot,P) := -F(P)^{-1}f^{(\cdot)}(\cdot,\kappa(P))$$

with a remainder $o(V(Q,P))$.

Proof. By a Taylor expansion,

$$\begin{aligned} 0 &= Q(f^{(i)}(\cdot,\kappa(Q))) \\ &= Q(f^{(i)}(\cdot,\kappa(P))) \\ &\quad + (\kappa_j(Q) - \kappa_j(P))\int_0^1 Q(f^{(ij)}(\cdot,(1-u)\kappa(P) + u\kappa(Q)))du . \end{aligned}$$

By the Lipschitz condition on $f^{(ij)}$,

$$|f^{(ij)}(x,(1-u)\kappa(P) + u\kappa(Q)) - f^{(ij)}(x,\kappa(P))|$$

$$\leq |\kappa(Q) - \kappa(P)|M(x) .$$

Hence

$$|Q(f^{(ij)}(\cdot,(1-u)\kappa(P)+u\kappa(Q)))-F_{ij}(P)|$$
$$\leq |\kappa(Q)-\kappa(P)|Q(M)+|F_{ij}(Q)-F_{ij}(P)|$$
$$= o(V(Q,P)^{o}) .$$

Therefore, with the Lipschitz condition for κ,

$$Q(f^{(\cdot)}(\cdot,\kappa(P)))+(\kappa(Q)-\kappa(P))'F(P) = o(V(Q,P)) .$$

Multiplication by $F(P)^{-1}$ yields

$$\kappa(Q)-\kappa(P) = -F(P)^{-1}Q(f^{(\cdot)}(\cdot,\kappa(P)))+o(V(Q,P)) ,$$

the assertion.

5.2.4. Remark. Of particular interest are contrast functions $f(x,t) = f_o(x-t)$ with $f_o\colon \mathbb{R}^k \to \mathbb{R}$. The pertaining minimum contrast functionals are location functionals, i.e., $\kappa(P*(x\to x+c)) = \kappa(P)+c$. In this case, if P admits a Lebesgue density p, then (5.2.2) implies identically in c

$$\int f_o^{(i)}(x-\kappa(P)+c)p(x+c)dx = 0, \qquad i=1,\dots,k,$$

hence

$$\int f_o^{(ij)}(x-\kappa(P))P(dx) = -\int f_o^{(i)}(x-\kappa(P))\ell^{(j)}(x,P)P(dx), \quad i,j=1,\dots,k,$$

so that

$$F_{ij}(P) = -\int f_o^{(i)}(x-\kappa(P))\ell^{(j)}(x,P)P(dx), \qquad i,j=1,\dots,k.$$

If f_o is bowl-shaped, symmetric about 0 and bounded, and if $P|\mathbb{B}^k$ admits a symmetric and unimodal Lebesgue density, then the minimum contrast functional $\kappa(P)$ is the center of symmetry of P. This follows easily from Anderson (1955, p. 170, Theorem 1).

Further examples of minimum contrast functionals can be found in Levit (1975, pp. 724f.) and Pfanzagl (1979, pp. 108f.).

5.3. Parameters

Let $P_\theta|\mathscr{A}$, $\theta \in \Theta$, be a parametrized family of p-measures with $\Theta \subset \mathbb{R}^k$, dominated by some σ-finite measure $\mu|\mathscr{A}$. Let $\kappa(P_\theta) := \theta$. This definition presumes that $P_\theta \neq P_\tau$ for $\theta \neq \tau$.

Whereas most other functionals are defined for all p-measures fulfilling certain regularity conditions, this functional is defined for a given family only.

To obtain the gradient of this functional, we remark that it may be considered as the minimum contrast functional pertaining to the contrast function $f(x,\theta) = -\log p(x,\theta)$, where $p(\cdot,\theta)$ is a μ-density of P_θ. Recall that in this case $t \to \int f(x,t)P_\theta(dx)$ has a unique minimum for $t = \theta$, because $Q \ll P$, $Q \neq P$, implies $P(\log q) < \log P(q) = 0$ for $q \in dQ/dP$, so that $-P_\theta(\log p(\cdot,\theta)) < -P_\theta(\log p(\cdot,t))$ for all $t \neq \theta$.

A similar proof as for Proposition 5.2.3, working this time with paths $P_{\theta+ta}$, $t \downarrow 0$, $a \in \mathbb{R}^k$, yields the following proposition.

5.3.1. Proposition. *Assume that at an inner point θ of Θ the second logarithmic derivatives of the density fulfill a Lipschitz condition*

$$|\ell^{(ij)}(x,\tau) - \ell^{(ij)}(x,\theta)| \leq |\tau-\theta| M(x)$$

with $P_\tau(M)$ bounded for $|\tau-\theta| < \varepsilon$, and that $P_\tau(\ell^{(\cdot\cdot)}(\cdot,\tau))$ equals $-L(\tau)$ and is nonsingular and continuous at θ. Then the functional $\kappa(P_\tau) := \tau$ has at θ the strong gradient

$$\kappa^{\boldsymbol{\cdot}}(x,P_\theta) = \Lambda(\theta)\ell^{(\cdot)}(x,\theta) \tag{5.3.2}$$

with a remainder $o(|\tau-\theta|)$.

<u>5.3.3. Remark</u>. Since the same family of p-measures can be parametrized in different ways, it remains to check whether the gradient remains unchanged under a change of the parametrization. To see this, let $\theta = g(\tau)$ for some sufficiently regular function $g: T \to \Theta$, $T \subset \mathbb{R}^k$. Let $\bar{P}_\tau := P_{g(\tau)}$. Then $\kappa(\bar{P}_\tau) := g(\tau)$ is the same functional as above. Hence invariance under change of parametrization requires (by (5.3.2))

$$\kappa^{\bullet}(\cdot,\bar{P}_\tau) = \Lambda(g(\tau))\ell^{(\cdot)}(\cdot,g(\tau)) \tag{5.3.4}$$

(where $\ell^{(\cdot)}$ and Λ have exactly the same meaning as above, i.e., they refer to the family $\{P_\theta: \theta \in \Theta\}$). To check whether this is, in fact, true, let $\bar{\ell}(\cdot,\tau) := \ell(\cdot,g(\tau))$. Furthermore, let $\bar{\kappa}(\bar{P}_\tau) = \tau$. Then (5.3.2), applied for the family $\{\bar{P}_\tau: \tau \in T\}$ and the functional $\bar{\kappa}$, yields

$$\bar{\kappa}^{\bullet}(\cdot,\bar{P}_\tau) = \bar{\Lambda}(\tau)\bar{\ell}^{(\cdot)}(\cdot,\tau) . \tag{5.3.5}$$

Let G denote the $k \times k$-matrix with elements

$$G_{jr} := \frac{\partial}{\partial \tau_j} g_r(\tau), \qquad j,r = 1,\dots,k.$$

(Since τ remains fixed, we suppress it in the notation for G.)

From

$$\bar{\ell}^{(\cdot)}(\cdot,\tau) = G\ell^{(\cdot)}(\cdot,g(\tau)) \tag{5.3.6}$$

we obtain

$$\bar{L}(\tau) = GL(g(\tau))G' ,$$

so that

$$\bar{\Lambda}(\tau) = (G')^{-1}\Lambda(g(\tau))G^{-1} . \tag{5.3.7}$$

From (5.3.5), (5.3.6) and (5.3.7),

$$\bar{\kappa}^{\bullet}(\cdot,\bar{P}_\tau) = (G')^{-1}\Lambda(g(\tau))\ell^{(\cdot)}(\cdot,g(\tau)) . \tag{5.3.8}$$

Since $\kappa(\bar{P}_\tau) = g(\bar{\kappa}(\bar{P}_\tau))$, we have (see Proposition 4.4.2),

$$\kappa^{\bullet}(\cdot,\bar{P}_\tau) = G'\bar{\kappa}^{\bullet}(\cdot,\bar{P}_\tau).$$

Together with (5.3.8) this implies (5.3.4).

5.4. Quantiles

Let $\beta \in (0,1)$ be fixed. For any p-measure $P|\mathbb{B}$ admitting a Lebesgue density let $\kappa_\beta(P)$ denote the β-quantile of P, defined by $P(-\infty,\kappa_\beta(P))=\beta$. To avoid technicalities we assume that $\kappa_\beta(P)$ is unique.

The β-quantile may also be considered as a minimum contrast functional, the pertaining contrast function being $f(x,t)=(x-t)(\beta-1_{(-\infty,t)}(x))$. Equivalently, $\kappa_\beta(P)$ is the solution in t of

$$\int(\beta - 1_{(-\infty,t)}(x))P(dx) = 0 . \tag{5.4.1}$$

Despite the fact that κ_β is a minimum contrast functional, the results of Proposition 5.2.3 are not applicable, since the regularity conditions of the proposition (differentiability of f, etc.) are not fulfilled in this case.

5.4.2. Proposition. *If* P *admits a Lebesgue density, say* p, *which is positive and continuous at* $\kappa_\beta(P)$, *then* κ_β *has at* P *the strong gradient*

$$\kappa_\beta^{\bullet}(\cdot,P) := p(\kappa_\beta(P))^{-1}(\beta - 1_{(-\infty,\kappa_\beta(P))})$$

with a remainder $o(\Delta(Q;P))$.

Proof. We have

$$|F_Q(x) - F_P(x)| = |Q(-\infty,x) - P(-\infty,x)|$$

$$\leq P(|\tfrac{q}{p} - 1|) \leq \Delta(Q;P) .$$

Since p is positive and continuous at $\kappa_\beta(P)$, the derivative of F_P is bounded from below in a neighborhood of $\kappa_\beta(P)$. Hence

$$\kappa_\beta(Q) - \kappa_\beta(P) = F_Q^{-1}(\beta) - F_P^{-1}(\beta) = O(\Delta(Q;P)) .$$

For notational convenience, we assume that $\kappa_\beta(Q) \geq \kappa_\beta(P)$. Since p is continuous in $\kappa(P)$, we obtain

$$P(\kappa_\beta(P),\kappa_\beta(Q)) = p(\kappa_\beta(P))(\kappa_\beta(Q) - \kappa_\beta(P)) + o(\kappa_\beta(Q) - \kappa_\beta(P))$$
$$= p(\kappa_\beta(P))(\kappa_\beta(Q) - \kappa_\beta(P)) + o(\Delta(Q;P)) \ .$$

From Hölder's inequality,

$$\beta - Q(-\infty,\kappa_\beta(P)) = Q(\kappa_\beta(P),\kappa_\beta(Q))$$
$$= P(\kappa_\beta(P),\kappa_\beta(Q)) + P((\tfrac{q}{p} - 1)1_{(\kappa_\beta(P),\kappa_\beta(Q))})$$
$$= P(\kappa_\beta(P),\kappa_\beta(Q)) + O(\Delta(Q;P)^{3/2}) \ .$$

Hence

$$(5.4.3) \qquad \kappa_\beta(Q) - \kappa_\beta(P) = p(\kappa_\beta(P))^{-1}P(\kappa_\beta(P),\kappa_\beta(Q)) + o(\Delta(Q;P))$$
$$= p(\kappa_\beta(P))^{-1}Q(\beta - 1_{(-\infty,\kappa_\beta(P))}) + o(\Delta(Q;P)) \ .$$

5.5. A location functional

Let K be a p-measure on the Borel algebra of $(0,1)$. For any p-measure $P|\mathbb{B}$ let F_P denote its distribution function. To avoid technicalities, we assume that P has a positive Lebesgue density, so that F_P is increasing everywhere. Hence F_P^{-1} is uniquely defined.

The functional κ is defined by

$$(5.5.1) \qquad \kappa(P) = \int_0^1 F_P^{-1}(u)K(du) \ .$$

The α-*trimmed mean* is obtained from $K(B) := (1-2\alpha)^{-1}\lambda(B \cap [\alpha,1-\alpha])$, the α-*Winsorized mean* from $K(B) := \lambda(B \cap [\alpha,1-\alpha]) + \alpha(1_B(\alpha) + 1_B(1-\alpha))$, the α-*quantile* from $K(B) := 1_B(\alpha)$.

If K has a Lebesgue density on $(0,1)$, say k, an alternative form of (5.5.1) is

(5.5.2) $\kappa(P) = \int \xi k(F_P(\xi))P(d\xi)$.

To see this, observe that

$$\kappa(P) = \int_0^1 F_P^{-1}(u)K(du) = \int xK*F_P^{-1}(dx) = \int xk(F_P(x))P(dx),$$

since $k \in dK/dE$ implies $k \circ F_P \in \frac{dK*F_P^{-1}}{dE*F_P^{-1}} = \frac{dK*F_P^{-1}}{dP}$, where E is the Lebesgue measure on (0,1).

Functionals of this type have been considered by Huber (1972) and Bickel and Lehmann (1975 b).

According to Bickel and Lehmann (1975b, p. 1052), κ may be interpreted as a location parameter. This interpretation is supported by the relation $\kappa(P*(x \to x+c)) = \kappa(P) + c$. Moreover, if K is symmetric about $\frac{1}{2}$, then $\kappa(P)$ is the center of symmetry for any symmetric p-measure P.

In fact, $\kappa(P)$, as defined by (5.5.1), is nothing else but the K-mixture of the quantiles of P. Hence a gradient of κ can formally be obtained as the K-mixture of the gradients of the quantiles given in Proposition 5.4.2, i.e.,

$$\int \kappa_u^{\bullet}(x,P)K(du) = \int p(F_P^{-1}(u))^{-1}(u - 1_{(-\infty,F_P^{-1}(u))}(x))K(du) .$$

By a change of the integration variable from u to $F_P(\xi)$ we obtain that the right side equals

$$\int (F_P(\xi) - 1_{(-\infty,\xi)}(x))k(F_P(\xi))d\xi$$
$$= \int F_P(\xi)k(F_P(\xi))d\xi - \int_x^\infty k(F_P(\xi))d\xi .$$

There is but one objection against this shortcut: The gradient of a K-mixture can be obtained as the K-mixture of the gradients only if the K-mixture of vanishing remainder terms vanishes. There is some freedom to place stronger regularity conditions on the measure K or to restrict the family of p-measures for which the gradient is to hold. To obtain a gradient valid for a rather general family, we assume that K has a compact support.

<u>5.5.3. Proposition</u>. *Assume that* $P|\mathbb{B}$ *admits a positive and continuous density with respect to the Lebesgue measure. If* K *has a Lebesgue density* k *vanishing outside an interval* $[\varepsilon,1-\varepsilon]$, *then the functional* $\kappa(P) := \int_0^1 F_P^{-1}(u)K(du)$ *has at* P *the strong gradient*

$$\kappa^{\cdot}(x,P) = \int F_P(\xi)k(F_P(\xi))d\xi - \int_x^\infty k(F_P(\xi))d\xi$$

with a remainder $o(\Delta(Q;P))$.

<u>Proof</u>. We have $|F_Q - F_P| \leq \Delta(Q;P)$. Let $\delta \in (0,\varepsilon)$. Since p is positive and continuous, it is bounded away from 0 on $F_P^{-1}[\delta,1-\delta]$. Hence uniformly for $u \in [\varepsilon,1-\varepsilon]$

$$F_Q^{-1}(u) - F_P^{-1}(u) = O(\Delta(Q;P)) .$$

Since p is uniformly continuous on $F_P^{-1}[\delta,1-\delta]$, we obtain from (5.4.3) (with $\kappa_\beta(Q)$ replaced by $F_P^{-1}(u)$) that uniformly for $u \in [\varepsilon,1-\varepsilon]$

$$\begin{aligned} u - Q(-\infty,F_Q^{-1}(u)) &= Q(F_P^{-1}(u),F_Q^{-1}(u)) \\ &= P(F_P^{-1}(u),F_Q^{-1}(u)) + P((q/p-1)1_{(F_P^{-1}(u),F_Q^{-1}(u))}) \\ &= P(F_P^{-1}(u),F_Q^{-1}(u)) + o(\Delta(Q;P)^{3/2}) . \end{aligned}$$

Hence uniformly for $u \in [\varepsilon,1-\varepsilon]$

$$\begin{aligned} F_Q^{-1}(u) - F_P^{-1}(u) &= p(F_P^{-1}(u))^{-1} P(F_P^{-1}(u),F_Q^{-1}(u)) + o(\Delta(Q;P)) \\ &= p(F_P^{-1}(u))^{-1}(u - Q(-\infty,F_P^{-1}(u))) + o(\Delta(Q;P)) . \end{aligned}$$

Integrating with respect to K, and using $k \circ F_P \in dK * F_P^{-1}/dP$, we obtain

$$\begin{aligned} \kappa(Q) - \kappa(P) &= \int (F_Q^{-1}(u) - F_P^{-1}(u))du \\ &= \int p(F_P^{-1}(u))^{-1}(u - Q(-\infty,F_P^{-1}(u))K(du) + o(\Delta(Q;P)) \\ &= \int F_P(\xi)k(F_P(\xi))d\xi - \int_x^\infty k(F_P(\xi))d\xi + o(\Delta(Q;P)) , \end{aligned}$$

the assertion.

5.5.4. Remark. The gradient given in Proposition 5.5.3 can be found in literature in various places. In Huber (1972, p. 1049), the gradient is written as

$$\Omega_F(x) = U(F_P(x)) - \int_0^1 U(t)\,dt$$

with

$$U(t) = \int_0^t p(F_P^{-1}(u))^{-1} k(u)\,du .$$

It is easy to see that $\Omega_F(x) = \kappa^{\bullet}(x,P)$.

According to Boos (1979, p. 956, Theorem 1),

$$\kappa(Q) - \kappa(P) = -\int k(F_P(x))(F_Q(x) - F_P(x))\,dx + o(\|F_Q - F_P\|_\infty). \tag{5.5.5}$$

Since

$$\int k(F_P(x))F_Q(x)\,dx = \int\int_x^\infty k(F_P(\xi))\,d\xi\,Q(dx) ,$$

relation (5.5.5) implies

$$\kappa(Q) - \kappa(P) = \int(\int_x^\infty k(F_P(\xi)\,d\xi)(Q-P)(dx) + o(\|F_Q - F_P\|_\infty) ,$$

in accordance with Proposition 5.5.3.

6. DISTANCE FUNCTIONS FOR PROBABILITY MEASURES

6.1. Some distance functions

Let $\mu|\mathscr{A}$ be a σ-finite measure and P, Q, P_1, P_2 p-measures with μ-densities p, q, p_1, p_2.

The *variational distance* (or *sup-distance*) is defined as

(6.1.1) $$V(Q,P) := \sup\{|Q(A) - P(A)| : A \in \mathscr{A}\}$$
$$= \tfrac{1}{2}\mu(|q-p|) = \tfrac{1}{2}P(|q/p - 1|) .$$

With Φ denoting the class of critical functions on $(X,\mathscr{A})$, we have

$$V(Q,P) = \sup\{|Q(\varphi) - P(\varphi)| : \varphi \in \Phi\}$$
$$= \sup\{Q(\varphi) - P(\varphi) : \varphi \in \Phi\}$$
$$= 1 - \inf\{Q(\varphi) + P(1-\varphi) : \varphi \in \Phi\} .$$

This provides us with another interpretation of the variational distance: If we use the test φ for testing Q against P, then $Q(\varphi)+P(1-\varphi)$ is the sum of the errors of the first and second kind, and $1 - V(Q,P)$ is the best possible value of this sum.

The *Hellinger distance* is defined as

(6.1.2) $$H(Q,P) := 2(\mu((q^{1/2} - p^{1/2})^2))^{1/2}$$
$$= 8^{1/2}(1 - \mu((qp)^{1/2}))^{1/2} = 8^{1/2}(1-P((q/p)^{1/2}))^{1/2} .$$

Usually, the Hellinger distance is defined without the factor 2. This factor is introduced here to make as. relations as in Proposition 6.2.2

and Theorem 8.4.1 simpler.

Moreover, we shall use

(6.1.3) $\Delta(Q_1,Q_2;P) := Q(((q_1-q_2)/p)^2)^{1/2}$.

Considered as a function of Q_1,Q_2, this is a distance, which turns out to be useful as a distance for p-measures 'near' P. To measure the distance of Q from P, we use

(6.14) $\Delta(Q;P)^2 := \Delta(Q,P;P)^2 = P((q/p-1)^2)$

$= P((q/p)^2) - 1$

$= Q(q/p) - 1$.

This is the distance measure introduced by K.Pearson (1900a,b). See also Lancaster (1969, Chapter VI).

A number of relations between these distances is known, so for instance the obvious relation

(6.1.5) $V(Q,P) \leq \frac{1}{2}\Delta(Q;P)$

and (see LeCam, 1970, p. 803)

(6.1.6) $\frac{1}{8}H(Q,P)^2 \leq V(Q,P) \leq \frac{1}{2}H(Q,P)(1-\frac{1}{16}H(Q,P)^2)^{1/2}$.

6.2. Asymptotic relations between distance functions

For our purposes, general relations between distance functions like (6.1.5) and (6.1.6) are of limited interest. What we need for an asymptotic theory are asymptotic relations between the different distance functions as Q and P approach a given p-measure.

The distance measure $(Q,P) \to \Delta(Q;P)$ appears in a natural way in connection with likelihood ratios. It is, however, neither symmetric nor does it fulfill the triangle inequality. The problem is to find distance f u n c t i o n s δ which are as. equivalent to Δ.

The following definition tries to give this a technically useful meaning.

<u>6.2.1. Definition</u>. A distance function δ on $\mathfrak{P}$ *approximates* Δ *at* P if uniformly for $Q_1, Q_2 \in \mathfrak{P}$

$$\delta(Q_1,Q_2)^2 = \Delta(Q_1,Q_2;P)^2 + o(\Delta(Q_1;P)^2) + o(\Delta(Q_2;P)^2).$$

Under suitable regularity conditions, this is, for instance, true for the Hellinger distance. Notice, however, that some widely used distances based on distribution functions, like the Cramér - von Mises and Kolmogorov-Smirnov distances, are not of this type (see Example 7.6.6 and Remark 15.2.8).

For a general theory, the condition $\Delta(Q;P) < \infty$ is, perhaps, too strong. In many cases, it is, however, possible to extract from $q/p - 1$ a square-integrable part g such that the remainder r, defined by $q/p - 1 = g+r$, is negligible in some technical sense. If $P((q/p-1)^2) = \infty$, this is, of course, always possible. The problem is to choose g such that $P(g^2)$ is as small as possible, and yet r is negligible compared with g. Below we formulate a set of conditions on g and r which tries to make this intuitive idea precise in a technically useful sense.

<u>6.2.2. Proposition</u>. *Let* N_i, $i = 0,1,2$, *be null-functions. Let* $P \in \mathfrak{P}$, *and for* $i = 1,2$ *let* $Q_i \in \mathfrak{P}$ *admit* P-*densities* $1 + g_i + r_i$ *such that*

(6.2.3) $\quad P(g_i^2 1_{\{|g_i| > c\|g_i\|\}}) \le \|g_i\|^2 N_o(c^{-1})$ *for all* $c > 0$,

(6.2.4) $\quad P(|r_i| 1_{\{|r_i| > 1\}}) \le \|g_i\|^2 N_1(\|g_i\|)$,

(6.2.5) $\quad P(r_i^2 1_{\{|r_i| \le 1\}}) \le \|g_i\|^2 N_2(\|g_i\|)$.

Then there exists a null-function N *such that*

$$|H(Q_1,Q_2)^2 - \|g_1-g_2\|^2| \le \|g_1\|^2 N(\|g_1\|) + \|g_2\|^2 N(\|g_2\|).$$

If (6.2.3) holds for $g_i = q_i/p - 1$, *then the assumptions hold with* $r_i = 0$, *and we obtain*

$$|H(Q_1,Q_2)^2 - \Delta(Q_1,Q_2;P)^2| \le \Delta(Q_1;P)^2 N(\Delta(Q_1;P)) + \Delta(Q_2;P)^2 N(\Delta(Q_2;P)) .$$

<u>Proof</u>. (i) In this proof, N denotes a generic null-function. Define

$$A_i := \{|g_i| \le N(\|g_i\|)\}, \qquad B_i := \{|r_i| \le 1\}.$$

By (6.2.3) and Lemma 19.1.1,

(6.2.6) $\quad P(A_i^c) \le \|g_i\|^2 N(\|g_i\|)$,

(6.2.7) $\quad P(|g_i| 1_{A_i^c}) \le \|g_i\|^2 N(\|g_i\|)$,

(6.2.8) $\quad P(g_i^2 1_{A_i^c}) \le \|g_i\|^2 N(\|g_i\|)$.

From (6.2.4) and (6.2.5) we obtain

(6.2.9) $\quad P(B_i^c) \le \|g_i\|^2 N(\|g_i\|)$,

(6.2.10) $\quad P(|r_i| 1_{B_i^c}) \le \|g_i\|^2 N(\|g_i\|)$,

(6.2.11) $\quad P(r_i^2 1_{B_i}) \le \|g_i\|^2 N(\|g_i\|)$.

We have

$$H(Q_1,Q_2)^2 = 8(1 - P(((1+g_1+r_1)(1+g_2+r_2))^{1/2})) .$$

On $D := A_1B_1A_2B_2$ we obtain an expansion

(6.2.12) $$((1+g_1+r_1)(1+g_2+r_2))^{1/2}$$
$$= 1 - \frac{1}{8}(g_1-g_2)^2 + \frac{1}{2}(g_1+r_1+g_2+r_2) + \frac{1}{2}(g_1r_2+g_2r_1+r_1r_2)$$
$$- \frac{1}{4}(g_1+g_2)(r_1+r_2+(g_1+r_1)(g_2+r_2))$$
$$- \frac{1}{8}(r_1+r_2+(g_1+r_1)(g_2+r_2))^2 + r$$

with

$$|r| \le \frac{1}{2}(g_1+r_1+g_2+r_2+(g_1+r_1)(g_2+r_2))^3 .$$

The assertion now follows by integrating (6.2.12) with respect to P and by applying the following relations to the remainder terms.

(ii) We have $A_j^c \cup B_j^c \subset A_i^c \cup A_i A_j^c \cup A_i B_j^c$. Hence by (6.2.7), (6.2.6) and (6.2.9),

(6.2.13) $$P(|g_i| 1_{A_j^c \cup B_j^c}) \leq \|g_1\|^2 N(\|g_1\|) + \|g_2\|^2 N(\|g_2\|) .$$

By (6.2.8), (6.2.6) and (6.2.9),

(6.2.14) $$P(g_i^2 1_{A_j^c \cup B_j^c}) \leq \|g_1\|^2 N(\|g_1\|) + \|g_2\|^2 N(\|g_2\|) .$$

Using $A_j^c \cup B_j^c \subset B_i^c \cup B_i A_j^c \cup B_i B_j^c$ we obtain from (6.2.10), (6.2.6) and (6.2.9)

(6.2.15) $$P(|r_i| 1_{A_j^c \cup B_j^c}) \leq \|g_1\|^2 N(\|g_1\|) + \|g_2\|^2 N(\|g_2\|) .$$

(iii) From (6.2.13) and (6.2.15), together with (6.2.6) and (6.2.9), we obtain

$$P(((1+g_1+r_1)(1+g_2+r_2))^{1/2} 1_{D^c})$$
$$\leq (P((1+g_1+r_1)1_{D^c}))^{1/2} (P((1+g_2+r_2)1_{D^c}))^{1/2}$$
$$\leq \|g_1\|^2 N(\|g_1\|) + \|g_2\|^2 N(\|g_2\|) .$$

From (6.2.14) we obtain

$$P((g_1-g_2)^2 1_{D^c}) \leq \|g_1\|^2 N(\|g_1\|) + \|g_2\|^2 N(\|g_2\|) .$$

(iv) It remains to show that we can neglect the P-integrals over D of the remaining terms in (6.2.12). From (6.2.13) and (6.2.15), together with $P(g_i) = P(r_i) = 0$, we obtain

$$P(g_i 1_D) \leq \|g_1\|^2 N(\|g_1\|) + \|g_2\|^2 N(\|g_2\|) ,$$

$$P(r_i 1_D) \leq \|g_1\|^2 N(\|g_1\|) + \|g_2\|^2 N(\|g_2\|) .$$

Furthermore, we apply (6.2.11) to

$$P(|g_i r_j| 1_D) \leq \|g_i\| \; \|r_j 1_{B_j}\| ,$$

$$P(|r_i r_j| 1_D) \leq \|r_i 1_{B_i}\| \; \|r_j 1_{B_j}\| .$$

For similar terms of higher order observe that on D the functions g_i are bounded by $N(\|g_i\|)$, and r_i by 1.

6.2.16. Remark. Let us now consider the case $Q_2 = P$. Writing Q for Q_1 and g for g_1 we obtain from Proposition 6.2.2 the existence of a null-function N such that for all p-measures Q fulfilling conditions (6.2.3) - (6.2.5),

(6.2.17) $\quad |H(Q,P)^2 - \|g\|^2| \leq \|g\|^2 N(\|g\|)$.

The interpretation: The decomposition of $q/p - 1$ into the sum $g + r$ with g and r fulfilling conditions (6.2.3) - (6.2.5) is to a certain extent arbitrary. But for all such decompositions leading to a function g with $\|g\|$ small, this amount agrees closely with the Hellinger distance $H(Q,P)$. This explains why the apparently arbitrary component g occurs through $\|g\|$ in results like Remark 8.4.5 on the slope of the as. envelope power function.

6.2.18. Proposition. *Assume that* $\overline{\mathfrak{P}}$ *is at* P *approximable by* $T_s(P,\overline{\mathfrak{P}})$, *and that* $\mathfrak{P}$ *is at* P *approximable by* $T_s(P,\mathfrak{P})$, *i.e. every* $Q \in \mathfrak{P}$ *admits a* P-*density* $1 + g + r$ *with* $g \in T_s(P,\mathfrak{P})$ *and* $\|r\| = o(\Delta(Q;P))$ *uniformly for* $Q \in \mathfrak{P}$. *Assume that* δ *approximates* Δ *at* P. *Then*

(6.2.19) $\quad \delta(Q,\overline{\mathfrak{P}})^2 \geq \|g-\bar{g}\|^2 + o(\Delta(Q;P)^2)$,

where $\bar{g}$ *denotes the projection of* g *into* $T_s(P,\overline{\mathfrak{P}})$.

Addendum. *If* $\overline{\mathfrak{P}}$ *fulfills (1.1.11), then equality holds in (6.2.19).*

Proof. (i) We have to show that uniformly for $P' \in \overline{\mathfrak{P}}$,

(6.2.20) $\quad \delta(Q,P')^2 \geq \|g-\bar{g}\|^2 + o(\Delta(Q;P)^2)$.

Since $\mathfrak{P}$ is at P approximable by $T_s(P,\mathfrak{P})$, every $Q \in \mathfrak{P}$ admits a P-density $1 + g + r$ with $g \in T_s(P,\mathfrak{P})$ and $\|r\| = o(\Delta(Q;P))$, uniformly for $Q \in \mathfrak{P}$. Since $\overline{\mathfrak{P}}$ is at P approximable by $T_s(P,\overline{\mathfrak{P}})$, every $P' \in \overline{\mathfrak{P}}$ admits a P-density $1 + g' + r'$ with $g' \in T(P,\overline{\mathfrak{P}})$ and $\|r'\| = o(\Delta(P';P))$. Since δ

approximates Δ at P and $\|g\| \geq \|g-\bar{g}\|$, we obtain uniformly for $Q \in \mathfrak{P}$

$$\delta(Q,P)^2 = \|g\|^2 + o(\Delta(Q;P)^2)$$
$$\geq \|g-\bar{g}\|^2 + o(\Delta(Q;P)^2) .$$

Hence (6.2.20) holds for $P' \in \bar{\mathfrak{P}}$ with $\delta(Q,P') \geq \delta(Q,P)$. If $\delta(Q,P') \leq \delta(Q,P)$, then $\delta(P',P) \leq 2\delta(Q,P)$. Since δ approximates Δ at P, we obtain uniformly for $Q \in \mathfrak{P}$

$$\delta(Q,P') = \Delta(Q,P';P)^2 + o(\Delta(Q;P)^2)$$
$$= \|g-g'\|^2 + o(\Delta(Q;P)^2)$$
$$\leq \|g-\bar{g}\|^2 + o(\Delta(Q;P)^2) .$$

(ii) Since $\bar{g} \in T_s(P,\bar{\mathfrak{P}})$, condition (1.1.10) implies the existence of $P^* \in \bar{\mathfrak{P}}$ with P-density $1+\bar{g}+r^*$ such that $\|r^*\| = o(\|\bar{g}\|) = o(\Delta(Q;P))$. This holds uniformly for $\bar{g} \in T(P,\bar{\mathfrak{P}})$ and, therefore, uniformly for $Q \in \mathfrak{P}$. Since δ approximates Δ at P and $\Delta(P^*;P) = O(\Delta(Q;P))$, we obtain uniformly for $Q \in \mathfrak{P}$

$$\delta(Q,P^*)^2 = \Delta(Q,P^*;P)^2 + o(\Delta(Q;P)^2) + o(\Delta(P^*;P)^2)$$
$$= \|g-\bar{g}\|^2 + o(\Delta(Q;P)^2) .$$

6.3. Distances in parametric families

For parametric families, an expression for the distance between p-measures in terms of the parameters will be useful.

<u>6.3.1. Proposition</u>. Let $\mathfrak{P} = \{P_\theta : \theta \in \Theta\}$, $\Theta \subset \mathbb{R}^k$. *Assume that the densities admit continuous partial derivatives such that for* $i = 1,\ldots,k$ *and for* $|\tau-\theta| < \varepsilon$, *say,*

$$|p^{(i)}(x,\tau) - p^{(i)}(x,\theta)| \leq |\tau-\theta| p(x,\theta) M(x,\theta) ,$$

where $M(\cdot,\theta)$ *and* $\ell^{(i)}(\cdot,\theta)$ *are* P_θ-square integrable and $L(\theta) := P_\theta(\ell^{(\cdot)}(\cdot,\theta)\ell^{(\cdot)}(\cdot,\theta)')$ *is nonsingular. Then the following relation holds uniformly for* $\tau', \tau'' \in \Theta$:

$$\Delta(P_{\tau'}, P_{\tau''}; P_\theta) = ((\tau'-\tau'')'L(\theta)(\tau'-\tau''))^{1/2} + |\tau'-\tau''|\,o(|\tau'-\theta| + |\tau''-\theta|) .$$

Proof. A Taylor expansion about $\tau = (\tau'+\tau'')/2$ yields

$$p(x,\tau') = p(x,\tau) + \tfrac{1}{2}(\tau'-\tau'')'\int_0^1 p^{(\cdot)}(x,(1+u)\tau'/2 + (1-u)\tau''/2)du$$

and

$$p(x,\tau'') = p(x,\tau) + \tfrac{1}{2}(\tau''-\tau')'\int_0^1 p^{(\cdot)}(x,(1+u)\tau''/2 + (1-u)\tau'/2)du,$$

hence

$$p(x,\tau')-p(x,\tau'') = (\tau'-\tau'')'\int_0^1 p^{(\cdot)}(x,(1+u)\tau'/2 + (1-u)\tau''/2)du$$

$$= (\tau'-\tau'')'p^{(\cdot)}(x,\theta) + (\tau'-\tau'')'r(x,\theta,\tau',\tau'')$$

with $r_i(x,\theta,\tau',\tau'') := \int_0^1 (p^{(i)}(x,(1+u)\tau'/2 + (1-u)\tau''/2) - p^{(i)}(x,\theta))du.$

Since $|r_i(x,\theta,\tau',\tau'')| \le \frac{3}{4}(|\tau'-\theta| + |\tau''-\theta|)p(x,\theta)M(x,\theta)$, the assertion follows easily.

6.4. Distances for product measures

Since our asymptotic results refer to product measures, it appears natural to consider how the distances between product measures depend on the distances between their components.

Using relations (6.1.4) and (6.1.2) we obtain

$$(6.4.1) \qquad 1 + \Delta(\mathop{\times}_{i=1}^{m} Q_i ; \mathop{\times}_{i=1}^{m} P_i)^2 = \prod_{i=1}^{m} (1 + \Delta(Q_i;P_i)^2)$$

and

$$(6.4.2) \qquad 1 - \tfrac{1}{8}H(\mathop{\times}_{i=1}^{m} Q_i , \mathop{\times}_{i=1}^{m} P_i)^2 = \prod_{i=1}^{m} (1 - \tfrac{1}{8}H(Q_i,P_i)^2).$$

From these relations we immediately obtain that both distances, Δ and H, fulfill the relations

$$(6.4.3) \qquad \delta(Q\times M, P\times M) = \delta(Q,P)$$

and

(6.4.4) $\delta(Q^2,P^2) = \sqrt{2}\,\delta(Q,P)(1+O(\delta(Q,P)^2))$.

(Since these relations hold true for $\delta = \Delta$, a n y distance function δ approximating Δ at P in the sense of Definition 6.2.1 fulfills relations (6.4.3) and (6.4.4) up to terms of order $o(\Delta(Q;P))$.)

Moreover, we mention that one obtains from (6.4.1) and (6.4.2) for sequences of p-measures Q_n with P-density $1+n^{-1/2}g+n^{-1/2}r_n$ under the assumption $P(r_n^2) = o(n^o)$ that

(6.4.5) $\Delta(Q_n^n;P^n)^2 = \exp[P(g^2)] - 1 + o(n^o)$

and

(6.4.6) $H(Q_n^n,P^n)^2 = 8(1-\exp[-\frac{1}{8}P(g^2)]) + o(n^o)$.

Though simple relations like (6.4.1) and (6.4.2) do not hold for the variational distance, Reiss (1980, p. 100) proves (for the case $r_n = 0$) a similar a s y m p t o t i c result, namely

(6.4.7) $V(Q_n^n,P^n) = 2\Phi(\frac{1}{2}P(g^2)^{1/2}) - 1 + O(n^{-1/2})$.

We mention these results for the following reason. The as. envelope power function (see Corollary 8.4.4) depends on the alternative Q through $n^{1/2}\delta(Q,\mathfrak{P}_o)$. For the alternatives Q_n defined above, we have $n^{1/2}\delta(Q_n,P) = P(g^2)^{1/2} + o(n^o)$. Hence $n^{1/2}\delta(Q_n,P)$ is of the same order of magnitude as $\delta(Q_n^n,P^n)$. This might suggest to consider the as. envelope power function as a function of $\delta(Q^n,\mathfrak{P}_o^n)$ (with $\mathfrak{P}_o^n = \{P^n: P\in\mathfrak{P}_o\}$ rather than $n^{1/2}\delta(Q,\mathfrak{P}_o)$. However, with relations (6.4.5) - (6.4.7) being rather complex, this would lead to a formula for the as. envelope power function which looks much more complicated than the one given in Corollary 8.4.4.

7. PROJECTIONS OF PROBABILITY MEASURES

7.1. Motivation

Our interest in projections results from the following problem occuring in estimation theory. Assume that we are given an estimator-sequence $\underline{x} \to P_n(\underline{x},\cdot)$, $n \in \mathbb{N}$, for p-measures in $\mathfrak{P}$ which is strict in the sense that $P_n(\underline{x},\cdot) \in \mathfrak{P}$ for $\underline{x} \in X^n$. If it is known that the true p-measure P belongs to some subfamily $\overline{\mathfrak{P}} \subset \mathfrak{P}$, is it then possible to obtain estimators for P which attain their values in $\overline{\mathfrak{P}}$ from the estimators P_n ? Given $P_n(\underline{x},\cdot) \in \mathfrak{P}$, it suggests itself to try an estimate $\overline{P}_n(\underline{x},\cdot)$ which is the projection of $P_n(\underline{x},\cdot)$ into $\overline{\mathfrak{P}}$ (i.e. that element of $\overline{\mathfrak{P}}$ which is 'closest' to $P_n(\underline{x},\cdot)$). In this way, we obtain at least estimators which are *strict* for $\overline{\mathfrak{P}}$. But we expect that, under appropriate regularity conditions, the projection does more than that, namely that it *improves* the estimators.

In the following sections we present some basic results on projections. These will be applied to estimation theory in Sections 10.4 - 10.6 and to testing theory in Sections 8.4 and 8.5.

7.2. The projection

Let $\overline{\mathfrak{P}}$ be a family of p-measures and Q a p-measure not in $\overline{\mathfrak{P}}$. Our problem is to define the projection of Q into $\overline{\mathfrak{P}}$. Since the intended applications are to asymptotic theory, our interest concentrates on p-measures Q close to $\overline{\mathfrak{P}}$. There are several possibilities to define a projection which appear to be asymptotically equivalent. The following definition has the advantage of not depending on any particular distance function.

7.2.1. Definition. $\overline{Q} \in \overline{\mathfrak{P}}$ is a *projection* of Q into $\overline{\mathfrak{P}}$ if $q/\overline{q} - 1$ is orthogonal to $T(\overline{Q},\overline{\mathfrak{P}})$, i.e.

$$\overline{Q}((q/\overline{q} - 1)g) = 0 \qquad \text{for all } g \in T(\overline{Q},\overline{\mathfrak{P}}) .$$

Equivalently,

(7.2.2) $$Q(g) = 0 \qquad \text{for all } g \in T(\overline{Q},\overline{\mathfrak{P}}) .$$

Notice that this definition presumes $(q/\overline{q} - 1)g$ to be $\overline{Q}$-integrable. This comes close to requiring that $(q/\overline{q} - 1)^2$ be $\overline{Q}$-integrable.

The projection will be unique only under additional regularity conditions. The existence of a projection is usually easy to see in any particular case, but difficult to establish as a general theorem.

Examples showing how the definition works will be given in Sections 7.6 - 7.8.

Some readers may be surprised by our efforts to achieve in the following sections results which hold u n i f o r m l y in the neighborhood of some fixed p-measure. Such a uniformity is required by applications in estimation theory where the 'true' p-measure is fixed, and where estimates are projected into the family of all p-measures belonging to the model (or a submodel).

7.3. Projections defined by distances

Some readers may find a definition of the 'projection' of Q through minimization of a distance $\delta(Q,P)$ for $P \in \overline{\mathfrak{P}}$ more natural. In fact, such a definition would serve our purposes equally well. But it appears that Definition 7.2.1 is technically somewhat easier to handle. Moreover, it has the advantage of not depending on a particular distance function.

Let δ be a distance function on $\mathfrak{P}$ which approximates Δ in the sense of Definition 6.2.1. In this section we shall show that the projection as defined by 7.2.1 minimizes the δ-distance from $\overline{\mathfrak{P}}$ approximately, and that, conversely, the p-measure minimizing the δ-distance is close to this projection. In view of Proposition 6.2.2, these relations hold in particular for the Hellinger distance. Similar relations hold for the distance measure $(Q,P) \to \Delta(Q;P)$, but they require severe regularity conditions and will, therefore, be omitted.

7.3.1. Lemma. *If* $\mathfrak{P}$ *is at* P *approximable by* $T_s(P,\mathfrak{P})$ *uniformly for* $P \in \overline{\mathfrak{P}}$, *then every* $Q \in \mathfrak{P}$ *admits a* $\overline{Q}$*-density* $1+g+r$ *with* $g \in T_s^{\perp}(\overline{Q};\overline{\mathfrak{P}},\mathfrak{P})$ *and* $\|r\|_{\overline{Q}} = o(\Delta(Q;\overline{Q}))$ *uniformly for* $Q \in \mathfrak{P}$.

Proof. Let $k := q/\overline{q} - 1$, and let g denote the projection of k into $T_s(\overline{Q},\mathfrak{P})$. Since $\overline{Q}$ is the projection of Q into $\overline{\mathfrak{P}}$, we have $k \perp T_s(\overline{Q},\overline{\mathfrak{P}})$, hence also $g \perp T_s(\overline{Q},\overline{\mathfrak{P}})$. Furthermore, since $\mathfrak{P}$ is at $\overline{Q}$ approximable by $T_s(\overline{Q},\mathfrak{P})$, we have $\|k-g\|_{\overline{Q}} = o(\Delta(Q;\overline{Q}))$ uniformly for $Q \in \mathfrak{P}$.

<u>7.3.2. Proposition</u>. *Assume that $\overline{\mathfrak{P}}$ is approximable by $T_s(\cdot,\overline{\mathfrak{P}})$, and that $\mathfrak{P}$ is at P approximable by $T_s(P,\mathfrak{P})$ uniformly for $P \in \overline{\mathfrak{P}}$. Let δ be a distance function on $\mathfrak{P}$ which approximates Δ at P uniformly for $P \in \overline{\mathfrak{P}}$. Then uniformly for $Q \in \mathfrak{P}$ and $P \in \overline{\mathfrak{P}}$*

$$(7.3.3) \qquad \delta(Q,P)^2 = \delta(Q,\overline{Q})^2 + \delta(P,\overline{Q})^2 + o(\Delta(Q;\overline{Q})^2) + o(\Delta(P;\overline{Q})^2) .$$

If, in addition, $\Delta(Q;P) = O(\delta(Q,P))$ uniformly for $Q \in \mathfrak{P}$ and $P \in \overline{\mathfrak{P}}$, then uniformly for $Q \in \mathfrak{P}$

$$(7.3.4) \qquad \delta(Q,\overline{Q}) = \delta(Q,\overline{\mathfrak{P}}) + o(\delta(Q,\overline{Q})) .$$

<u>Proof</u>. (i) By Lemma 7.3.1, every $Q \in \mathfrak{P}$ admits a $\overline{Q}$-density $1+g+r$ with $g \in T_s^{\perp}(\overline{Q};\overline{\mathfrak{P}},\mathfrak{P})$ and $\|r\|_{\overline{Q}} = o(\Delta(Q;\overline{Q}))$ uniformly for $Q \in \mathfrak{P}$. Since $\overline{\mathfrak{P}}$ is approximable by $T_s(\cdot,\overline{\mathfrak{P}})$, every $P \in \overline{\mathfrak{P}}$ admits a $\overline{Q}$-density $1+h+s$ with $h \in T_s(\overline{Q},\overline{\mathfrak{P}})$ and $\|s\|_{\overline{Q}} = o(\Delta(P;\overline{Q}))$ uniformly for $Q \in \mathfrak{P}$ and $P \in \overline{\mathfrak{P}}$. Since $g \perp h$, we have $\|g-h\|^2_{\overline{Q}} = \|g\|^2_{\overline{Q}} + \|h\|^2_{\overline{Q}}$. Since δ approximates Δ, we obtain uniformly for $Q \in \mathfrak{P}$

$$\begin{aligned} \delta(Q,P)^2 &= \Delta(Q,P;\overline{Q})^2 + o(\Delta(Q;\overline{Q})^2) + o(\Delta(P;\overline{Q})^2) \\ &= \Delta(Q;\overline{Q})^2 + \Delta(P;\overline{Q})^2 + o(\Delta(Q;\overline{Q})^2) + o(\Delta(P;\overline{Q})^2) \\ &= \delta(Q,\overline{Q})^2 + \delta(P,\overline{Q})^2 + o(\Delta(Q;\overline{Q})^2) + o(\Delta(P;\overline{Q})^2) . \end{aligned}$$

(ii) In order to prove (7.3.4), we have to show for all $P \in \overline{\mathfrak{P}}$

$$\delta(Q,P) \geq \delta(Q,\overline{Q}) + o(\delta(Q,\overline{Q})) .$$

The case $\delta(Q,P) \geq \delta(Q,\overline{Q})$ is trivial. If $\delta(Q,P) < \delta(Q,\overline{Q})$, then

$$\delta(P,\overline{Q}) \leq \delta(Q,\overline{Q}) + \delta(Q,P) < 2\delta(Q,\overline{Q}) .$$

Hence (7.3.3) implies

$$\begin{aligned} \delta(Q,P)^2 &= \delta(Q,\overline{Q})^2 + \delta(P,\overline{Q})^2 + o(\delta(Q,\overline{Q})^2) \\ &\geq \delta(Q,\overline{Q})^2 + o(\delta(Q,\overline{Q})^2) . \end{aligned}$$

Assertion (7.3.4) follows immediately.

Let $\widetilde{Q} \in \overline{\mathfrak{P}}$ denote a p-measure which minimizes the distance $\delta(Q,P)$ for $P \in \overline{\mathfrak{P}}$, i.e.

$$\delta(Q,\tilde{Q}) = \delta(Q,\overline{\mathfrak{P}}) .$$

The following proposition shows that $\tilde{Q}$ coincides approximately with the projection of Q into $\overline{\mathfrak{P}}$, if δ approximates Δ.

7.3.5. Proposition. *Under the assumptions of Proposition 7.3.2 we have* $\delta(\tilde{Q},\overline{Q}) = o(\delta(Q,\overline{Q}))$ *uniformly for* $Q \in \mathfrak{P}$.

Proof. By (7.3.3) we have uniformly for $Q \in \mathfrak{P}$

$$\delta(Q,\tilde{Q})^2 = \delta(Q,\overline{Q})^2 + \delta(\tilde{Q},\overline{Q})^2 + o(\delta(Q,\overline{Q})^2) + o(\delta(\tilde{Q},\overline{Q})^2) .$$

By definition of $\tilde{Q}$ we have $\delta(Q,\tilde{Q}) = \delta(Q,\overline{\mathfrak{P}})$ and

$$\delta(\tilde{Q},\overline{Q}) \leq \delta(Q,\tilde{Q}) + \delta(Q,\overline{Q}) \leq 2\delta(Q,\overline{Q}) .$$

Moreover, $\delta(Q,\overline{Q}) = \delta(Q,\overline{\mathfrak{P}}) + o(\delta(Q,\overline{Q}))$ uniformly for $Q \in \mathfrak{P}$ by (7.3.4). The assertion follows immediately from these relations.

7.3.6. Remark. Let $\kappa: \mathfrak{P} \to \mathbb{R}^k$ be a k-dimensional functional and $\overline{\mathfrak{P}} := \{P \in \mathfrak{P}: \kappa(P) = c\}$. Assume that

(i) $\mathfrak{P}$ is at P approximable by $T_s(P,\mathfrak{P})$ uniformly for $P \in \overline{\mathfrak{P}}$,

(ii) κ is strongly differentiable in the sense that uniformly for $Q \in \mathfrak{P}$ and $P \in \overline{\mathfrak{P}}$

$$\kappa(Q) - \kappa(P) = Q(\kappa^{\bullet}(\cdot,P)) + o(\Delta(Q;P)) ,$$

(iii) $T_s(P,\overline{\mathfrak{P}}) = \{g \in T_s(P,\mathfrak{P}): P(g\kappa^{\bullet}(\cdot,P)) = 0\}$ for $P \in \overline{\mathfrak{P}}$ (see Proposition 4.5.1),

(iv) $\Sigma(P) := P(\kappa^{\bullet}(\cdot,P)\kappa^{\bullet}(\cdot,P)')$ and its inverse, say $D(P)$, are bounded for $P \in \overline{\mathfrak{P}}$.

Then uniformly for $Q \in \mathfrak{P}$

$$(\kappa(Q) - c)'D(\overline{Q})(\kappa(Q) - c) = \Delta(Q;\overline{Q})^2 + o(\Delta(Q;\overline{Q})^2) . \tag{7.3.7}$$

Addendum. If D is continuous in the sense that uniformy for $Q \in \mathfrak{P}$ and $P \in \overline{\mathfrak{P}}$,

$$D(Q) - D(P) = o(\Delta(Q;P)^{\circ}) ,$$

then (7.3.7) holds with $D(\overline{Q})$ replaced by $D(Q)$.

Proof. By Proposition 7.3.1, Q has a $\bar{Q}$-density $1+g+r$ with $g \in T_s^{\perp}(\bar{Q};\bar{\mathfrak{P}},\mathfrak{P})$ and $\|r\|_{\bar{Q}} = o(\Delta(Q;\bar{Q}))$. Strong differentiability of κ implies

$$\kappa(Q) - \kappa(\bar{Q}) = \bar{Q}((q/\bar{q} - 1)\kappa^{\bullet}(\cdot,\bar{Q})) + o(\Delta(Q;\bar{Q}))$$
$$= \bar{Q}(g\kappa^{\bullet}(\cdot,\bar{Q})) + o(\Delta(Q;\bar{Q})) .$$

Assumption (iii) implies

$$T_s^{\perp}(\bar{Q};\bar{\mathfrak{P}},\mathfrak{P}) = [\kappa_1^{\bullet}(\cdot,\bar{Q}),\ldots,\kappa_k^{\bullet}(\cdot,\bar{Q})] .$$

Hence $g = a'\kappa^{\bullet}(\cdot,\bar{Q})$, and

$$\kappa(Q) - \kappa(\bar{Q}) = a'\Sigma(\bar{Q}) + o(\Delta(Q;\bar{Q})).$$

This implies

$$a = D(\bar{Q})(\kappa(Q) - \kappa(\bar{Q})) + o(\Delta(Q;\bar{Q})) .$$

The assertion now follows from

$$\Delta(Q;\bar{Q})^2 = \bar{Q}(g^2)(1 + o(\Delta(Q;\bar{Q})^{\circ})) = a'\Sigma(\bar{Q})a(1 + o(\Delta(Q;\bar{Q})^{\circ}))$$

and assumption (iv), since

$$|\kappa(Q) - \kappa(\bar{Q})| \leq \Delta(Q;\bar{Q})\|\kappa^{\bullet}(\cdot,\bar{Q})\|_{\bar{Q}} = O(\Delta(Q;\bar{Q})) .$$

7.4. Projections of measures - projections of densities

Let $P \in \bar{\mathfrak{P}}$ be fixed, and let Q be a p-measure with P-density $1 + k_Q$, say. Let $\bar{Q}$ denote the projection of Q into $\bar{\mathfrak{P}}$, and $1 + k_{\bar{Q}}$ the P-density of $\bar{Q}$. (For an intuitive interpretation, think of P as the true p-measure, Q as an estimator, and $\bar{Q}$ as an 'improved' estimator obtained by projecting Q into $\bar{\mathfrak{P}}$.)

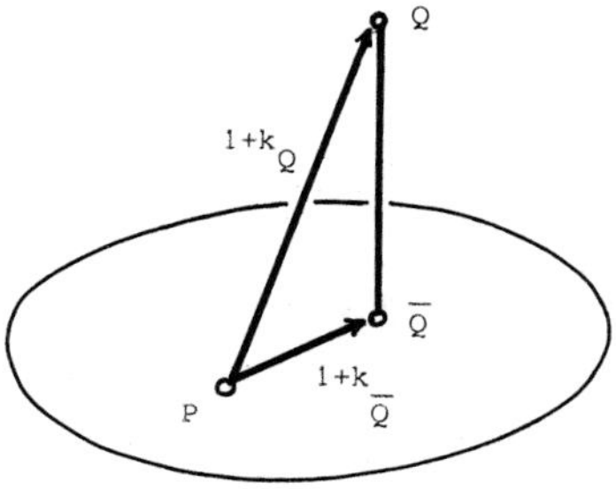

7.4.1. Lemma. *Assume that* $T_s(\cdot,\overline{\mathfrak{P}})$ *is continuous at* P *(see Definition 1.1.12). Then uniformly for* $Q \in \mathfrak{P}$ and $f \in T_s(P,\overline{\mathfrak{P}})$,

$$|P((k_Q - k_{\overline{Q}})f)| = \|f\|(\Delta(Q;P) + \Delta(\overline{Q};P))o(\Delta(\overline{Q};P)^{0}) .$$

Proof. By definition of $\overline{Q}$ we have $Q(\overline{f}) = 0$ for all $\overline{f} \in T_s(\overline{Q},\overline{\mathfrak{P}})$. Hence for $f \in T_s(P,\overline{\mathfrak{P}})$, $\overline{f} \in T_s(\overline{Q},\overline{\mathfrak{P}})$,

$$P((k_Q - k_{\overline{Q}})f) = P((1+k_Q)f) - P((1+k_{\overline{Q}})f)$$
$$= Q(f) - \overline{Q}(f) = Q(f - \overline{Q}(f) - \overline{f}) .$$

Since $T_s(\cdot,\overline{\mathfrak{P}})$ is continuous at P, for every $f \in T_s(P,\overline{\mathfrak{P}})$ there exists $f^* \in T_s(\overline{Q},\overline{\mathfrak{P}})$ such that $r := f - \overline{Q}(f) - f^*$ fulfills uniformly for $Q \in \mathfrak{P}$ and $f \in T_s(P,\overline{\mathfrak{P}})$

$$\|r\| = \|f\| o(\|k_{\overline{Q}}\|^{0}) .$$

We have $\overline{Q}(r) = 0$, i.e. $P((1+k_{\overline{Q}})r) = 0$, hence $P(r) = -P(k_{\overline{Q}}r)$, and therefore

$$|P(r)| \le \|k_{\overline{Q}}\| \, \|r\| = \|f\| \, o(\|k_{\overline{Q}}\|) .$$

Hence

$$|P((k_Q - k_{\overline{Q}})f)| = |Q(r)| = |P((1+k_Q)r)| \le |P(r)| + |P(k_Q r)|$$
$$\le \|f\|(\|k_{\overline{Q}}\| + \|k_Q\|)o(\|k_{\overline{Q}}\|^{0}) .$$

This is the assertion.

<u>7.4.2. Corollary</u>. *If* $\overline{k}_Q$ *and* $\overline{k}_{\overline{Q}}$ *denote the projections into* $T_s(P,\overline{\mathfrak{P}})$ *of* k_Q *and* $k_{\overline{Q}}$, *respectively, then uniformly for* $Q \in \mathfrak{P}$

$$\|\overline{k}_Q - \overline{k}_{\overline{Q}}\| = (\Delta(Q;P) + \Delta(\overline{Q};P))o(\Delta(\overline{Q};P)^o) .$$

<u>Proof</u>. Since $\overline{k}_Q$ and $\overline{k}_{\overline{Q}}$ are the projections of k_Q and $k_{\overline{Q}}$, respectively, we have $P(\overline{k}_Q f) = P(k_Q f)$ and $P(\overline{k}_{\overline{Q}} f) = P(k_{\overline{Q}} f)$ for $f \in T_s(P,\overline{\mathfrak{P}})$. Hence Lemma 7.4.1 implies uniformly for $Q \in \mathfrak{P}$ and $f \in T_s(P,\overline{\mathfrak{P}})$

$$P((\overline{k}_Q - \overline{k}_{\overline{Q}})f) = \|f\|(\|k_Q\| + \|k_{\overline{Q}}\|)o(\|k_{\overline{Q}}\|^o) .$$

The assertion follows by applying this relation for $f = \overline{k}_Q - \overline{k}_{\overline{Q}}$.

<u>7.4.3. Remark</u>. If the projections fulfill $\Delta(\overline{Q};P) = O(\Delta(Q;P))$ uniformly for $Q \in \mathfrak{P}$, then the bounds in Theorem 7.4.1 and Corollary 7.4.2 can be simplified to $\|f\|o(\Delta(Q;P))$ and $o(\Delta(Q;P))$, respectively.

In applications to estimation theory, P is the (unknown) true p-measure and Q an estimator. It is, therefore, desirable to replace $\Delta(\overline{Q};P) = O(\Delta(Q;P))$ by a property like $\Delta(Q;\overline{Q}) = O(\Delta(Q;\overline{\mathfrak{P}}))$. This is possible if $\mathfrak{P}$ is sufficiently regular:

Assume that there exists a distance function δ which approximates Δ at P uniformly for $P \in \overline{\mathfrak{P}}$ and fulfills $\Delta(Q;P) = O(\delta(Q,P))$ uniformly for $Q \in \mathfrak{P}$ and $P \in \overline{\mathfrak{P}}$. If the projection is chosen such that $\Delta(Q;\overline{Q}) = O(\Delta(Q;\overline{\mathfrak{P}}))$ uniformly for $Q \in \mathfrak{P}$, then $\Delta(\overline{Q};P) = O(\Delta(Q;P))$ uniformly for $Q \in \mathfrak{P}$ and $P \in \overline{\mathfrak{P}}$.

(We have $\Delta(\overline{Q};P) = O(\delta(\overline{Q},P))$ and $\delta(\overline{Q},P) \le \delta(Q,P) + \delta(Q,\overline{Q})$. Furthermore,

$$\delta(Q,\overline{Q}) = \Delta(Q;\overline{Q}) + o(\Delta(Q;\overline{Q})) = O(\Delta(Q;\overline{\mathfrak{P}})) .$$

Since $\Delta(Q;\overline{\mathfrak{P}}) \le \Delta(Q;P)$, the assertion follows.)

<u>7.4.4. Proposition</u>. *Assume that* $\overline{\mathfrak{P}}$ *is at* P *approximable by* $T_s(P,\overline{\mathfrak{P}})$, *and that* $T_s(\cdot,\overline{\mathfrak{P}})$ *is continuous at* P. *If* $\overline{Q}$ *fulfills* $\Delta(\overline{Q};P) = O(\Delta(Q;P))$ *uniformly for* $Q \in \mathfrak{P}$, *then uniformly for* $Q \in \mathfrak{P}$

$$\|k_{\overline{Q}} - \overline{k}_Q\| = o(\Delta(Q;P)) .$$

In a nutshell: If we characterize each p-measure M by its 'reduced' P-density $k_M = m/p - 1$, then we can say that the reduced density of the projection can be closely approximated by the projection of the reduced density.

Proof. Since $\overline{\mathfrak{P}}$ is at P approximable by $T_s(P,\overline{\mathfrak{P}})$, we obtain uniformly for $Q \in \mathfrak{P}$

$$\| k_{\overline{Q}} - \overline{k}_{\overline{Q}} \| = o(\Delta(Q;P)) .$$

By Corollary 7.4.2 we have uniformly for $Q \in \mathfrak{P}$

$$\| \overline{k}_Q - \overline{k}_{\overline{Q}} \| = o(\Delta(Q;P)) .$$

These two relations imply the assertion.

7.5. Iterated projections

If H is a Hilbert space and $H_o \subset H_1 \subset H$ closed subspaces, then the projection into H_o of the projection into H_1 of any element of H equals the projection of this element into H_o. Since locally our families of p-measures behave like Hilbert spaces, we expect a corresponding relationship.

Let $\mathfrak{P}_o \subset \mathfrak{P}_1$ and let Q be a p-measure not necessarily belonging to $\mathfrak{P}_1$. Let Q_i denote the projection of Q into $\mathfrak{P}_i$, i = 0,1. The following theorem asserts that the projection $Q_{1,o}$ of Q_1 into $\mathfrak{P}_o$ agrees closely with Q_o.

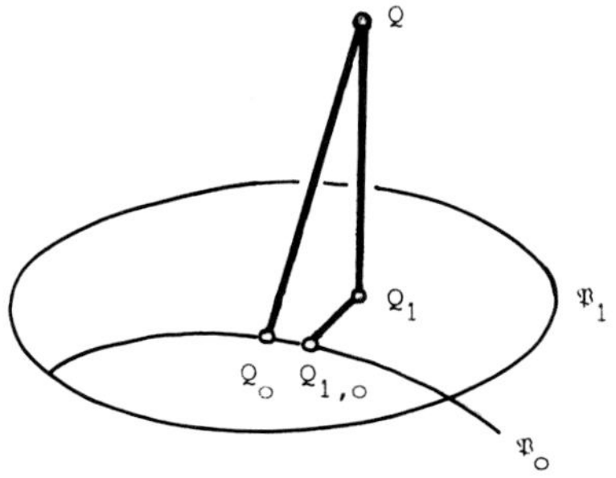

7.5.1. Theorem. *If $\mathfrak{P}_o$ is approximable by $T_s(\cdot,\mathfrak{P}_o)$, $T_s(\cdot,\mathfrak{P}_o)$ is continuous on $\mathfrak{P}_o$, and $T_s(\cdot,\mathfrak{P}_1)$ is continuous at P uniformly for $P \in \mathfrak{P}_o$, then uniformly for $Q \in \mathfrak{P}$,*

$$\Delta(Q_{1,o};Q_o) + o(\Delta(Q_{1,o};Q_o)) = \Delta(Q;Q_o)o(\Delta(Q_1;Q_o)^o)$$
$$+ \Delta(Q_1;Q_o)o(\Delta(Q_{1,o};Q_o)^o) .$$

If, in addition, $\Delta(Q_1;Q_o) = O(\Delta(Q;Q_o))$ and $\Delta(Q_{1,o};Q_o) = O(\Delta(Q_1;Q_o))$, then uniformly for $Q \in \mathfrak{P}$,

$$\Delta(Q_{1,o};Q_o) = o(\Delta(Q;Q_o)) .$$

Proof. By (7.2.2) we have

(7.5.2) $\quad Q(g) = 0 \quad$ for $g \in T_s(Q_o,\mathfrak{P}_o)$,

(7.5.3) $\quad Q(g) = 0 \quad$ for $g \in T_s(Q_1,\mathfrak{P}_1)$,

(7.5.4) $\quad Q_1(g) = 0 \quad$ for $g \in T_s(Q_{1,o},\mathfrak{P}_o)$.

In the following, let $\| \; \| = \| \; \|_{Q_o}$. Since $\mathfrak{P}_o$ is approximable by $T_s(\cdot,\mathfrak{P}_o)$, every projection $Q_{1,o}$ admits a Q_o-density $1+h+r$ with $h \in T_s(Q_o,\mathfrak{P}_o)$, $r \perp h$, and $\|r\| = o(\Delta(Q_{1,o};Q_o))$ uniformly for $Q \in \mathfrak{P}$. In particular,

(7.5.5) $\quad \Delta(Q_{1,o};Q_o) = \|h+r\| = \|h\| + o(\Delta(Q_{1,o};Q_o))$.

Since $T_s(\cdot,\mathfrak{P}_1)$ is continuous on $\mathfrak{P}_o$, there exist $h_1 \in T_s(Q_1,\mathfrak{P}_1)$ such that $r_1 := h - Q_1(h) - h_1$ fulfills uniformly for $Q \in \mathfrak{P}$

(7.5.6) $\quad \|r_1\| = \|h\| o(\Delta(Q_1;Q_o)^o)$.

Since $T_s(\cdot,\mathfrak{P}_o)$ is continuous on $\mathfrak{P}_o$, there exist $h_{1,o} \in T_s(Q_{1,o},\mathfrak{P}_o)$ such that $r_{1,o} := h - Q_{1,o}(h) - h_{1,o}$ fulfills

(7.5.7) $\quad \|r_{1,o}\| = \|h\| o(\Delta(Q_{1,o};Q_o)^o)$.

From (7.5.3) and (7.5.4) we obtain

$$0 = Q(h_1) \quad = Q(h) - Q_1(h) - Q(r_1) ,$$
$$0 = Q_1(h_{1,o}) = Q_1(h) - Q_{1,o}(h) - Q_1(r_{1,o}) .$$

Since $Q(h) = 0$ by (7.5.2), these relations imply

$$Q_{1,o}(h) = Q(r_1) + Q_1(r_{1,o}) .$$

Since $r \perp h$, this implies

(7.5.8) $\quad Q_o(h^2) = Q(r_1) + Q_1(r_{1,o})$.

By (7.5.6),

$$|Q(r_1)| = |Q_o((\frac{q}{q_o} - 1)r_1)| \leq \Delta(Q;Q_o)\|r_1\| = \|h\|\Delta(Q;Q_o)o(\Delta(Q_1;Q_o)^o).$$

Similarly, by (7.5.7),

$$|Q_1(r_{1,o})| = |Q_o((\frac{q_1}{q_o} - 1)r_{1,o}| \leq \Delta(Q_1;Q_o)\|r_{1,o}\|$$

$$= \|h\|\Delta(Q_1;Q_o)o(\Delta(Q_{1,o};Q_o)^o) .$$

Together with (7.5.8) this implies

$$\|h\| = \Delta(Q;Q_o)o(\Delta(Q_1;Q_o)^o) + \Delta(Q_1;Q_o)o(\Delta(Q_{1,o};Q_o)^o) .$$

he assertion now follows from (7.5.5).

7.6. Projections into a parametric family

Let $\mathfrak{P} = \{P_\theta : \theta \in \Theta\}$, $\Theta \subset \mathbb{R}^k$, be a parametric family with tangent space $T(P_\theta,\mathfrak{P}) = [\ell^{(1)}(\cdot,\theta),\ldots,\ell^{(k)}(\cdot,\theta)]$. Let Q be a p-measure not belonging to $\mathfrak{P}$. By Definition 7.2.1, the projection of Q into $\mathfrak{P}$, say $P_{\underline{\theta}}$, is determined by

(7.6.1) $\quad Q(\ell^{(\cdot)}(\cdot,\underline{\theta})) = 0 .$

Let now $\theta \in \Theta$ be fixed, and consider p-measures Q close to P_θ. (The interpretation: P_θ is the true p-measure, and Q an estimate.) Since

$$\ell^{(\cdot)}(x,\underline{\theta})' = \ell^{(\cdot)}(x,\theta)' + (\underline{\theta} - \theta)'\int_0^1 \ell^{(\cdot\cdot)}(x,(1-u)\theta + u\underline{\theta})du ,$$

we obtain from (7.6.1) under suitable regularity conditions that

(7.6.2) $\quad \underline{\theta} - \theta = c(\theta,Q) + O(\Delta(Q;P_\theta)^2) ,$

where

(7.6.3) $\quad c(\theta,Q) := \Lambda(\theta)Q(\ell^{(\cdot)}(\cdot,\theta)) .$

Corollary 7.4.2 asserts that the projections of $k_Q := q/p(\cdot,\theta) - 1$ and $k_{\underline{\theta}} := p(\cdot,\underline{\theta})/p(\cdot,\theta) - 1$ into $T(P_\theta,\mathfrak{P})$, say $\overline{k}_Q$ and $\overline{k}_{\underline{\theta}}$, are almost identical. To check this for our particular case, we use that (see Proposition 4.2.3)

$$\overline{k}_Q = c(\theta,Q)'\ell^{(\cdot)}(\cdot,\theta) \quad \text{and} \quad \overline{k}_{\underline{\theta}} = c(\theta,P_{\underline{\theta}})'\ell^{(\cdot)}(\cdot,\theta).$$

Since

$$P_{\underline{\theta}}(\ell^{(\cdot)}(\cdot,\theta)) = L(\theta)(\underline{\theta}-\theta) + O(|\underline{\theta}-\theta|^2)$$
$$= L(\theta)(\underline{\theta}-\theta) + O(\Delta(Q;P_\theta)^2) ,$$

we obtain from (7.6.2) and (7.6.3) that

$$P_{\underline{\theta}}(\ell^{(\cdot)}(\cdot,\theta)) = L(\theta)\Lambda(\theta)Q(\ell^{(\cdot)}(\cdot,\theta)) + O(\Delta(Q;P_\theta)^2)$$
$$= Q(\ell^{(\cdot)}(\cdot,\theta)) + O(\Delta(Q;P_\theta)^2) ,$$

so that

$$c(\theta,P_{\underline{\theta}}) = c(\theta,Q) + O(\Delta(Q,P_\theta)^2) .$$

Hence

$$\|\bar{k}_Q - \bar{k}_{\underline{\theta}}\| = O(\Delta(Q;P_\theta)^2) .$$

To illustrate Theorem 7.5.1, let now $\mathfrak{P}_1 = \{P_{(\tau,\eta)}: \tau \in T,\ \eta \in H\}$ with $T \subset \mathbb{R}^q$, $H \subset \mathbb{R}^{k-q}$, and $\mathfrak{P}_o = \{P_{(\tau,\eta_o)}: \tau \in T\}$, with $\eta_o \in H$ fixed. Let $\theta = (\tau,\eta_o)$. According to (7.6.2), the projection of Q into $\mathfrak{P}_1$ is P_{θ_1} with

(7.6.4) $\quad \theta_1 - \theta = \Lambda(\theta)Q(\ell^{(\cdot)}(\cdot,\theta)) + O(\Delta(Q;P_\theta)^2)$;

and the projection of Q into $\mathfrak{P}_o$ is P_{θ_o} with $\theta_o = (\tau_o,\eta_o)$ and

(7.6.5) $\quad \tau_o - \tau = \Lambda^*(\theta)Q(\ell^{(*)}(\cdot,\theta)) + O(\Delta(Q;P_\theta)^2)$,

where Λ^* is the inverse of $(L_{\alpha,\beta})_{\alpha,\beta=1,\ldots,q}$, and $\ell^{(*)}$ is the vector with components $\ell^{(1)},\ldots,\ell^{(q)}$.

Theorem 7.5.1 asserts that the projection of P_{θ_1} into $\mathfrak{P}_o$, say $P_{\theta_{1,o}}$, has distance $o(\Delta(Q;P_\theta))$ from P_{θ_o}. To verify this for our particular case, we proceed as follows.

Using (7.6.4) we obtain

$$P_{\theta_1}(\ell^{(*)}(\cdot,\theta)) = P_\theta(\ell^{(*)}(\cdot,\theta)\ell^{(\cdot)}(\cdot,\theta)')\Lambda(\theta)Q(\ell^{(\cdot)}(\cdot,\theta)) + O(\Delta(Q;P_\theta)^2)$$
$$= Q(\ell^{(*)}(\cdot,\theta)) + O(\Delta(Q;P_\theta)^2) .$$

Hence we obtain for $\theta_{1,o} = (\tau_{1,o},\eta_o)$ from (7.6.5), applied for $Q = P_{\theta_1}$,

$$\tau_{1,o} - \tau = \Lambda^*(\theta)P_{\theta_1}(\ell^{(*)}(\cdot,\theta)) + O(\Delta(Q;P_\theta)^2)$$
$$= \Lambda^*(\theta)Q(\ell^{(*)}(\cdot,\theta)) + O(\Delta(Q;P_\theta)^2)$$
$$= \tau_o - \tau + O(\Delta(Q;P_\theta)^2) ,$$

so that $\tau_{1,o} = \tau_o + O(\Delta(Q;P_\theta)^2)$. Since $\theta_{1,o} = (\tau_{1,o},\eta_o)$ and $\theta_o = (\tau_o,\eta_o)$, this implies $\theta_{1,o} = \theta_o + O(\Delta(Q;P_\theta)^2)$.

Proposition 7.3.2 asserts that $P_{\underline{\theta}}$ minimizes $\delta(Q,P_\theta)$ for $P_\theta \in \mathfrak{P}$, up to a term of order $o(\delta(Q,P_\theta))$. We verify this in our particular case for the distance Δ. With $\underline{\theta}$ fulfilling (7.6.2), we have

$$\Delta(Q;P_\tau)^2 = Q(\frac{q}{p_\tau}) - 1 = P_\theta(\frac{q}{p_\theta}\frac{q}{p_\tau}) - 1$$

$$\doteq P_\theta((1+k_Q)^2(1-(\tau-\theta)'\ell^{(\cdot)}(\cdot,\theta))) - 1$$

$$\doteq P_\theta((k_Q - (\tau-\theta)'\ell^{(\cdot)}(\cdot,\theta))^2) .$$

The 'approximate distance', $\|k_Q - (\tau-\theta)'\ell^{(\cdot)}(\cdot,\theta)\|$, attains its minimum for

$$\tau = \theta + \Lambda(\theta)P_\theta(k_Q\ell^{(\cdot)}(\cdot,\theta)) ,$$

which, by (7.6.2), differs from the projection as defined by (7.6.1) only by an amount of order $O(\Delta(Q;P_\theta)^2)$.

<u>7.6.6. Remark</u>. If we use for the projection distances of a different nature, for instance distances based on the distribution function, the resulting parameter values will differ from $\underline{\theta}$ (as defined by (7.6.1)) by an amount of order $\Delta(Q;P_\theta)$.

As an example, consider a distance of Cramér-von Mises type, $\int(F_Q(t) - F_P(t))^2 w(t)dt$. In our case,

$$F_Q(t) := \int_{-\infty}^{t} Q(d\xi) = \int_{-\infty}^{t} (1+k(\xi))P_\theta(d\xi) ,$$

$$F_\tau(t) := \int_{-\infty}^{t} P_\tau(d\xi) = \int_{-\infty}^{t} (1+(\tau-\theta)'\ell^{(\cdot)}(\xi,\theta) + r(\xi;\theta,\tau))P_\theta(d\xi).$$

With

$$K(t) := \int_{-\infty}^{t} k(\xi)P_\theta(d\xi)$$

we obtain

$$F_Q(t) - F_\tau(t) = K(t) - (\tau-\theta)'F_\theta^{(\cdot)}(t) + o(|\tau-\theta|) .$$

Hence the projection is approximately determined by the parameter value τ for which

$$\int (K(t) - (\tau-\theta)' F_\theta^{(\cdot)}(t))^2 w(t)dt = \min .$$

This value τ differs in general by an amount of order $\|k\|$ from the value τ minimizing $\|k - (\tau-\theta)'\ell^{(\cdot)}(\cdot,\theta)\|$.

7.7. Projections into a family of product measures

Assume that the basic space is a product space, say $(X_1 \times X_2, \mathscr{A}_1 \times \mathscr{A}_2)$, and let $\mathfrak{P}$ be a family of product measures on $\mathscr{A}_1 \times \mathscr{A}_2$, say

$$\mathfrak{P} = \{P_1 \times P_2 | \mathscr{A}_1 \times \mathscr{A}_2 : P_i \in \mathfrak{P}_i, \ i = 1,2\} .$$

In Proposition 2.4.1 it was shown that

$$T(P_1 \times P_2, \mathfrak{P}) = \{(x,y) \to g_1(x) + g_2(y) : g_i \in T(P_i, \mathfrak{P}_i), \ i = 1,2\}.$$

Let $Q|\mathscr{A}_1 \times \mathscr{A}_2$ be arbitrary. Our problem is to determine the projection of Q into $\mathfrak{P}$. According to (7.2.2), this projection, say $\underline{Q}_1 \times \underline{Q}_2$, is determined by

(7.7.1) $\quad \int (g_1(x) + g_2(y)) Q(d(x,y)) = 0 \qquad$ for all $g_i \in T(\underline{Q}_i, \mathfrak{P}_i)$, $i = 1,2$.

Let Q_i denote the i-th marginal of Q. Then (7.7.1) is equivalent to

(7.7.2) $\quad \int g_1(x) Q_1(dx) + \int g_2(y) Q_2(dy) = 0 \qquad$ for all $g_i \in T(\underline{Q}_i, \mathfrak{P}_i)$, $i = 1,2$,

which implies

(7.7.3) $\quad \int g_i(z) Q_i(dz) = 0 \qquad$ for all $g_i \in T(\underline{Q}_i, \mathfrak{P}_i)$, $i = 1,2$.

Trivial solutions of these equations are $\underline{Q}_i = Q_i$, $i = 1,2$ (and these will be the only ones if $T(P_i, \mathfrak{P}_i)$ is sufficiently large).

To summarize: *The projection of an arbitrary p-measure into the family of product measures is the product of its marginals.*

Consider now the case $X_1 = X_2$, $\mathscr{A}_1 = \mathscr{A}_2$, and the smaller family $\mathfrak{P}^2 = \{P^2 : P \in \mathfrak{P}\}$. According to Proposition 2.4.1,

$$T(P^2,\mathfrak{P}^2) = \{(x,y) \to g(x)+g(y): g \in T(P,\mathfrak{P})\}.$$

In this case, the projection, say $\underline{Q}^2$, is determined by

$$\int (g(x)+g(y))Q(d(x,y)) = 0 \qquad \text{for } g \in T(\underline{Q},\mathfrak{P}), \tag{7.7.4}$$

equivalently:

$$\int g(z)Q_1(dz) + \int g(z)Q_2(dz) = 0 \qquad \text{for } g \in T(\underline{Q},\mathfrak{P}). \tag{7.7.5}$$

If $\mathfrak{P}$ contains with Q_1,Q_2 also the p-measure $\frac{1}{2}Q_1 + \frac{1}{2}Q_2$, then $\underline{Q} = \frac{1}{2}Q_1 + \frac{1}{2}Q_2$ is the trivial solution of (7.7.5). If $\mathfrak{P}$ is loc. as. convex (see Definition 1.3.1), then there exists at least an approximate projection for Q close to $\mathfrak{P}^2$.

7.8. Projections into a family of symmetric distributions

Let $\mathfrak{P}$ denote the family of all p-measures on $\mathbb{B}$ with symmetric and positive Lebesgue density. Our problem is to determine the projection of an arbitrary p-measure $Q|\mathbb{B}$ into $\mathfrak{P}$. According to Propositions 2.3.1 and 2.3.3,

$$T(P,\mathfrak{P}) = [\ell'(\cdot,P)] \oplus \Psi(P),$$

where $\Psi(P)$ is the class of all functions in $\mathscr{L}_*(P)$ which are symmetric about the median of P. According to (7.2.2), the projection $\overline{Q}$ of Q into $\mathfrak{P}$ is determined by

$$Q(a\ell'(\cdot,\overline{Q}) + \psi) = 0 \qquad \text{for all } a \in \mathbb{R},\ \psi \in \Psi(\overline{Q}). \tag{7.8.1}$$

This relation implies

$$Q(\ell'(\cdot,\overline{Q})) = 0, \tag{7.8.2}$$

$$Q(\psi) = 0 \qquad \text{for all } \psi \in \Psi(\overline{Q}). \tag{7.8.3}$$

Relation (7.8.3) can be written as

$$\int \psi(\xi)q(\xi)d\xi = 0 \qquad \text{for all } \psi \in \Psi(\overline{Q}). \tag{7.8.4}$$

Let M denote the center of symmetry of $\overline{Q}$. Since ψ is symmetric about M, this implies

(7.8.5) $\qquad \int\psi(\xi)q(2M-\xi)d\xi = 0 \qquad$ for all $\psi \in \Psi(\bar{Q})$.

Let $q_M(\xi) := \frac{1}{2}q(\xi) + \frac{1}{2}q(2M-\xi)$. By (7.8.4) and (7.8.5), $\int\psi(\xi)q_M(\xi)d\xi = 0$ for all ψ which are symmetric about M and fulfill $\int\psi(\xi)\bar{q}(\xi)d\xi = 0$. Since both, q_M and $\bar{q}$, are symmetric about M, this implies $\bar{q} = q_M$ Lebesgue-a.e. In other words, the projection $\bar{Q}$ has Lebesgue density $x \rightarrow \frac{1}{2}q(x) + \frac{1}{2}q(2M-x)$. Finally, condition (7.8.2) can be used to determine M. We can write (7.8.2) as

$$\int \frac{q'(x)-q'(2M-x)}{q(x)+q(2M-x)} q(x)dx = 0 ,$$

equivalently,

(7.8.6) $$\int \frac{q(x)-q(2M-x)}{q(x)+q(2M-x)} q'(x)dx = 0 .$$

We expect that the function

$$M \rightarrow \int \frac{q(x)-q(2M-x)}{q(x)+q(2M-x)} q'(x)dx$$

will be negative for some M smaller than the median of Q and positive for some M larger than the median of Q if q is unimodal. Observe that the derivative of this function is

$$M \rightarrow -4\int \frac{q'(2M-x)q'(x)}{(q(x)+q(2M-x))^2} q(x)dx .$$

Hence a solution of (7.8.6) will exist.

As an alternative, $\bar{Q}$ could be determined as the measure in $\mathfrak{P}$ which minimizes a certain distance function. For Q an estimate, this was carried through by Beran (1978, pp. 295ff.) for the Hellinger distance.

8. ASYMPTOTIC BOUNDS FOR THE POWER OF TESTS

8.1. Hypotheses and co-spaces

Let $(X,\mathscr{A})$ be a measurable space, and $\mathfrak{P}$ the basic family of p-measures $Q|\mathscr{A}$. Let $\mathfrak{P}_0 \subset \mathfrak{P}$ be a subfamily, interpreted as a *hypothesis* which is to be tested against alternatives from $\mathfrak{P} - \mathfrak{P}_0$.

For each p-measure $P \in \mathfrak{P}_0$, we consider two tangent spaces, $T(P,\mathfrak{P})$ and the subspace $T(P,\mathfrak{P}_0)$. Whether an asymptotic solution for a testing problem is simple or not depends essentially on the relationship between these two tangent spaces. To explain this, we introduce for $P \in \mathfrak{P}_0$ the *co-space* $T^{\perp}(P;\mathfrak{P}_0,\mathfrak{P})$ as the orthogonal complement of $T(P,\mathfrak{P}_0)$ in $T(P,\mathfrak{P})$, i.e. we represent $T(P,\mathfrak{P})$ as

$$T(P,\mathfrak{P}) = T(P,\mathfrak{P}_0) \oplus T^{\perp}(P;\mathfrak{P}_0,\mathfrak{P}).$$

In other words, we represent each function $g \in T(P,\mathfrak{P})$ as the sum of its projection $g_0 \in T(P,\mathfrak{P}_0)$ and a function $k \in T^{\perp}(P;\mathfrak{P}_0,\mathfrak{P})$ which is orthogonal to $T(P,\mathfrak{P}_0)$, i.e.

$$g = g_0 + k .$$

This representation is unique.

8.1.1. Example. Let $\mathfrak{P}$ denote a full family of p-measures on $\mathbb{B}$ which are equivalent to the Lebesgue measure, and $\mathfrak{P}_0$ the subfamily of all measures in $\mathfrak{P}$ which are symmetric about 0. Under suitable regularity conditions (see Propositions 2.3.1 and 2.3.3) we have $T(P,\mathfrak{P}) = \mathscr{L}_*(P)$ and $T(P,\mathfrak{P}_o) = \Psi(P)$ (the class of all functions in $\mathscr{L}_*(P)$ which are symmetric about the median of P). Then $T^{\perp}(P;\mathfrak{P}_o,\mathfrak{P})$ consists of all functions in $\mathscr{L}_*(P)$ which are skew-symmetric about 0.

If $\mathfrak{Q} \subset \mathfrak{P}$ denotes the subfamily of all symmetric p-measures, then for $P \in \mathfrak{Q}$ we have (see Propositions 2.3.1 and 2.3.3) $T(P,\mathfrak{Q})$ $= [\ell'(\cdot,P)] \oplus \Psi(P)$ and therefore $T^{\perp}(P;\mathfrak{P}_o,\mathfrak{Q}) = [\ell'(\cdot,P)]$.

8.1.2. Example. Let $\mathfrak{P}$ be a family of p-measures with tangent space $T(P,\mathfrak{P})$. Given a functional $\kappa: \mathfrak{P} \to \mathbb{R}$, let $\mathfrak{P}_o$ be the hypothesis that κ attains a prescribed value c_o, i.e. $\mathfrak{P}_o = \{P \in \mathfrak{P}: \kappa(P) = c_o\}$. If κ has a gradient $\kappa^{\cdot}(\cdot,P) \in T(P,\mathfrak{P})$, then by Proposition 4.5.1

$$T(P,\mathfrak{P}_o) = \{g \in T(P,\mathfrak{P}): P(g\kappa^{\cdot}(\cdot,P)) = 0\}.$$

Correspondingly,

$$T^{\perp}(P;\mathfrak{P}_o,\mathfrak{P}) = [\kappa^{\cdot}(\cdot,P)].$$

8.1.3. Example. Let $\mathfrak{P} = \{P_\theta: \theta \in \Theta\}$, $\Theta \subset \mathbb{R}^k$, be a family of mutually absolutely continuous p-measures, and $\mathfrak{P}_o = \{P_\theta: \theta \in \Theta,\ F(\theta) = 0\}$, where the function $F: \Theta \to \mathbb{R}$, occuring in the side condition, is differentiable. For the functional $\kappa(P_\theta) = F(\theta)$ we obtain from Propositions 2.2.1, 4.4.2 and 5.3.1 the canonical gradient

$$\kappa^{\cdot}(\cdot,P_\theta) = F^{(\cdot)}(\theta)'\Lambda(\theta)\ell^{(\cdot)}(\cdot,\theta).$$

From Example 8.1.2 we obtain

$$T^{\perp}(P_\theta;\mathfrak{P}_o,\mathfrak{P}) = [F^{(\cdot)}(\theta)'\Lambda(\theta)\ell^{(\cdot)}(\cdot,\theta)].$$

8.1.4. Example. Let $\mathfrak{P}_o = \{P_\theta: \theta \in \Theta\}$, $\Theta \subset \mathbb{R}^k$, be a parametric family of mutually absolutely continuous p-measures, and $\mathfrak{P}$ the family of all p-measures which are mutually absolutely continuous with $\mathfrak{P}_o$. Then

$$T(P_\theta,\mathfrak{P}_o) = [\ell^{(1)}(\cdot,\theta),\dots,\ell^{(k)}(\cdot,\theta)],$$

$$T(P_\theta,\mathfrak{P}) = \mathscr{L}_*(P_\theta),$$

and $\quad T^{\perp}(P_\theta;\mathfrak{P}_o,\mathfrak{P}) = \{g \in \mathscr{L}_*(P_\theta): P(g\ell^{(i)}(\cdot,\theta)) = 0 \quad \text{for } i=1,\ldots,k\}$.
In this case, the co-space is infinite-dimensional.

8.1.5. Example. Let $\mathfrak{P}$ denote a full family of mutually absolutely continuous p-measures on $\mathscr{A}_1 \times \mathscr{A}_2$, and $\mathfrak{P}_o$ the subfamily of all product measures. We have

$$T(P,\mathfrak{P}) = \mathscr{L}_*(P)$$

and by Proposition 2.4.1,

$$T(P_1 \times P_2,\mathfrak{P}_o) = \{(x,y) \to g_1(x) + g_2(y): g_i \in \mathscr{L}_*(P_i),\ i=1,2\}.$$

$T^{\perp}(P_1 \times P_2;\mathfrak{P}_o,\mathfrak{P})$ consists of all functions $k \in \mathscr{L}_*(P_1 \times P_2)$ such that

$$\int\int k(x,y)(g_1(x)+g_2(y))P_1(dx)P_2(dy) = 0 \quad \text{for all } g_i \in \mathscr{L}_*(P_i),\ i=1,2,$$

i.e.,

$$\int k(x,\eta)P_2(d\eta) = 0 \qquad \text{for } P_1\text{-a.a. } x \in \mathbb{R}$$

and

$$\int k(\xi,y)P_1(d\xi) = 0 \qquad \text{for } P_2\text{-a.a. } y \in \mathbb{R}.$$

Thus the co-space $T^{\perp}(P_1 \times P_2;\mathfrak{P}_o,\mathfrak{P})$ has an intuitive interpretation: p-measures which differ from $P_1 \times P_2$ in a direction belonging to $T^{\perp}(P_1 \times P_2;\mathfrak{P}_o,\mathfrak{P})$ have locally the same marginals P_1 and P_2. An arbitrary $g \in \mathscr{L}_*(P_1 \times P_2)$ can be represented as

$$g(x,y) = g_1(x) + g_2(y) + k(x,y)$$

with $((x,y) \to g_1(x)+g_2(y)) \in T(P_1 \times P_2,\mathfrak{P}_o)$ and $k \in T^{\perp}(P_1 \times P_2;\mathfrak{P}_o,\mathfrak{P})$, where

$$g_1(x) = \int g(x,\eta)P_2(d\eta)\ ,$$
$$g_2(y) = \int g(\xi,y)P_1(d\xi)\ ,$$
$$k(x,y) = g(x,y) - (g_1(x)+g_2(y))\ .$$

8.1.6. Example. For any $Q|\mathbb{B}$, let $Q_a := Q*(x \to x+a)$. Let $\mathfrak{Q}$ denote a full family of p-measures on $\mathbb{B}$ with positive Lebesgue density, and let

$$\mathfrak{P} = \{Q \times Q_a: Q \in \mathfrak{Q},\ a \in \mathbb{R}\}$$
$$\mathfrak{P}_o = \{Q \times Q: Q \in \mathfrak{Q}\}\ .$$

From Proposition 2.4.1 and Example 2.2.4 we obtain

$T(Q \times Q, \mathfrak{P}_o) = \{(x,y) \to g(x) + g(y) : g \in \mathscr{L}_*(Q)\},$

$T(Q \times Q, \mathfrak{P}) = \{(x,y) \to g(x) + g(y) + a\ell'(y,Q) : g \in \mathscr{L}_*(Q),\ a \in \mathbb{R}\}.$

The condition that $(x,y) \to g_o(x) + g_o(y) + a_o\ell'(y,Q)$ be orthogonal to $T(Q \times Q, \mathfrak{P}_o)$ is equivalent to

$$\int (g_o(x)+g_o(y)+a_o\ell'(y,Q))(g(x)+g(y))Q(dx)Q(dy) = 0 \text{ for all } g \in \mathscr{L}_*(Q).$$

Equivalently,

$$\int [2g_o(y) + a_o\ell'(y,Q)]g(y)Q(dy) = 0 \qquad \text{for all } g \in \mathscr{L}_*(Q)$$

and therefore

$$2g_o(y) + a_o\ell'(y,Q) = c \qquad \text{for } \lambda\text{-a.a. } y \in \mathbb{R}.$$

Since g_o and $\ell'(\cdot,Q)$ have expectation 0, this implies

$$g_o(y) = -\tfrac{1}{2}a_o\ell'(y,Q) \qquad \text{for } \lambda\text{-a.a. } y \in \mathbb{R}.$$

Hence

$$g_o(x) + g_o(y) + a_o\ell'(y,Q) = \tfrac{1}{2}a_o(\ell'(y,Q) - \ell'(x,Q)) \text{ for } \lambda^2\text{-a.a. } (x,y).$$

Therefore, $T^{\perp}(Q \times Q; \mathfrak{P}_o, \mathfrak{P})$ is the one-dimensional space generated by the function $(x,y) \to \ell'(y,Q) - \ell'(x,Q)$.

8.1.7. Remark. For $i = 1,\dots,k$ let $\mathfrak{P}_i \subset \mathfrak{P}$ be a subfamily with tangent space $T(P,\mathfrak{P}_i)$. For $P \in \bigcap_{i=1}^{k} \mathfrak{P}_i$ let

$$T(P,\mathfrak{P}) = T(P,\mathfrak{P}_i) \oplus T^{\perp}(P;\mathfrak{P}_i,\mathfrak{P}).$$

If the hypothesis is $\mathfrak{P}_o = \bigcap_{i=1}^{k} \mathfrak{P}_i$, then

$$T(P,\mathfrak{P}_o) = \bigcap_{i=1}^{k} T(P,\mathfrak{P}_i)$$

and $T^{\perp}(P;\mathfrak{P}_o,\mathfrak{P})$ is the linear space spanned by $T^{\perp}(P;\mathfrak{P}_i,\mathfrak{P})$, $i = 1,\dots,k$.

8.1.8. Corollary. If we are given k differentiable functionals $\kappa_i : \mathfrak{P} \to \mathbb{R}$ with gradients $\dot\kappa_i(\cdot,P) \in T(P,\mathfrak{P})$, $i = 1,\dots,k$, and the hypothesis is $\mathfrak{P}_o = \{P \in \mathfrak{P} : \kappa_i(P) = c_i,\ i = 1,\dots,k\}$, then

$$T(P,\mathfrak{P}_o) = \{g \in T(P,\mathfrak{P}) : P(g\dot\kappa_i(\cdot,P)) = 0,\ i = 1,\dots,k\}$$

and

$$T^{\perp}(P;\mathfrak{P}_o,\mathfrak{P}) = [\dot\kappa_1(\cdot,P),\dots,\dot\kappa_k(\cdot,P)].$$

8.1.9. Remark. The co-space may be trivial (i.e. $T^{\perp}(P;\mathfrak{P}_o,\mathfrak{P}) = \{o\}$) in degenerate cases. (See Example 2.2.5, where $T(P,\mathfrak{P}_o) = T(P,\mathfrak{P})$ despite the fact that $\mathfrak{P}_o$ is a genuine subfamily of $\mathfrak{P}$.)

8.2. The dimension of the co-space

The dimension of the co-space of a hypothesis can be anything between zero and infinity. The following sections are for the most part devoted to one-dimensional co-spaces, because multidimensional co-spaces pose problems of an essentially different nature.

If the co-space is multidimensional, the alternative may deviate from the hypothesis in different directions. Deviations in different directions will, in general, differ in the likelihood of their occurence, and in their consequences. On the other hand, we are free to allocate the power of our test over the different directions. Without any knowledge about the likelihood of occurence and without weighting the consequences, we have no basis for the allocation of testing power.

As an example, consider a production process which renders items with a quality characteristic which is normally distributed. Let (μ_o,σ_o) specify the current state of the production process. The purpose of the test is to discover deviations from the current state. Statisticians usually formulate this as the problem of testing the hypothesis $(\mu,\sigma) = (\mu_o,\sigma_o)$ against the alternatives $(\mu,\sigma) \neq (\mu_o,\sigma_o)$. This formulation as a mathematical problem neglects essential aspects of the practical problem, for instance: That changes of μ occur much more frequently than changes of σ. That σ increases (but never decreases) if it changes at all. Usually a change of σ is much more momentous than a change of μ. A change of μ may mean that the machine has to be reset, an increase of σ may mean that the machine has to be replaced.

It is not surprising that the formulation as a testing problem which brushes aside such aspects leaves us wondering what to do. The surprising thing is that statisticians try to 'solve' such indeterminate testing problems.

What are the 'solutions' offered in literature? Without knowing anything about likelihood and relevance of the alternatives, there is no other possibility than to choose a test which looks plausible, and to accept the allocation of power over the different alternatives which this test brings about. If the problem is a parametric one, most statisticians may find it appropriate to use the likelihood ratio test (or any other as. equivalent test). Knowing that the as. power function of this test is constant on ellipses $(\mu-\mu_o)^2 + 2(\sigma-\sigma_o)^2$ = const, one may question whether such a test is adequate for the practical problem, which gave rise to the testing problem $(\mu,\sigma) = (\mu_o,\sigma_o)$ against $(\mu,\sigma) \neq (\mu_o,\sigma_o)$.

The careless way in which such 'omnibus' tests are presented in textbooks is in remarkable contrast to the care which Neyman devotes to the discussion of the choice of alternatives (see, e.g., Neyman, 1941, section 5).

If the model is nonparametric, the situation becomes worse. In such situations it is popular to suggest a certain test, and let other people find out what the test really does. Tests based on the distance between the empirical distribution function and the hypothesis $N(\mu_o,\sigma_o)$ are certainly plausible. Yet it took about forty years to get a rough idea of what tests based on the Cramér-von Mises distance really do to the different alternatives. With the persuading suggestion of Durbin and Knott (1972) to use orthogonal components of the Cramér-von Mises statistic for testing a hypothesis like $N(\mu_o, \sigma_o)$, we are, finally, back to a procedure in the spirit of Neyman's smooth tests (see Neyman 1937).

Many authors find it delightful to prove that certain tests are as. most powerful against certain (contiguous) alternatives. This is a popular game, because almost all tests occuring in the literature turn out to be as. most powerful against some alternatives. Surprisingly enough, being as. most powerful is considered as a virtue, neglecting the fact that a test which is as. most powerful against certain alternatives is bound to have as. power zero against certain other alternatives, if the co-space is multidimensional (see Section 8.5).

Another favorite resort for statisticians confronted with an indeterminate testing problem is to switch tacitly to another testing problem. Consider as an example the problem of testing for independence. Many authors try to 'solve' this problem by replacing the hypothesis of independence by the hypothesis that a certain measure of dependence (like the correlation coefficient) is zero. Formally, this amounts to replacing the original hypothesis by a larger one, thereby reducing the dimension of the co-space from infinity to one.

If the measure of dependence is adequate for the problem, then this procedure is legitimate. (The objection remains that then the original hypothesis was inadequate.) In most cases, however, the measure of dependence comes to the original problem as some sort of 'deus ex machina'. Considered as a test for the original hypothesis, such a test is of objectionable performance if the measure of dependence attains the value 0 also in certain cases of dependence.

Tests for independence based on the distance between the two-dimensional empirical distribution function and the product of its marginals are free from this defect. But again: Depending on the distance function chosen, they allocate their power somehow over the different kinds of dependence - and nobody knows how.

For these reasons, the following considerations place particular emphasis upon tests with one-dimensional co-space.

8.3. The concept of asymptotic power functions

The *power function* of a test φ_n is $Q \to Q^n(\varphi_n)$. The *asymptotic power function* is a comparatively complex construct, referring to a s e q u e n c e of tests, evaluated under a sequence of contiguous alternatives, converging to an element of the hypothesis.

Consider as an example the simplest case, that of a two-parameter family $\mathfrak{P} = \{P_{\theta,\tau}: (\theta,\tau) \in \Theta \times T\}$ and the hypothesis $\mathfrak{P}_o = \{P_{\theta_o,\tau}: \tau \in T\}$ Then it is common to consider a sequence of alternatives $P_{\theta_o+n^{-1/2}s,\tau}$, $n \in \mathbb{N}$.

In the general case of an arbitrary family $\mathfrak{P}$ and an arbitrary hypothesis $\mathfrak{P}_o$, the analogue to sequences $P_{\theta_o+n^{-1/2}s,\tau}$, $n \in \mathbb{N}$, are sequences $P_{n^{-1/2},tg}$, $n \in \mathbb{N}$, with $P \in \mathfrak{P}_o$ and $g \in T(P,\mathfrak{P})$. Under appropriate regularity conditions we find for reasonable test-sequences φ_n, $n \in \mathbb{N}$, that $P^n_{n^{-1/2},tg}(\varphi_n)$, $n \in \mathbb{N}$, converges to $\Phi(N_\alpha + ta_P(g))$. If this is the case, we may take $\Phi(N_\alpha + ta_P(g))$ as the asymptotic power function of the sequence φ_n, $n \in \mathbb{N}$. The factor $a_P(g)$ is the slope of the as. power function against alternatives $P_{n^{-1/2},tg}$. Without such a concise expression any statement about as. envelope power functions would necessarily involve the level α, which is usually irrelevant in comparisons of as. envelope power functions.

Notice that this definition of the slope incorporates an arbitrary element: The factorization tg is arbitrary, and may also be written as $(t/r)rg$, say. Hence, if $P^n_{n^{-1/2},tg}(\varphi_n)$ converges for all $t \in \mathbb{R}$ and $g \in T(P,\mathfrak{P})$ to $\Phi(N_\alpha + ta_P(g))$, then a_P is necessarily a homogeneous function of g, i.e. $a_P(rg) = ra_P(g)$. The homogeneity of a_P suggests to use a standardized version, say $g \to a_P(g/\|g\|)$. For assessing the as. efficiency of a test-sequence it is the r a t i o of the slope of its as. power function to the slope of the as. envelope power function which counts. Hence any non-standardized version of a_P serves this purpose equally well.

What are the possible uses of as. power functions? First of all, to provide numerical approximations to the true power functions, because these are usually difficult to compute. Second, to reveal the general structure of power functions, i.e., to make transparent how the power depends on the alternative. If numerical computations were the only base for comparing power functions, it would, probably, be difficult to arrive at general conclusions (such as: tests constructed by method A are in regular cases approximately optimal). What we need are approximations to the true power functions which are, for one thing, transparent enough to allow general conclusions, and which are, for another thing, accurate enough so that the general conclusions derived from these approximations reflect the relationships between the true power functions with sufficient accuracy.

It is not obvious to us that 'asymptotic power functions' in the usual sense serve these purposes. In the first place, we have to disentangle ourselves from the idea - almost universally accepted in literature - of a sequence of alternatives moving towards the hypothesis. Nothing moves in reality. We have a certain alternative, say Q, and we wish to approximate $Q^n(\varphi_n)$. Considering this alternative as an element of a sequence of alternatives converging to the hypothesis brings an arbitrary aspect to our consideration which can in no way contribute to our problem of approximating $Q^n(\varphi_n)$.

Among the infinity of sequences into which Q can be embedded, there is, perhaps, one natural choice: to consider Q as an element of the sequence which converges to $\overline{Q}$, the projection of Q into $\mathfrak{P}_o$, i.e. a sequence converging to $\mathfrak{P}_o$ from a direction orthogonal to $T(\overline{Q},\mathfrak{P}_o)$.

However: The sequence $P_{\theta_o+n^{-1/2}s,\tau}$, $n \in \mathbb{N}$, which appeared so natural to us at the beginning of this section, is not of this type! Orthogonality requires sequences $P_{\theta_o+n^{-1/2}s,\tau+n^{-1/2}ta(\tau)}$, $n \in \mathbb{N}$, with $a_i(\tau) = \Lambda_{oo}(\theta_o,\tau)^{-1}\Lambda_{oi}(\theta_o,\tau)$, $i = 1,\dots,k$.

In the following sections, the reader will find approximations to power functions of the following type:

(8.3.1) $$Q^n(\varphi_n) = \Phi(N_\alpha + n^{1/2}\delta(Q,\mathfrak{P}_o)) + R_n(P) ,$$

where $\delta(Q,\mathfrak{P}_o)$ measures the distance of Q from the hypothesis $\mathfrak{P}_o$.

The alternatives Q for which the approximation of $Q^n(\varphi_n)$ by $\Phi(N_\alpha + n^{1/2}\delta(Q,\mathfrak{P}_o))$ is useful will depend on the sample size n: They will be those alternatives for which $\Phi(N_\alpha + n^{1/2}\delta(Q,\mathfrak{P}_o))$ is neither too close to 0 nor too close to 1, say alternatives Q for which $n^{1/2}\delta(Q,\mathfrak{P}_o)$ is between 0 and $2|N_\alpha|$. If the sample size increases, this set of 'relevant alternatives' shrinks towards the hypothesis. At the same time, the approximation of $Q^n(\varphi_n)$ by $\Phi(N_\alpha + n^{1/2}\delta(Q,\mathfrak{P}_o))$, applied to the relevant alternatives Q, becomes more and more accurate.

How can we describe this second effect connected with an increasing sample size? It is meaningless, of course, to state that $R_n(Q)$ converges to zero as n tends to infinity, since the *relevant* alternatives Q differ for different n. What we need is that R_n, applied to alternatives relevant for the sample size n, attains values which tend to zero as n tends to infinity. In other words: A formula like (8.3.1) is meaningful if

(8.3.2) $$\sup\{R_n(Q): Q \in \mathfrak{P},\ \delta(Q,\mathfrak{P}_o) \le cn^{-1/2}\} = o(n^o) .$$

Such a relation will certainly hold true for regular parametric families, at least as long as Q remains bounded away from the boundary of $\mathfrak{P}$. In the general case of a hypothesis with infinite-dimensional co-space, the situation may be different. If we think of Q as represented by its *distance* from the hypothesis $\mathfrak{P}_o$ and the *direction* in which Q deviates from $\mathfrak{P}_o$, then we cannot take for granted that $\delta(Q,\mathfrak{P}_o) \downarrow 0$ implies that $R_n(Q) \downarrow 0$ uniformly over all directions. This effect does not occur for parametric families, because here the co-space is finite-dimensional.

Even in regular cases, approximations like (8.3.1), based on the normal distribution, have an error term $R_n(Q)$ for which $\sup\{R_n(Q): Q \in \mathfrak{P}, \delta(Q,\mathfrak{P}_o) \leq cn^{-1/2}\}$ decreases like $n^{-1/2}$, a rather slow rate. Numerical computations show that in most cases the error of such approximations is not at all negligible, even for sample sizes of several hundred. For moderate sample sizes, a formula like (8.3.1) can, therefore, hardly be used as a satisfactory approximation. Its real meaning is that of a prototype, refinements of which may eventually become useful for practical purposes.

8.4. The asymptotic envelope power function

Let $\mathfrak{P}$ be a family of p-measures, and $\mathfrak{P}_o \subset \mathfrak{P}$ the hypothesis. The following theorem gives an as. bound for the power of tests for the hypothesis $\mathfrak{P}_o$, expressed in terms of the distance of the alternative from $\mathfrak{P}_o$. A meaningful interpretation of such a bound requires an error term which becomes negligible uniformly over a class of alternatives which shrinks towards the hypothesis. (See Section 8.3 for a discussion of this approach.) Hence certain assumptions of the theorem are to hold uniformly over alternatives.

For regular parametric families, these assumptions will usually be fulfilled. In the general case, they will hold uniformly only for certain subfamilies of $\mathfrak{P}$. The assertion of the following theorem is to be interpreted in the sense that the sequences s_n and δ_n depend only on the bounds in the assumptions on $\mathfrak{P}$ and $T^{\perp}(\cdot;\mathfrak{P}_o,\mathfrak{P})$.

The same reservations have to be made for the uniformity of the level required by (8.4.2). What is really needed is uniformity over those p-measures $\bar{Q} \in \mathfrak{P}_o$ which occur as projections as Q varies in a certain subfamily of $\mathfrak{P}$.

For each p-measure $Q \in \mathfrak{P}$ let $\bar{Q}$ denote its projection into $\mathfrak{P}_o$ (see Definition 7.2.1). To keep the conditions of the following theorem transparent, we assume that $\Delta(Q;\bar{Q}) < \infty$. Let $k_Q := q/\bar{q} - 1$, and let g_Q denote the projection of k_Q into $T(\bar{Q},\mathfrak{P})$. Since k_Q is orthogonal to $T(\bar{Q},\mathfrak{P})$ by definition of $\bar{Q}$, g_Q shares this property. The theorem assumes $\|k_Q - g_Q\|_{\bar{Q}} = o(\Delta(Q;\bar{Q}))$ uniformly for $Q \in \mathfrak{P}$. By Proposition 7.3.1, this is always true if $\mathfrak{P}$ is at P approximable by $T_s(P,\mathfrak{P})$ uniformly for $P \in \bar{\mathfrak{P}}$. Intuitively, the condition $\|k_Q - g_Q\|_{\bar{Q}} = o(\Delta(Q;\bar{Q}))$ is much weaker, since it requires this approximability only for directions orthogonal to the tangent space of the hypothesis.

8.4.1. Theorem. *Assume that*

(i) $\|k_Q - g_Q\|_{\bar{Q}} = o(\Delta(Q;\bar{Q}))$ *uniformly for* $Q \in \mathfrak{P}$;

(ii) The class of functions $g/\|g\|$, $g \in T_s^{\perp}(P;\mathfrak{P}_o,\mathfrak{P})$, *is uniformly* P*-square integrable, uniformly for* $P \in \mathfrak{P}_o$.

Let φ_n, $n \in \mathbb{N}$, *be a test-sequence of as. level* α *for* $\mathfrak{P}_o$ *in the sense that*

(8.4.2) $\quad \alpha_n := \max\{\alpha, \sup\{P^n(\varphi_n) : P \in \mathfrak{P}_o\}\} = \alpha + o(n^o)$.

Then there exist sequences $s_n \uparrow \infty$ *and* $\delta_n \downarrow 0$ *such that for all* $n \in \mathbb{N}$ *and all* $Q \in \mathfrak{P}$ *with* $\Delta(Q;\bar{Q}) \leq n^{-1/2}s_n$

$$Q^n(\varphi_n) \leq \Phi(N_\alpha + n^{1/2}\Delta(Q;\bar{Q})) + \delta_n + (\alpha_n - \alpha)/\alpha \ .$$

The form of the error term is chosen to separate the influence of the approximation error, bounded by δ_n, which depends on the regularity of the family $\mathfrak{P}$, and the error resulting from the level of the test-sequence exceeding α.

Proof. Define

$$c_n := -n^{1/2}\|g_Q\|_{\bar{Q}} N_\alpha - \tfrac{1}{2}n\|g_Q\|_{\bar{Q}}^2 - \varepsilon_n ,$$

$$C_n := \{\sum_{\nu=1}^{n} \log(q(x_\nu)/\bar{q}(x_\nu)) > c_n\}.$$

(We abstain from expressing the dependence of c_n and C_n on Q.) By Corollary 19.2.9, there exist sequences $s_n \uparrow \infty$, $\delta_n \downarrow 0$ and $\varepsilon_n \downarrow 0$ such that for all $n \in \mathbb{N}$ and all $Q \in \mathfrak{P}$ with $\|g_Q\|_{\bar{Q}} \le n^{-1/2}s_n$,

$$\alpha \le \bar{Q}^n(C_n) \le \alpha + \delta_n ,$$

$$|Q^n(C_n) - \Phi(N_\alpha + n^{1/2}\|g_Q\|_{\bar{Q}})| \le \delta_n .$$

The test $\psi_n := \alpha_n^{-1}\alpha\varphi_n$ is of level α. Since C_n is a most powerful critical region for testing $\bar{Q}^n$ against Q^n, we obtain from the Neyman-Pearson Lemma $Q^n(\psi_n) \le Q^n(C_n)$ and therefore

(8.4.3) $$Q^n(\varphi_n) = \alpha^{-1}\alpha_n Q^n(\psi_n) \le \alpha^{-1}\alpha_n Q^n(C_n)$$
$$\le Q^n(C_n) + (\alpha_n-\alpha)/\alpha \le \Phi(N_\alpha + n^{1/2}\|g_Q\|_{\bar{Q}}) + \delta_n + (\alpha_n-\alpha)/\alpha .$$

By assumption,

$$\|g_Q\|_{\bar{Q}} = \Delta(Q;\bar{Q}) + o(\Delta(Q;\bar{Q})) .$$

The assertion now follows from (8.4.3) (with modified sequences δ_n and s_n).

8.4.4. Corollary. *Assume that*

(i) $\mathfrak{P}_0$ *is approximable by* $T_s(\cdot,\mathfrak{P}_0)$,

(ii) $\mathfrak{P}$ *is at* P *approximable by* $T_s(\cdot,\mathfrak{P})$, *uniformly for* $P \in \mathfrak{P}_0$,

(iii) The class of functions $g/\|g\|$, $g \in T_s^{\perp}(P;\mathfrak{P}_0,\mathfrak{P})$, *is uniformly* P*-square integrable uniformly for* $P \in \mathfrak{P}_0$,

(iv) δ *approximates* Δ *at* P *uniformly for* $P \in \mathfrak{P}_0$,

(v) $\Delta(Q;P) = O(\delta(Q,P))$ *uniformly for* $Q \in \mathfrak{P}$ *and* $P \in \mathfrak{P}_0$,

(vi) The projections are chosen such that $\delta(Q,\bar{Q}) = O(\delta(Q,\mathfrak{P}_0))$ *uniformly for* $Q \in \mathfrak{P}$.

If φ_n, $n \in \mathbb{N}$, *is of as. level* α *in the sense of (8.4.2), then there exist* $s_n \uparrow \infty$ *and* $\delta_n \downarrow 0$ *such that for all* $n \in \mathbb{N}$ *and all* $Q \in \mathfrak{P}$ *with* $\delta(Q,\mathfrak{P}_0) \leq n^{-1/2} s_n$

$$Q^n(\varphi_n) \leq \Phi(N_\alpha + n^{1/2}\delta(Q,\mathfrak{P}_0)) + \delta_n + (\alpha_n - \alpha)/\alpha.$$

Proof. By Proposition 7.3.1, the assumptions of Theorem 8.4.1 are fulfilled. Proposition 7.3.2 implies uniformly for $Q \in \mathfrak{P}$

$$\delta(Q,\overline{Q}) = \delta(Q,\mathfrak{P}_0) + o(\delta(Q,\overline{Q})) = \delta(Q,\mathfrak{P}_0) + o(\delta(Q,\mathfrak{P}_0)) .$$

The function $Q \to \Phi(N_\alpha + n^{1/2}\delta(Q,\mathfrak{P}_0))$ will be called *as. envelope power function.* This is justified by the fact that test-sequences attaining this power function asymptotically exist in regular cases (see Section 8.5).

8.4.5. Remark. The difficulties with the uniformity of the assumptions smooth down if we consider sequences of alternatives, converging to a fixed p-measure $P \in \mathfrak{P}_0$, say $P_{n^{-1/2},tg}$ with $g \in T_s(P,\mathfrak{P})$. Though this is in conflict with our position explained in Section 8.3, there are occasions where it is persuadingly convenient to think in terms of sequences of alternatives. To prepare for such a misuse, we remind the reader (see Proposition 6.2.18) that under strong regularity conditions, $\delta(P_{n^{-1/2},tg},\mathfrak{P}_0) = n^{-1/2}t\|g-g_0\| + o(n^{-1/2}t)$, where g_0 denotes the projection of g into $T(P,\mathfrak{P}_0)$. Under weaker regularity conditions (compare Theorem 8.5.3), the function $t \to \Phi(N_\alpha + t\|g-g_0\|)$ can be interpreted as an as. envelope power function for alternatives $P_{n^{-1/2},tg}$ in the sense that any test-sequence φ_n, $n \in \mathbb{N}$, of as. level α for a sequence $P_{n^{-1/2},tg_0}$ in the hypothesis fulfills

$$P^n_{n^{-1/2},tg}(\varphi_n) \leq \Phi(N_\alpha + t\|g-g_0\|) + o(n^0) .$$

In agreement with the terminology introduced in Section 8.3 we speak of $\|g-g_o\|$ as the *slope* of the as. envelope power function against alternatives $P^n_{n^{-1/2},tg}$.

8.4.6. Remark. Consider the particular case of a hypothesis determined by a k-dimensional functional $\kappa: \mathfrak{P} \to \mathbb{R}^k$, say

$$\mathfrak{P}_o = \{P \in \mathfrak{P}: \kappa(P) = c\}.$$

By Remark 7.3.6 we have uniformly for $Q \in \mathfrak{P}$

$$\Delta(Q;\overline{Q})^2 = (\kappa(Q) - c)'D(Q)(\kappa(Q) - c) + o(\Delta(Q;\overline{Q})^2)$$

where $D(Q)$ is the inverse of $\Sigma(Q) := Q(\kappa^{\cdot}(\cdot,Q)\kappa^{\cdot}(\cdot,Q)')$. For every test-sequence φ_n, $n \in \mathbb{N}$, of as. level α we obtain from Theorem 8.4.1 uniformly for $Q \in \mathfrak{P}$ with $|\kappa(Q) - c| \leq n^{-1/2}s_n$

$$Q^n(\varphi_n) \leq \Phi(N_\alpha + n^{1/2}((\kappa(Q)-c)'D(Q)(\kappa(Q)-c))^{1/2}) + o(n^o).$$

For $P \in \mathfrak{P}_o$ and paths $P_{t,g}$, $t \downarrow 0$, with strong derivative $g \in T_s(P,\mathfrak{P})$ we have

$$\kappa(P_{t,g}) - c = \kappa(P_{t,g}) - \kappa(P) = tP(g\kappa^{\cdot}(\cdot,P)) + o(t).$$

Hence (see Remark 8.4.5) the as. envelope power function against alternatives $P_{n^{-1/2},tg}$, $n \in \mathbb{N}$, has the slope

(8.4.7) $$(P(g\kappa^{\cdot}(\cdot,P))'D(P)P(g\kappa^{\cdot}(\cdot,P)))^{1/2}.$$

In the case of a real-valued functional, the slope becomes

(8.4.8) $$P(g\kappa^{\cdot}(\cdot,P))/P(\kappa^{\cdot}(\cdot,P)^2)^{1/2}.$$

8.4.9. Remark. Let $\kappa: \mathfrak{P} \to \mathbb{R}$ be a functional. Since $T^{\perp}(P;\mathfrak{P}_o,\mathfrak{P}) = [\kappa^{\cdot}(\cdot,P)]$ in the case of a hypothesis $\mathfrak{P}_o = \{P \in \mathfrak{P}: \kappa(P) = c\}$, the alternatives $P_{n^{-1/2},tg}$ with $g \in [\kappa^{\cdot}(\cdot,P)]$ are the natural ones. If we choose $g_o = \kappa^{\cdot}(\cdot,P)/P(\kappa^{\cdot}(\cdot,P)^2)$, we have

$$\lim_{t\to 0} t^{-1}(\kappa(P_{t,g_o}) - \kappa(P)) = 1.$$

Hence the length of g_o is related in a natural way to the change of κ. From (8.4.8) we obtain that the as. envelope power function for alternatives $P_{n^{-1/2},tg_o}$ has slope $P(\kappa^{\bullet}(\cdot,P)^2)^{-1/2}$. The expression $P(\kappa^{\bullet}(\cdot,P)^2)$ occurs in Theorem 9.2.2 as the as. variance bound for estimators of $\kappa(P)$. This generalizes a relationship well known for k-parameter families: The slope of the as. envelope power function for tests of the hypothesis $\{P_\theta: \theta \in \Theta,\ \theta_1 = \theta_1^o\}$ against alternatives with parameter $(\theta_1^o + n^{-1/2}t, \theta_2,\ldots,\theta_k)$ is $\Lambda_{11}(\theta_1^o,\theta_2,\ldots,\theta_k)^{-1/2}$; the as. variance bound for estimators of θ_1 is $\Lambda_{11}(\theta_1,\ldots,\theta_k)$. (See also the following Remark 8.4.10.)

8.4.10. Remark. We conclude this section with an application of Theorem 8.4.1 to parametric families. Let $\mathfrak{P} = \{P_\theta: \theta \in \Theta\}$ with $\Theta \subset \mathbb{R}^k$ and, for an $m < k$,

$$\mathfrak{P}_o = \{P_\theta: \theta \in \Theta,\ \theta_i = \theta_i^o,\ i = 1,\ldots,m\}.$$

Corresponding to the subdivision of the vector θ into $\theta(1) = (\theta_1,\ldots,\theta_m)$ and $\theta(2) = (\theta_{m+1},\ldots,\theta_k)$ we partition the matrix $L(\theta)$ into

$$L = \begin{pmatrix} L(1,1) & L(1,2) \\ L(2,1) & L(2,2) \end{pmatrix}.$$

By Proposition 5.3.1, the m-dimensional functional $\kappa(P_\theta) = \theta(1)$ has the strong gradient $\kappa^{\bullet}(\cdot,P_\theta)$ with

$$\kappa_i^{\bullet}(\cdot,P_\theta) = \Lambda_{ij}(\theta)\ell^{(j)}(\cdot,\theta), \qquad i = 1,\ldots,m.$$

Since

$$A := L(1,1) - L(1,2)L(2,2)^{-1}L(2,1)$$

is the inverse of

$$\Lambda(1,1) = (P_\theta(\kappa_i^{\bullet}(\cdot,P_\theta)\kappa_j^{\bullet}(\cdot,P_\theta)))_{i,j=1,\ldots,m}$$

we obtain from Remark 7.3.6 uniformly for $\theta \in \Theta$

$$H(P_\theta,\overline{P}_\theta)^2 = (\theta(1) - \theta^o(1))'A(\theta)(\theta(1) - \theta^o(1)) + o(\Delta(P_\theta;\overline{P}_\theta)^2).$$

As in Remark 8.4.6 we obtain for every test-sequence φ_n, $n \in \mathbb{N}$, of as. level α, uniformly for $\theta \in \Theta$ with $\Delta(P_\theta;\overline{P}_\theta) \le n^{-1/2}s_n$

$$P_\theta^n(\varphi_n) \le \Phi(N_\alpha + n^{1/2}((\theta(1)-\theta^o(1))'A(\theta)(\theta(1)-\theta^o(1)))^{1/2}) + o(n^o).$$

The as. envelope power function against alternatives $P_{\theta^o(1)+n^{-1/2}ta,\theta(2)}$, $n \in \mathbb{N}$, with $a \in \mathbb{R}^m$, has the slope $(a'A(\theta^o(1),\theta(2))a)^{1/2}$.

Of particular interest is the case $m = 1$, i.e. the test for the hypothesis $\theta_1 = \theta_1^o$ against alternatives $\theta_1 > \theta_1^o$, say. In this case, the matrix $A(\theta)$ reduces to a real number, which equals $\Lambda_{11}(\theta)^{-1}$. Hence we obtain a bound of the type $\Phi(N_\alpha + n^{1/2}\Lambda_{11}(\theta)^{-1/2}(\theta_1 - \theta_1^o))$. Rescaling θ_1 as $\theta_1^o + n^{-1/2}s$, and using the continuity of $\theta \to \Lambda_{11}(\theta)$, we arrive at the form in which this as. envelope power function is usually written, namely $\Phi(N_\alpha + s\Lambda_{11}(\theta_1^o,\theta_2,\ldots,\theta_k)^{-1/2})$.

8.5. The power function of asymptotically efficient tests

Let φ_n, $n \in \mathbb{N}$, be a test-sequence of as. level α for the hypothesis $\mathfrak{P}_o$. Let δ be a distance function on $\mathfrak{P} \supset \mathfrak{P}_o$ which approximates Δ in the sense of Definition 6.2.1. It was shown in Corollary 8.4.4 that for all $c > 0$ and uniformly for $Q \in \mathfrak{P}$ with $\delta(Q,\mathfrak{P}_o) \leq cn^{-1/2}$,

(8.5.1) $$Q^n(\varphi_n) \leq \Phi(N_\alpha + n^{1/2}\delta(Q,\mathfrak{P}_o)) + o(n^o).$$

Under suitable regularity conditions (see also Chapter 12) there exist test-sequences for which equality holds in (8.5.1) for all alternatives in a subfamily of $\mathfrak{P}$ in which $\mathfrak{P}_o$ has co-dimension one. It is the existence of such test-sequences which justifies the interpretation of $Q \to \Phi(N_\alpha + n^{1/2}\delta(Q,\mathfrak{P}_o))$ as the *as. envelope power function*. We call a test-sequence φ_n, $n \in \mathbb{N}$, *as. efficient* for $\mathfrak{Q} \subset \mathfrak{P}$ if for all $c > 0$ and uniformly for $Q \in \mathfrak{Q}$ with $\delta(Q,\mathfrak{P}_o) \leq cn^{-1/2}$, equality holds in (8.5.1). This definition presumes regularity conditions under which Corollary 8.4.4 holds true.

At first sight, this definition seems quite different from the usual definitions based on sequences of alternatives. To make the connection more explicit, we consider a sequence of alternatives converging to a fixed p-measure $P \in \mathfrak{P}_o$, say $P_{n^{-1/2},tg}$ with $g \in T(P,\mathfrak{P})$. By Remark 8.4.5, we have for all $c > 0$

$$(8.5.2) \qquad P^n_{n^{-1/2},tg}(\varphi_n) \leq \Phi(N_\alpha + t\|g-g_o\|) + o(n^o) ,$$

uniformly for $t \in [0,c]$. The definition of as. efficiency given above becomes in this case: A test-sequence φ_n, $n \in \mathbb{N}$, is as. efficient for $P_{n^{-1/2},tg}$ if for all $c > 0$ equality holds in (8.5.2), uniformly for $t \in [0,c]$.

This definition differs from the usual one only by requiring uniformity in $t \in [0,c]$. It is, however, clear that equality in (8.5.2) for all $t \geq 0$ entails in regular cases that the equality holds uniformly on compact subsets of $[0,\infty)$.

Even more holds true: If equality holds true in (8.5.2) for a certain t_o, then it holds uniformly in t on compact subsets of $[0,\infty)$. This is a particular consequence of the following Theorem 8.5.3 (which generalizes a parametric version given in Pfanzagl, 1974, p.31, Theorem 6).

The theorem shows more generally, that the as. power function of a test-sequence which is as. efficient for a certain sequence of alternatives $P_{n^{-1/2},g}$, is by this fact uniquely determined for all alternatives. It is as. efficient for the alternatives $P_{n^{-1/2},tg}$ with $t > 0$, but as. inefficient for all other alternatives. For $P_{n^{-1/2},th}$ its as. power function has the slope $P(h(g-g_o))/\|g-g_o\|$, compared to the slope of the as. envelope power function, $\|h-h_o\|$ (with h_o the projection of h into $T(P,\mathfrak{P}_o)$). Hence the as. efficiency is $P(h(g-g_o))^2/\|g-g_o\|^2\|h-h_o\|^2$. Since g_o is the projection of g into $T(P,\mathfrak{P}_o)$, we have $P(h(g-g_o)) = P((h-h_o)(g-g_o))$. Hence the as. efficiency is always less than one, unless $h-h_o$ is proportional to $g-g_o$.

In particular: Unless $T^\perp(P;\mathfrak{P}_o,\mathfrak{P}) = [g-g_o]$, there will always be alternatives for which the test-sequence is of as. efficiency zero.

8.5.3. Theorem. *Let* $P \in \mathfrak{P}_o$, *and let* φ_n, $n \in \mathbb{N}$ *be a test-sequence which is of as. level* α *for* $\mathfrak{P}_o$ *in the sense that for each* $f \in T_w(P,\mathfrak{P}_o)$ *and each path* $P_{t,f}$, $t \downarrow 0$, *with weak derivative* f,

$$(8.5.4) \qquad P^n_{n^{-1/2},f}(\varphi_n) \le \alpha + o(n^o).$$

Let $P_{t,g}$, $t \downarrow 0$, *be a path with weak derivative* $g \in T_w(P,\mathfrak{P}) - T_w(P,\mathfrak{P}_o)$, *and assume that* φ_n, $n \in \mathbb{N}$, *is as. efficient for the sequence of alternatives* $P_{n^{-1/2},g}$, $n \in \mathbb{N}$, *i.e.,*

$$(8.5.5) \qquad P^n_{n^{-1/2},g}(\varphi_n) = \Phi(N_\alpha + \|g-g_o\|) + o(n^o),$$

where g_o *is the projection of* g *into* $T_w(P,\mathfrak{P}_o)$.

Then for each $h \in T_w(P,\mathfrak{P})$ *and each path* $P_{t,h}$, $t \downarrow 0$, *with weak derivative* h,

$$(8.5.6) \qquad P^n_{n^{-1/2},h}(\varphi_n) = \Phi(N_\alpha + P(h(g-g_o))/\|g-g_o\|) + o(n^o).$$

In particular, φ_n, $n \in \mathbb{N}$, *is as. efficient for all sequences of alternatives* $P_{n^{-1/2},tg}$, $n \in \mathbb{N}$, *with* $t > 0$.

Proof. Since $k := g - g_o \in T_w(P,\mathfrak{P})$, there exists a path $P_{t,k} \in \mathfrak{P}$, $t \downarrow 0$, with weak derivative k at P. For $\rho_n \downarrow 0$ with $n^{1/2}\rho_n \uparrow \infty$ define

$$g_n := g1_{\{|g| \le n^{1/2}\rho_n\}},$$

$$g_{o,n} := g_o 1_{\{|g_o| \le n^{1/2}\rho_n\}},$$

$$k_n := g_n - g_{o,n},$$

$$C_n := \{\tilde{k}_n > -\|k\| N_\alpha\}$$

By Corollary 19.2.9, for each $h \in T_w(P,\mathfrak{P})$ and each path $P_{t,h} \in \mathfrak{P}$, $t \downarrow 0$, with weak derivative h we can choose $\rho_n \downarrow 0$ such that

$$(8.5.7) \qquad P^n_{n^{-1/2},h}(C_n) = \Phi(N_\alpha + P(hk)/\|k\|) + o(n^o).$$

Since $P(g_o k) = 0$, we obtain in particular

$$P^n_{n^{-1/2},g_o}(C_n) = \alpha + o(n^o),$$

$$P^n_{n^{-1/2},g}(C_n) = \Phi(N_\alpha + \|k\|) + o(n^o).$$

With assumptions (8.5.4) and (8.5.5) we obtain for $\Delta_n := 1_{C_n} - \varphi_n$

(8.5.8) $$P^n_{n^{-1/2},g_o}(\Delta_n) \geq o(n^o) ,$$

(8.5.9) $$P^n_{n^{-1/2},g}(\Delta_n) = o(n^o)$$

Define

$$G_n := \{|\tilde{g}_n - \|g\|^2/2| \leq s_n\} ,$$

$$G_{o,n} := \{|\tilde{g}_{o,n} - \|g_o\|^2/2| \leq s_n\} .$$

By Čebyšev's inequality, $s_n \uparrow \infty$ implies

(8.5.10) $$P^n(G_n^c) = o(n^o), \qquad P^n_{n^{-1/2},g_o}(G_n^c) = o(n^o),$$

(8.5.11) $$P^n(G_{o,n}^c) = o(n^o), \qquad P^n_{n^{-1/2},g}(G_{o,n}^c) = o(n^o) .$$

Define $H_n := G_n G_{o,n}$. By Corollary 19.2.26 and (8.5.10), relation (8.5.8) implies for $s_n \uparrow \infty$ slowly enough,

$$P^n(\exp[\tilde{g}_{o,n} - \|g_o\|^2/2]\Delta_n 1_{H_n})$$
$$= P^n(\exp[\tilde{g}_{o,n} - \|g_o\|^2/2]\Delta_n 1_{G_{o,n}})$$
$$- P^n(\exp[\tilde{g}_{o,n} - \|g_o\|^2/2]\Delta_n 1_{G_n^c G_{o,n}})$$
$$= P^n_{n^{-1/2},g_o}(\Delta_n) - P^n_{n^{-1/2},g_o}(\Delta_n 1_{G_n^c}) + o(n^o) \geq o(n^o) .$$

Similarly, using (8.5.11) instead of (8.5.10), we obtain from (8.5.9)

$$P^n(\exp[\tilde{g}_n - \|g\|^2/2]\Delta_n 1_{H_n}) = o(n^o) .$$

Hence

(8.5.12) $$P^n((\exp[\tilde{g}_n] - \exp[\tilde{g}_{o,n} - \|k\|N_\alpha])\Delta_n 1_{H_n}) \leq o(n^o) .$$

By definition of C_n, the integrand in (8.5.12) is nonnegative, so that (8.5.12) holds with $\leq$ replaced by $=$. Let $\varepsilon_n \downarrow 0$ and define

$$A_n := \{|\exp[\tilde{g}_n] - \exp[\tilde{g}_{o,n} - \|k\|N_\alpha]| > \varepsilon_n\} .$$

Since $|u-v| < \varepsilon_n |e^u - e^v|$ for $u,v > \log \varepsilon_n$, we obtain from $\tilde{k}_n = \tilde{g}_n - \tilde{g}_{o,n}$ that

$$A_n^c \subset \{|\tilde{k}_n - \|k\| N_\alpha| < \varepsilon_n^2\} \cup \{\tilde{g}_n \leq \log \varepsilon_n\} \cup \{\tilde{g}_{o,n} - \|k\| N_\alpha \leq \log \varepsilon_n\}.$$

Hence

$$P^n(A_n^c) = o(n^o) .$$

Choosing $\varepsilon_n \downarrow 0$ slowly enough, we obtain from (8.5.12) that $P^n(|\Delta_n| 1_{H_n}) = o(n^o)$. Together with (8.5.10) and (8.5.11) this implies for each $h \in T_w(P,\mathfrak{P})$ and each path $P_{t,h} \in \mathfrak{P}$, $t \downarrow 0$, with weak derivative h,

$$P^n_{n^{-1/2},h}(|\Delta_n|) = o(n^o)$$

and therefore

$$P^n_{n^{-1/2},h}(\varphi_n) = P^n_{n^{-1/2},h}(c_n) + o(n^o) .$$

The assertion now follows from (8.5.7).

8.6. Restrictions of the basic family

In the preceding sections we discussed the problem of testing a hypothesis $\mathfrak{P}_o$ in a basic family $\mathfrak{P}$. Now we consider the problem of how the as. envelope power can be increased if the prior knowledge becomes more precise, i.e., if the basic family $\mathfrak{P}$ is replaced by a smaller family, say $\mathfrak{Q}$.

The restriction from $\mathfrak{P}$ to $\mathfrak{Q}$ brings about a corresponding restriction of the hypothesis from $\mathfrak{P}_o$ to $\mathfrak{Q}_o := \mathfrak{P}_o \cap \mathfrak{Q}$.

Optimal tests for the hypothesis $\mathfrak{Q}_o$ will usually have higher power against alternatives in $\mathfrak{Q}$ than tests for the larger hypothesis $\mathfrak{P}_o$. There are, however, exceptional cases in which this is not true.

A familiar example is that of testing the hypothesis of symmetry about 0 against alternatives in the family of all symmetric distributions, $\mathfrak{P}$. In this case, the restriction to $\mathfrak{Q} = \{N(\mu,1): \mu \in \mathbb{R}\}$ (or any other location parameter family) does not increase the as. envelope power. In this section we try to explain this phenomenon in terms of tangent spaces.

In all natural examples we find that - corresponding to the relation $\mathfrak{Q}_o = \mathfrak{P}_o \cap \mathfrak{Q}$ - we have

(8.6.1) $\quad T(P,\mathfrak{Q}_o) = T(P,\mathfrak{P}_o) \cap T(P,\mathfrak{Q}) \qquad$ for all $P \in \mathfrak{Q}_o$.

Relation (8.6.1) is trivially fulfilled if $\mathfrak{P}_o \subset \mathfrak{Q}$, for then $T(P,\mathfrak{P}_o) \subset T(P,\mathfrak{Q})$ and $\mathfrak{Q}_o = \mathfrak{P}_o$, so that $T(P,\mathfrak{Q}_o) = T(P,\mathfrak{P}_o) = T(P,\mathfrak{P}_o) \cap T(P,\mathfrak{Q})$. The following example illustrates a situation in which relation (8.6.1) fails to be true.

<u>8.6.2. Example</u>. Let $\mathfrak{P} = \{P_\theta: \theta \in \Theta\}$ with $\Theta \subset \mathbb{R}^3$ be a parametric family, $\mathfrak{P}_o = \{P_\theta: \theta \in \mathbb{R}^3, \theta_3 = 0\}$, and $\mathfrak{Q} = \{P_\theta: \theta \in \mathbb{R}^3, \theta_3 = \theta_2^2\}$. Then $\mathfrak{Q}_o = \{P_\theta: \theta \in \mathbb{R}^3, \theta_2 = \theta_3 = 0\}$.

For $\theta \in \Theta$, $T(P_\theta,\mathfrak{P})$ is the space spanned by $\ell^{(i)}(\cdot,\theta)$, $i = 1,2,3$, i.e. the set of all functions $c_i\ell^{(i)}(\cdot,\theta)$, $(c_1,c_2,c_3) \in \mathbb{R}^3$ (see Proposition 2.2.1). For $T(P_\theta,\mathfrak{P}_o)$ the triple (c_1,c_2,c_3) is subject to the condition $c_3 = 0$, for $T(P_\theta,\mathfrak{Q})$ to the condition $2\theta_2 c_2 - c_3 = 0$ (see Remark 2.2.7(ii)). For $P_\theta \in \mathfrak{Q}_o$, we have $\theta_2 = \theta_3 = 0$, so that both side conditions become identical, i.e.,

$$P_\theta \in \mathfrak{Q}_o \text{ implies } T(P_\theta,\mathfrak{P}_o) = T(P_\theta,\mathfrak{Q}) = [\ell^{(1)}(\cdot,\theta),\ell^{(2)}(\cdot,\theta)].$$

On the other hand,

$$T(P_\theta,\mathfrak{Q}_o) = [\ell^{(1)}(\cdot,\theta)] .$$

<u>8.6.3. Proposition</u>. *For* $P \in \mathfrak{Q}_o$ *let* $S(P)$ *denote the projection of* $T^\perp(P;\mathfrak{P}_o,\mathfrak{P})$ *into* $T(P,\mathfrak{Q})$. *Then*

$$T(P,\mathfrak{Q}) = (T(P,\mathfrak{P}_o) \cap T(P,\mathfrak{Q})) \oplus S(P) .$$

<u>8.6.4. Addendum</u>. *If (8.6.1) is fulfilled, then* $T^\perp(P;\mathfrak{Q}_o,\mathfrak{Q})$ *is the projection of* $T^\perp(P;\mathfrak{P}_o,\mathfrak{P})$ *into* $T(P,\mathfrak{Q})$.

Proof. We have $T(P,\mathfrak{Q}) = S(P) \oplus S(P)^{\perp}$, where

$$S(P)^{\perp} = \{g \in T(P,\mathfrak{Q}) : g \perp S(P)\}.$$

By definition of $S(P)$, $g \in T(P,\mathfrak{Q})$ implies $P(gh) = P(gh_o)$ if $h \in T^{\perp}(P;\mathfrak{P}_o,\mathfrak{P})$, and if h_o is the projection of h into $T(P,\mathfrak{Q})$. Hence for $g \in T(P,\mathfrak{Q})$ we have $g \perp S(P)$ iff $g \perp T^{\perp}(P;\mathfrak{P}_o,\mathfrak{P})$, i.e.,

$$\begin{aligned} S(P)^{\perp} &= \{g \in T(P,\mathfrak{Q}) : g \perp T^{\perp}(P;\mathfrak{P}_o,\mathfrak{P})\} \\ &= T(P,\mathfrak{Q}) \cap \{g \in T(P,\mathfrak{P}) : g \perp T^{\perp}(P;\mathfrak{P}_o,\mathfrak{P})\} \\ &= T(P,\mathfrak{Q}) \cap T(P,\mathfrak{P}_o). \end{aligned}$$

8.6.5. Proposition. *The reduction of the basic family* $\mathfrak{P}$ *to a subfamily* $\mathfrak{Q}$ *does not increase the as. envelope power function iff*

(8.6.6) $\quad T^{\perp}(P;\mathfrak{Q}_o,\mathfrak{Q}) \subset T^{\perp}(P;\mathfrak{P}_o,\mathfrak{P}) \qquad$ *for all* $P \in \mathfrak{Q}_o$.

An intuitive equivalent formulation is that under condition (8.6.6), $\delta(Q,\mathfrak{Q}_o)$ equals $\delta(Q,\mathfrak{P}_o)$ up to a term of order $o(\delta(Q,\mathfrak{Q}_o))$.

Proof. Let $k \in T^{\perp}(P;\mathfrak{Q}_o,\mathfrak{Q})$ be arbitrary. By Remark 8.4.5 the slope of the as. envelope power function for tests of the hypothesis $\mathfrak{Q}_o$ against alternatives $P_{n^{-1/2},tk}$ is $\|k\|$ (since the projection of k into $T(P,\mathfrak{Q}_o)$ is zero). The slope of the as. envelope power function for tests of the hypothesis $\mathfrak{P}_o$ against alternatives $P_{n^{-1/2},tk}$ is $\|k-k_1\|$, where k_1 is the projection of k into $T(P,\mathfrak{P}_o)$. The relation $\|k\| \geq \|k-k_1\|$ reflects the fact that the restriction to $\mathfrak{Q}$ usually increases the as. power. If this is n o t the case, we have $k_1 = 0$, i.e. $k \in T^{\perp}(P;\mathfrak{P}_o,\mathfrak{P})$, so that $T^{\perp}(P;\mathfrak{Q}_o,\mathfrak{Q}) \subset T^{\perp}(P;\mathfrak{P}_o,\mathfrak{P})$.

Assume, conversely, that $T^{\perp}(P;\mathfrak{Q}_o,\mathfrak{Q}) \subset T^{\perp}(P;\mathfrak{P}_o,\mathfrak{P})$. Let $g \in T(P,\mathfrak{Q})$ be arbitrary. Then the slope of the as. envelope power function for tests of the hypothesis $\mathfrak{Q}_o$ against alternatives $P_{n^{-1/2},tg}$ is $\|g-g_o\|$, where g_o is the projection of g into $T(P,\mathfrak{Q}_o)$. By definition of g_o we have $g-g_o \in T^{\perp}(P;\mathfrak{Q}_o,\mathfrak{Q})$ and therefore $g-g_o \in T^{\perp}(P;\mathfrak{P}_o,\mathfrak{P})$. Since

$g_o \in T(P,\mathfrak{Q}_o) \subset T(P,\mathfrak{P}_o)$, this implies that g_o is also the projection of g into $T(P,\mathfrak{P}_o)$, so that the slope of the as. envelope power function for tests of the hypothesis $\mathfrak{P}_o$ is $\|g-g_o\|$, too.

The phenomenon that a restriction of the basic family $\mathfrak{P}$ to a subfamily $\mathfrak{Q}$ leaves the envelope power function on $\mathfrak{Q}$ unchanged, is particularly striking if $\mathfrak{P}$ is a large (nonparametric) family, and $\mathfrak{Q}$ a small parametric family. In the following we give a few examples. Further examples of this phenomenon can be found in Section 13.1, Remark 14.2.7, Section 15.1, Remark 17.3.7, and Sections 18.4, 18.5 and 18.7.

<u>8.6.7. Example</u>. Let $\mathfrak{P}$ denote the family of all symmetric p-measures on $\mathbb{B}$ which are equivalent to the Lebesgue measure. The hypothesis $\mathfrak{P}_o$ is the subfamily of all p-measures in $\mathfrak{P}$ which are symmetric about o. By Example 8.1.1 we have $T(P,\mathfrak{P}_o) = \Psi(P)$ (the class of all functions in $\mathscr{L}_*(P)$ which are symmetric about o), and $T^{\perp}(P;\mathfrak{P}_o,\mathfrak{P}) = [\ell'(\cdot,P)]$.

Let $\mathfrak{Q} := \{Q_o * (x \to x-\mu) : \mu \in \mathbb{R}\}$, with $Q_o \in \mathfrak{P}_o$ fixed. Then $\mathfrak{Q}_o = \{Q_o\}$, $T(Q_o,\mathfrak{Q}_o) = \{o\}$, and $T^{\perp}(Q_o;\mathfrak{Q}_o,\mathfrak{Q}) = T(Q_o,\mathfrak{Q}) = [\ell'(\cdot,Q_o)]$. Hence $T^{\perp}(Q_o;\mathfrak{P}_o,\mathfrak{P}) = T^{\perp}(Q_o;\mathfrak{Q}_o,\mathfrak{Q})$, and the restriction from the nonparametric family $\mathfrak{P}$ to the one-parametric family $\mathfrak{Q}$ does not increase the as. envelope power function for $\mathfrak{Q}$.

<u>8.6.8. Example</u>. Let $\mathfrak{P}$ be the family of two-dimensional normal distributions with parameter $(\mu_1,\mu_2,\sigma_1,\sigma_2,\rho)$, and let $\mathfrak{P}_o$ be the family with $\rho = \rho_o$. Then $T^{\perp}(N(\mu_1,\mu_2,\sigma_1,\sigma_2,\rho_o);\mathfrak{P}_o,\mathfrak{P})$ is spanned by

$$k_1(x_1,x_2;\mu_1,\mu_2,\sigma_1,\sigma_2,\rho_o) = \frac{x_1-\mu_1}{\sigma_1}\,\frac{x_2-\mu_2}{\sigma_2} - \frac{\rho_o}{2}\Big(\Big(\frac{x_1-\mu_1}{\sigma_1}\Big)^2 + \Big(\frac{x_2-\mu_2}{\sigma_2}\Big)^2\Big).$$

Let $\mathfrak{Q}$ be the subfamily with $\sigma_1 = \sigma_2$. Then $T^{\perp}(N(\mu_1,\mu_2,\sigma,\sigma,\rho_o);\mathfrak{Q}_o,\mathfrak{Q})$ is spanned by

$$k_o(x_1,x_2;\mu_1,\mu_2,\sigma,\sigma,\rho_o) = \frac{x_1-\mu_1}{\sigma}\,\frac{x_2-\mu_2}{\sigma} - \frac{\rho_o}{2}\Big(\Big(\frac{x_1-\mu_1}{\sigma}\Big)^2 + \Big(\frac{x_2-\mu_2}{\sigma}\Big)^2\Big).$$

Since $k_o(\cdot;\mu_1,\mu_2,\sigma,\sigma,\rho_o)$ equals $k_1(\cdot;\mu_1,\mu_2,\sigma,\sigma,\rho_o)$, relation (8.6.6) holds true. Hence the restriction to the subfamily of normal distributions with equal variances is ineffective for testing a hypothesis about the correlation coefficient.

(It is also easy to see $k_1(\cdot;\mu_1,\mu_2,\sigma,\sigma,\rho) \in T(N(\mu_1,\mu_2,\sigma,\sigma,\rho),\mathfrak{Q})$ directly. $T(N(\mu_1,\mu_2,\sigma,\sigma,\rho),\mathfrak{Q})$ is spanned by $\ell^{(1)},\ell^{(2)},\ell^{(3)}+\ell^{(4)},\ell^{(5)}$. This space contains $k_1 = \Lambda_{5i}\ell^{(i)}$, since $\sigma_1 = \sigma_2$ implies $\Lambda_{53} = \Lambda_{54}$.)

8.6.9. Remark. The question whether a reduction of the basic family leads to a higher as. envelope power, admits a particularly simple answer if the hypothesis is of the type $\mathfrak{P}_o = \{P \in \mathfrak{P}: \kappa(P) = c_o\}$, for some differentiable functional κ. Let $\kappa^{\cdot}(\cdot,P) \in T(P,\mathfrak{P})$ be a gradient. Since $T^{\perp}(P;\mathfrak{P}_o,\mathfrak{P}) = [\kappa^{\cdot}(\cdot,P)]$ by Example 8.1.2, we have $T^{\perp}(P;\mathfrak{Q}_o,\mathfrak{Q}) \subset T^{\perp}(P;\mathfrak{P}_o,\mathfrak{P})$ iff $\kappa^{\cdot}(\cdot,P) \in T(P,\mathfrak{Q})$.

Hence (see Proposition 8.6.5) *the reduction from $\mathfrak{P}$ to $\mathfrak{Q}$ is ineffective iff the canonical gradient of κ in $T(P,\mathfrak{P})$ belongs to $T(P,\mathfrak{Q})$.*

There are a few natural conditions which imply (8.6.6), for instance

(8.6.10) $\quad T^{\perp}(P;\mathfrak{Q},\mathfrak{P}) \subset T(P,\mathfrak{P}_o) \quad$ for all $P \in \mathfrak{Q}_o$.

Since $T(P,\mathfrak{P}) = T(P,\mathfrak{Q}) \oplus T^{\perp}(P;\mathfrak{Q},\mathfrak{P})$, relation (8.6.10) is equivalent to $T^{\perp}(P;\mathfrak{P}_o,\mathfrak{P}) \subset T(P,\mathfrak{Q})$. According to Addendum 8.6.4, this implies $T^{\perp}(P;\mathfrak{P}_o,\mathfrak{P}) = T^{\perp}(P;\mathfrak{Q}_o,\mathfrak{Q})$, provided condition (8.6.1) is fulfilled.

Another sufficient condition for (8.6.6) is $\mathfrak{P}_o \subset \mathfrak{Q}$. Since $\mathfrak{Q} \subset \mathfrak{P}$, we have $T^{\perp}(P;\mathfrak{P}_o,\mathfrak{Q}) \subset T^{\perp}(P;\mathfrak{P}_o,\mathfrak{P})$. Since $\mathfrak{P}_o \subset \mathfrak{Q}$ implies $\mathfrak{P}_o = \mathfrak{Q}_o$, this is (8.6.6).

8.6.11. Example. Let $\mathfrak{P}$ be an arbitrary family, and $\mathfrak{P}_o = \{P \in \mathfrak{P}: \kappa_i(P) = c_i$ for $i = 1,\ldots,k\}$, where κ_i is a differentiable functional with gradient $\kappa_i^{\cdot}(\cdot,P) \in T(P,\mathfrak{P})$. By Corollary 8.1.8,

$$T(P,\mathfrak{P}_o) = \{g \in T(P,\mathfrak{P}): P(g\kappa_i^{\bullet}(\cdot,P)) = 0 \quad \text{for } i = 1,\dots,k\}.$$

Assume that the family $\mathfrak{P}$ is restricted to $\mathfrak{Q} = \{P \in \mathfrak{P}: \kappa_o(P) = 0\}$, where κ_o is a differentiable functional with gradient $\kappa_o^{\bullet}(\cdot,P) \in T(P,\mathfrak{P})$. Since $T^{\perp}(P;\mathfrak{Q},\mathfrak{P}) = [\kappa_o^{\bullet}(\cdot,P)]$ by Example 8.1.2, relation (8.6.10) is fulfilled iff $P(\kappa_o^{\bullet}(\cdot,P)\kappa_i^{\bullet}(\cdot,P)) = 0$ for $i = 1,\dots,k$. In this case the knowledge of the value of κ_o does not help to increase the as. envelope power for tests of a hypothesis about κ_i, $i = 1,\dots,k$.

As an example let $\mathfrak{P} = \{P_\theta: \theta \in \Theta\}$, $\Theta \subset \mathbb{R}^k$, $\mathfrak{P}_o = \{P_\theta: \theta \in \Theta: \theta_i = \theta_i^o$, $i = 1,\dots,m\}$ for some $m < k$, and $\mathfrak{Q} = \{P_\theta: \theta \in \Theta,\ F(\theta_{m+1},\dots,\theta_k) = 0\}$, where F is real-valued and differentiable. If we apply the result obtained above for $\kappa_i(\theta) = \theta_i$, $i = 1,\dots,m$, and $\kappa_o(\theta) = F(\theta_{m+1},\dots,\theta_k)$, we have $\kappa_i^{\bullet}(\cdot,P_\theta) = \Lambda_{ij}(\theta)\ell^{(j)}(\cdot,\theta)$, for $i = 1,\dots,m$, and $\kappa_o^{\bullet}(\cdot,P_\theta)$ $= F^{(\alpha)}(\theta)\Lambda_{\alpha j}(\theta)\ell^{(j)}(\cdot,\theta)$, where the summation over α extends from $m+1$ to k. We have $P_\theta(\kappa_o^{\bullet}(\cdot,P_\theta)\kappa_i^{\bullet}(\cdot,P_\theta)) = \Lambda_{i\alpha}(\theta)F^{(\alpha)}(\theta)$. Hence $\kappa_o^{\bullet}(\cdot,P_\theta)$ is orthogonal to $\kappa_i^{\bullet}(\cdot,P_\theta)$ for $i = 1,\dots,m$ if F includes among the arguments $\theta_{m+1},\dots,\theta_k$ only such θ_α for which $\Lambda_{i\alpha}(\theta) = 0$ for $\alpha = 1,\dots,m$. Parameters θ_i,θ_α with $\Lambda_{i\alpha}(\theta) = 0$ are called *unrelated*. (For examples of unrelated parameters see 9.5.1.)

With this terminology, our result can be states as follows: *A restriction of the parametric family is ineffective if the parameters subject to this restriction are unrelated to the parameters fixed by the hypothesis.*

<u>8.6.12. Remark</u>. Let $\mathfrak{P}$ be an arbitrary family and $\mathfrak{P}_o \subset \mathfrak{P}$ the hypothesis. Let $\mathfrak{Q}_o \subset \mathfrak{P}_o$ be a subfamily, $P \in \mathfrak{Q}_o$ and $k \in T(P,\mathfrak{P})$.

The reduction of $\mathfrak{P}_o$ to $\mathfrak{Q}_o$ does not reduce the as. envelope power function in direction k iff the projection of k into $T(P,\mathfrak{P}_o)$ belongs to $T(P,\mathfrak{Q}_o)$.

This follows immediately from Remark 8.4.5, according to which the slope of the as. envelope power function for the hypothesis $\mathfrak{P}_o$

in direction k is $\|k-k_o\|$, where k_o is the projection of k into $T(P,\mathfrak{P}_o)$. If $k_o \in T(P,\mathfrak{Q}_o)$, then k_o is, at the same time, the projection of k into $T(P,\mathfrak{Q}_o)$.

8.7. Asymptotic envelope power functions using the Hellinger distance

In Theorem 8.4.1, the as. envelope power function was given as $\Phi(N_\alpha + n^{1/2}\Delta(Q;\bar{Q}))$, under the assumption that $\Delta(Q;\bar{Q}) < \infty$. This was done only to keep the assumptions more transparent. If we use Assumption 19.2.1 on $\bar{Q}$ and Q, then we obtain Theorem 8.4.1 with an as. variance bound $\Phi(N_\alpha + n^{1/2}\|g\|_{\bar{Q}})$, where g is the square-integrable part occuring in the decomposition of $q/\bar{q}-1$ described in Assumption 19.2.1. By Proposition 6.2.2, we have under this assumption $H(Q,\bar{Q}) = \|g\|_{\bar{Q}}(1+o(\|g\|_{\bar{Q}}^o))$, so that $n^{1/2}\|g\|_{\bar{Q}} \le c$ implies $n^{1/2}H(Q,\bar{Q}) = n^{1/2}\|g\|_{\bar{Q}} + o(n^o)$. Hence the as. envelope power function may under this assumption also be written as

(8.7.1) $\quad \Phi(N_\alpha + n^{1/2}H(Q,\bar{Q}))$.

Assumption 19.2.1 which implies that the as. envelope power function can be represented by means of the Hellinger distance, places restrictions on the deviation of $q/\bar{q}$ from 1. In the following theorem we try to achieve this goal under weaker conditions of the same nature. These conditions require that the Q-probability is small for the set on which $q/\bar{q}$ is large, and that the $\bar{Q}$-probability is small where $\bar{q}/q$ is large. For notational convenience we write P instead of $\bar{Q}$.

8.7.2. Theorem. *There exists a universal constant* c *such that for all* $\varepsilon > 0$, $\delta > 0$, *and all* P,Q *with* $n^{1/2}H(P,Q) \le k$,

$$|P^n\{\underline{x}\in X^n: \sum_{\nu=1}^{n} \log q(x_\nu)/p(x_\nu) < t\} - \Phi((t+nH(P,Q)^2/2)/n^{1/2}H(P,Q))|$$
$$\leq c\Big(H(P,Q)^2 + \frac{k^3}{n^{1/2}H(P,Q)}[\delta + H(P,Q)^{-2}\, P\{x\in X: p(x)/q(x) > 1+\delta\}$$
$$+ \varepsilon + H(P,Q)^{-2}\, Q\{x\in X: q(x)/p(x) > 1+\varepsilon\}]\Big).$$

This bound is, of course, not the most suitable one for the purpose of numerical computations. It was chosen to make transparent the essential aspects.

Using Theorem 8.7.2 with P,Q interchanged, we obtain the following corollary about the envelope power function for (exact) level-α-tests of P^n against Q^n.

8.7.3. Corollary. *There exists a universal constant* c *such that for all* $\varepsilon,\delta > 0$, *and all* P,Q *with* $n^{1/2}H(P,Q) \leq k$,

$$|\sup_{\varphi_n}\{Q^n(\varphi_n): P^n(\varphi_n) \leq \alpha\} - \Phi(N_\alpha + n^{1/2}H(P,Q))|$$
$$\leq c\Big(H(P,Q)^2 + \frac{k^3}{n^{1/2}H(P,Q)}[\delta + H(P,Q)^{-2}\, P\{x\in X: p(x)/q(x) > 1+\delta\}$$
$$+ \varepsilon + H(P,Q)^{-2}\, Q\{x\in X: q(x)/p(x) > 1+\varepsilon\}]\Big).$$

This corollary gives conditions under which $\Phi(N_\alpha + n^{1/2}H(P,Q))$ is a useful approximation to the envelope power function for the sample size n. This will be the case for all alternatives Q with $n^{1/2}H(P,Q)$ between $|N_\alpha|$ and $2|N_\alpha|$, say (corresponding to rejection probabilities between $\frac{1}{2}$ and $(1-\alpha)$), for which

(8.7.4') $\inf\{\delta + H(P,Q)^{-2}P\{p/q > 1+\delta\}: \delta \in (0,\frac{1}{2})\} = o(H(P,Q)^0)$,

(8.7.4") $\inf\{\varepsilon + H(P,Q)^{-2}P\{q/p > 1+\varepsilon\}: \varepsilon \in (0,\frac{1}{2})\} = o(H(P,Q)^0)$.

8.7.5. Example. Let Q denote alternatives with P-density $1 + tg$, where g runs through a class of uniformly square-integrable functions with $P(g^2) = 1$. We have

$$P\{1+tg<\frac{1}{1+\delta}\} = P\{g<-t^{-1}\delta/(1+\delta)\}$$

$$\leq t^2\delta^{-2}(1+\delta)^2\ P(g^2 1_{\{g<-t^{-1}\delta/(1+\delta)\}}) .$$

Since $H(P,Q) = t+o(t)$ (see Remark 6.2.16), we obtain

$$\delta + H(P,Q)^{-2}P\{1+tg<\frac{1}{1+\delta}\} \leq \delta+\delta^{-2}(1+\delta)^2\ P(g^2 1_{\{g<-t^{-1}\delta(1+\delta)\}}) + o(t^o).$$

Uniform integrability ascertains (see Lemma 19.1.1) that the infimum over δ is of the order $o(t^o)$, and therefore of the order $o(H(P,Q)^o)$. The other inequality follows similarly from

$$Q\{1+tg>1+\varepsilon\} = Q\{g>t^{-1}\varepsilon\} \leq t^2(\varepsilon^{-2}+\varepsilon^{-1})P(g^2 1_{\{g>t^{-1}\varepsilon\}}) .$$

The following example illustrates a boundary case.

<u>8.7.6. Example</u>. For $\alpha \in [0,1]$ and $\theta \in [0,\infty)$ let $Q_{\alpha,\theta}$ denote the p-measure with Lebesgue density

$$x \to (1-\alpha)\exp[-x] + \alpha\theta\exp[-\theta x], \qquad x > 0 .$$

For each $\theta > 0$ fixed, $Q_{\alpha,\theta} \to Q_{0,0} =: P$ as $\alpha \to 0$.

If $\theta > \frac{1}{2}$, the situation is regular in the sense that the Δ-distance is finite. We have

$$\Delta(Q_{\alpha,\theta};P) = \alpha|1-\theta|(2\theta-1)^{-1/2}$$

and

$$H(Q_{\alpha,\theta},P) = \alpha|1-\theta|(2\theta-1)^{-1/2} + o(\alpha) .$$

With θ fixed, the family $Q_{\alpha,\theta}$ is of the type discussed above, with P-density $1+\alpha g$, $\alpha \in [0,1]$, with $g(x) = \theta\exp[(1-\theta)x]-1$ and $P(g^2) < \infty$.

For $\theta = \frac{1}{2}$ we have $\Delta(Q_{\alpha,1/2};P) = \infty$ for all $\alpha \in (0,1]$, whereas $H(Q_{\alpha,1/2},P) = \alpha|\log\alpha|^{1/2}/\sqrt{2}$. To apply Corollary 8.7.3, we have to prove (8.7.4') and (8.7.4"). Since $q_{\alpha,1/2}(x)/p(x) = 1+\alpha(\frac{1}{2}\exp[x/2]-1)$, relation (8.7.4') is trivially fulfilled with $\delta = \alpha/(1-\alpha)$. An elementary computation shows that

$$Q_{\alpha,1/2}\{1+\alpha(\tfrac{1}{2}\exp[x/2]-1) > 1+\varepsilon\} \leq \alpha^2(1+2\varepsilon)/4\varepsilon^2.$$

Hence

$$\varepsilon+H(P,Q_{\alpha,1/2})^{-2}Q_{\alpha,1/2}\{1+\alpha(\tfrac{1}{2}\exp[x/2]-1)>1+\varepsilon\} \leq \varepsilon + \frac{1+2\varepsilon}{2\varepsilon^2|\log\alpha|} .$$

Choosing $\varepsilon = |\log \alpha|^{-1/3}$, we obtain a bound of order $|\log \alpha|^{-1/3}$ which is of the order $|\log H(Q_{\alpha,1/2},P)|^{-1/3} = o(H(Q_{\alpha,1/2},P)^{o})$. This proves (8.7.4").

The following example demonstrates that, without conditions restricting the deviations of q/p from 1, an approximation of the envelope power function by the Hellinger distance is not possible any more.

8.7.7. Example. Let P be the uniform distribution over (0,1). For $a \in (0,\frac{1}{2})$ and $b \in (0,1)$ let $Q_{a,b}$ denote the p-measure with P-density

$$q_{a,b}(x) = \begin{cases} 1+b & 0 < x < a \\ 1 & a \leq x \leq \frac{1}{2} \\ 1-2ab & \frac{1}{2} < x < 1 . \end{cases}$$

For a fixed, we have

$$H(Q_{a,b},P) = a^{1/2}(1+2a)^{1/2}b + o(b^2) .$$

Relations (8.7.4') and (8.7.4") can be fulfilled by choosing $\delta = 2ab$ and $\varepsilon = b$, which leads to

$$\inf_{\delta}[\delta + H(Q_{a,b},P)^{-2}P\{p/q_{a,b} > 1+\delta\}] \leq 2ab \leq b,$$

$$\inf_{\varepsilon}[\varepsilon + H(Q_{a,b},P)^{-2}Q_{a,b}\{q_{a,b}/p > 1+\varepsilon\}] \leq b .$$

Hence the bound provided by Corollary 8.7.3 is of the order $H(Q_{a,b},P)$, which is $n^{-1/2}$ for the interesting alternatives.

If we keep b fixed, we have

$$H(Q_{a,b},P) = \sqrt{8}\, a^{1/2}[1 + \tfrac{b}{2} - \sqrt{1+b}]^{1/2} + o(a) .$$

In this case, condition (8.7.4') becomes

$$\inf_{\delta}[\delta + H(Q_{a,b},P)^{-2}P\{p/q_{a,b} > 1+\delta\}] \leq 2ab.$$

Difficulties arise with condition (8.7.4"). We have

$$\varepsilon + H(Q_{a,b},P)^{-2}Q\{q_{a,b}/p > 1+\varepsilon\}$$

$$= \begin{cases} \varepsilon & \text{for } \varepsilon \geq b \\ \varepsilon + \dfrac{(1+b)}{8[1 + \frac{b}{2} - \sqrt{1+b}]} + o(a^{1/2}) & \text{for } \varepsilon < b , \end{cases}$$

so that

$$\inf_{\varepsilon}[\varepsilon+H(Q_{a,b},P)^{-2}Q\{q_{a,b}/p > 1+\varepsilon\}] = \min\{b, \frac{1+b}{8[1+\frac{b}{2}-\sqrt{1+b}]} + O(a^{1/2})\}.$$

Hence for b fixed, the corollary does not lead to an error bound of the envelope power function tending to zero. In fact, it turns out that the as. envelope power function is of a different type in this case.

If we consider the sequence of alternatives with

$$a_n = t^2 n^{-1},$$

we have

$$n^{1/2} H(Q_{a_n,b},P) = \sqrt{8}\, t[1+\frac{b}{2}-\sqrt{1+b}]^{1/2} + o(n^o).$$

The as. envelope power function is, however,

$$\Phi(N_\alpha(1+b)^{-1/2} + tb(1+b)^{-1/2}).$$

This follows easily from

$$P^n*(\underline{x} \to \sum_{\nu=1}^{n} \log r_n(x_\nu)) \Rightarrow N(t^2(\log(1+b)-b), t^2[\log(1+b)]^2)$$

and

$$Q^n_{a_n,b}*(\underline{x} \to \sum_{\nu=1}^{n} \log r_n(x_\nu)) \Rightarrow N(t^2((1+b)\log(1+b)-b), t^2(1+b)[\log(1+b)]^2).$$

The following proof uses certain relations occuring in Oosterhoff and van Zwet (1979).

Proof of Theorem 8.7.2. For notational convenience, let $r := q/p$. In the following, δ and ε denote numbers in $(0,\frac{1}{2})$. Define

(8.7.8') $A_\delta := \{p/q > 1+\delta\}$,

(8.7.8") $B_\delta := \{q/p > 1+\delta\}$.

We have

(8.7.9') $P(B_\delta) = Q(\frac{p}{q} 1_{B_\delta}) \le (1+\delta)^{-1} Q(B_\delta) \le Q(B_\delta)$,

(8.7.9") $Q(A_\delta) = P(\frac{q}{p} 1_{A_\delta}) \le (1+\delta)^{-1} P(A_\delta) \le P(A_\delta)$.

Let $M_{\delta,\varepsilon} := A^c_\delta \cap B^c_\varepsilon$. Using (8.7.9') and (8.7.9") we obtain the following relations:

(8.7.10') $P(M^c_{\delta,\varepsilon}) \leq P(A_\delta) + Q(B_\varepsilon)$,

(8.7.10") $Q(M^c_{\delta,\varepsilon}) \leq P(A_\delta) + Q(B_\varepsilon)$,

and

(8.7.11) $|P((r-1)1_{M_{\delta,\varepsilon}})| \leq P(A_\delta) + Q(B_\varepsilon)$.

(Hint: $P((r-1)1_M) = -P((r-1)1_{M^c}) = P(M^c) - Q(M^c)$ implies $-Q(M^c) \leq P((r-1)1_M) \leq P(M^c)$.)

Moreover, we need the following relations:

(8.7.12) $x \in M_{\delta,\varepsilon}$ implies $-\delta < r(x) - 1 \leq \varepsilon$

and $-\delta < \sqrt{r(x)} - 1 \leq \varepsilon$,

(8.7.13) $P((\sqrt{r} - 1)^2 1_{M^c_{\delta,\varepsilon}}) \leq P(A_\delta) + Q(B_\varepsilon)$.

This follows from

(8.7.13') $P((\sqrt{r} - 1)^2 1_{A_\delta}) \leq P(A_\delta)$,

(8.7.13") $P((\sqrt{r} - 1)^2 1_{B_\varepsilon}) \leq P(r 1_{B_\varepsilon}) = Q(B_\varepsilon)$.

Furthermore,

(8.7.14) $x \in M_{\delta,\varepsilon}$ implies $-2\delta < \log r(x) \leq \varepsilon$.

(Proof: We have

$$\log(1+u) \begin{cases} \leq u & \text{for } u \geq 0 \\ > 2u & \text{for } -\frac{1}{2} < u < 0 . \end{cases}$$

Using (8.7.12) we obtain for $x \in M_{\delta,\varepsilon}$

$$\log r(x) = \log(1+(r(x)-1)) \begin{cases} \leq r(x)-1 \leq \varepsilon \\ > -2|r(x)-1| \geq -2\delta .) \end{cases}$$

In the following expansion

(8.7.15) $\log r = 2(\sqrt{r} - 1) - (\sqrt{r} - 1)^2(1+R)$,

(8.7.15') $x \in M_{\delta,\varepsilon}$ implies $|R(x)| \leq 6 \max\{\delta,\varepsilon\}$.

(Proof: $\log(1+u) = u - \frac{u^2}{2}(1 - \frac{2}{3} \frac{u}{(1+\theta u)^3})$ with $\theta \in [0,1]$. We have

$$\frac{u}{(1+\theta u)^3} \begin{cases} \leq u & \text{for } u \geq 0 \\ \geq 8u & \text{for } -\frac{1}{2} < u < 0 . \end{cases}$$

Applying these relations for $u = \sqrt{r(x)} - 1$ we obtain

$$R(x) = -\frac{2}{3} \frac{\sqrt{r(x)} - 1}{(1+\theta(\sqrt{r(x)} - 1))^3}$$

and, from (8.7.12),

$$-8\delta \leq \frac{\sqrt{r(x)} - 1}{(1+\theta(\sqrt{r(x)} - 1))^3} \leq \varepsilon .$$

From this, (8.7.15') follows immediately.)

Since

$$2(\sqrt{r} - 1) = -(\sqrt{r} - 1)^2 + (r-1) ,$$

we obtain from (8.7.15)

(8.7.16) $\log r = -2(\sqrt{r} - 1)^2 + (r-1) - (\sqrt{r} - 1)^2 R .$

Using (8.7.15'), (8.7.13) and (8.7.11) we obtain

$$|P((\log r)1_{M_{\delta,\varepsilon}}) + 2P((\sqrt{r} - 1)^2)|$$
$$\leq 2P((\sqrt{r} - 1)^2 1_{M^c_{\delta,\varepsilon}}) + |P((r-1)1_{M_{\delta,\varepsilon}})| + 6 \max\{\delta,\varepsilon\}P((\sqrt{r} - 1)^2)$$
$$\leq 3(P(A_\delta) + Q(B_\varepsilon)) + 6 \max\{\delta,\varepsilon\}P((\sqrt{r} - 1)^2) ,$$

i.e.

(8.7.17) $|P((\log r)1_{M_{\delta,\varepsilon}}) + \frac{1}{2}H(P,Q)^2|$

$$\leq 3(P(A_\delta) + Q(B_\varepsilon)) + \frac{3}{2} \max\{\delta,\varepsilon\}H(P,Q)^2 .$$

Moreover, (8.7.15), (8.7.12) and (8.7.15') imply

$$|P((\log r)^2 1_{M_{\delta,\varepsilon}}) - 4P((\sqrt{r} - 1)^2)|$$
$$\leq 4P((\sqrt{r} - 1)^2 1_{M^c_{\delta,\varepsilon}}) + 24 \max\{\delta,\varepsilon\}P((\sqrt{r} - 1)^2) .$$

Using (8.7.13) we obtain

(8.7.18) $|P((\log r)^2 1_{M_{\delta,\varepsilon}}) - H(P,Q)^2|$

$$\leq 4(P(A_\delta) + Q(B_\varepsilon)) + 6 \max\{\delta,\varepsilon\}H(P,Q)^2 .$$

For bounded functions f with $P(f) = 0$ and $P(f^2) = 1$ we have for all $n \in \mathbb{N}$

(8.7.19) $|P\{\tilde{f} < t\} - \Phi(t)| \leq c\|f\|_\infty n^{-1/2}$

(where c is a universal constant; see Theorem 19.1.2). This version of the central limit theorem will be applied to

$$(8.7.20)\qquad f = ((\log r)1_{M_{\delta,\varepsilon}} - \mu_{\delta,\varepsilon})/\sigma_{\delta,\varepsilon}$$

with $\mu_{\delta,\varepsilon} := P((\log r)1_{M_{\delta,\varepsilon}})$, $\sigma^2_{\delta,\varepsilon} := P((\log r)^2 1_{M_{\delta,\varepsilon}}) - \mu^2_{\delta,\varepsilon}$.

Since $|\log r| 1_{M_{\delta,\varepsilon}} \le 2\max\{\delta,\varepsilon\}$ by (8.7.14), we obtain

$$(8.7.21)\qquad \|f\|_\infty \le 2\max\{\delta,\varepsilon\}/\sigma_{\delta,\varepsilon}.$$

By (8.7.10'),

$$(8.7.22)\qquad |P^n\{\sum_{\nu=1}^n \log r(x_\nu) < t\} - P^n\{\sum_{\nu=1}^n (\log r(r_\nu))1_{M_{\delta,\varepsilon}}(x_\nu) < t\}|$$
$$\le nP(M^c_{\delta,\varepsilon}) \le n(P(A_\delta) + Q(B_\varepsilon)).$$

Moreover, by (8.7.20),

$$P^n\{\sum_{\nu=1}^n (\log r(x_\nu))1_{M_{\delta,\varepsilon}}(x_\nu) < t\} = P^n\{\tilde{f} < (t-n\mu_{\delta,\varepsilon})/n^{1/2}\sigma_{\delta,\varepsilon}\},$$

so that by (8.7.19) and (8.7.21)

$$(8.7.23)\qquad |P^n\{\sum_{\nu=1}^n (\log r(x_\nu))1_{M_{\delta,\varepsilon}}(x_\nu) < t\} - \Phi((t-n\mu_{\delta,\varepsilon})/n^{1/2}\sigma_{\delta,\varepsilon})|$$
$$\le 2c\max\{\delta,\varepsilon\}\sigma^{-1}_{\delta,\varepsilon} n^{-1/2}.$$

Finally, for all $t \in \mathbb{R}$,

$$(8.7.24')\qquad |\Phi(t+a) - \Phi(t)| \le |a|/\sqrt{2\pi}, \qquad a \in \mathbb{R},$$

$$(8.7.24'')\qquad |\Phi(t/(1+b)) - \Phi(t)| \le |b|\sqrt{2/\pi e}, \qquad |b| \le \tfrac{1}{2}.$$

Given P and Q, let

$$c'(\delta) := \delta + P(A_\delta)/H(P,Q)^2,$$
$$c''(\varepsilon) := \varepsilon + Q(B_\varepsilon)/H(P,Q)^2.$$

Relation (8.7.17) implies

$$(8.7.25)\qquad \mu_{\delta,\varepsilon} = -\tfrac{1}{2}H(P,Q)^2 + a_{\delta,\varepsilon}$$

with

$$(8.7.25')\qquad |a_{\delta,\varepsilon}| \le 3H(P,Q)^2(c'(\delta) + c''(\varepsilon)).$$

Relations (8.7.17) and (8.7.18) imply

$$(8.7.26)\qquad \sigma^2_{\delta,\varepsilon} = H(P,Q)^2[1 + b_{\delta,\varepsilon}]$$

with

$$(8.7.26')\qquad |b_{\delta,\varepsilon}| \le 6(c'(\delta) + c''(\varepsilon)) + \mu^2_{\delta,\varepsilon}/H(P,Q)^2.$$

We remark that

$$\mu^2_{\delta,\varepsilon}/H(P,Q)^2 \le \tfrac{1}{2}H(P,Q)^2 + 2a^2_{\delta,\varepsilon}/H(P,Q)^2 \le H(P,Q)^2[\tfrac{1}{2} + 18(c'(\delta)+c''(\varepsilon))].$$

Since $|u-1| \le |u^2-1|$ for $u \ge 0$, we have

(8.7.27) $$\sigma_{\delta,\varepsilon} = H(P,Q)(1+\hat{b}_{\delta,\varepsilon})$$

with

(8.7.27') $$|\hat{b}_{\delta,\varepsilon}| \le |b_{\delta,\varepsilon}| .$$

We have

$$(t-n\mu_{\delta,\varepsilon})/n^{1/2}\sigma_{\delta,\varepsilon} = (t+\tfrac{n}{2}H(P,Q)^2 - na_{\delta,\varepsilon})/n^{1/2}H(P,Q)(1+\hat{b}_{\delta,\varepsilon})$$

$$= \frac{t+\frac{n}{2}H(P,Q)^2}{n^{1/2}H(P,Q)} \frac{1}{1+\hat{b}_{\delta,\varepsilon}} - \frac{n^{1/2}a_{\delta,\varepsilon}}{H(P,Q)(1+\hat{b}_{\delta,\varepsilon})}$$

and therefore, by (8.7.24') and (8.7.24"),

(8.7.28) $$|\Phi((t-n\mu_{\delta,\varepsilon})/n^{1/2}\sigma_{\delta,\varepsilon}) - \Phi((t+\tfrac{n}{2}H(P,Q)^2)/n^{1/2}H(P,Q))|$$

$$\le \frac{n^{1/2}|a_{\delta,\varepsilon}|}{\sqrt{2\pi}H(P,Q)(1+\hat{b}_{\delta,\varepsilon})} + \sqrt{2/\pi e}\,|\hat{b}_{\delta,\varepsilon}| .$$

Throughout the following we assume that $n^{1/2}H(P,Q) \le k$ and $k \ge 1$. Relations (8.7.22), (8.7.23), (8.7.27) and (8.7.28) together imply the existence of a universal constant c such that

(8.7.29) $$|P^n\{\sum_{\nu=1}^{n} \log r(x_\nu) < t\} - \Phi((t+\tfrac{n}{2}H(P,Q)^2)/n^{1/2}H(P,Q))|$$

$$\le n(P(A_\delta)+Q(B_\varepsilon)) + c\big(\max\{\delta,\varepsilon\}/n^{1/2}H(P,Q)$$

$$+ k[c'(\delta)+c''(\varepsilon)] + H(P,Q)^2\big)$$

since for $c'(\delta)+c''(\varepsilon) > \frac{1}{12}$ or $H(P,Q)^2 > \frac{1}{12}$, say, the inequality becomes trivial by choosing $c \ge 12$. Moreover,

$$nP(A_\delta) + ck^2\delta/n^{1/2}H(P,Q) \le ck^3c'(\delta)/n^{1/2}H(P,Q) ,$$

$$nQ(B_\varepsilon) + ck^2\varepsilon/n^{1/2}H(P,Q) \le ck^3c''(\varepsilon)/n^{1/2}H(P,Q) .$$

Hence we obtain from (8.7.29) that for some universal constant c

$$|P^n\{\sum_{\nu=1}^{n} \log r(x_\nu) < t\} - \Phi((t+\tfrac{n}{2}H(P,Q)^2)/n^{1/2}H(P,Q))|$$

$$\le \frac{ck^3}{n^{1/2}H(P,Q)}[c'(\delta)+c''(\varepsilon)] + cH(P,Q)^2 .$$

This is the assertion.

9. ASYMPTOTIC BOUNDS FOR THE CONCENTRATION OF ESTIMATORS

Let $\mathfrak{P}$ be an arbitrary family of p-measures on a measurable space $(X,\mathcal{A})$ and $\kappa: \mathfrak{P} \to \mathbb{R}^k$ a functional. By an *estimator* for κ for the sample size n we mean a measurable map $\kappa^n: X^n \to \mathbb{R}^k$. The goal of this chapter is to obtain standards for the evaluation of particular estimators by providing bounds for the concentration of estimators (with certain properties). This program can be carried through with success if the sample size is large.

As. bounds for the concentration of estimators are well-known for parametric families, and have been carried over to special nonparametric problems by a number of authors. A general result for differentiable functionals was obtained by Koshevnik and Levit (1976, p. 745, Theorem 2), who obtain a lower bound of the local as. minimax type, i.e.,

$$\underline{\lim}_{n\to\infty} \inf_{\kappa^n} \sup_{P\in\mathcal{U}} P^n(L(n^{1/2}(\kappa^n-\kappa(P)))) \geq \sup_{P\in\mathcal{U}} \int L(u)N(0,\Sigma(P))(du) \quad ,$$

where $\mathcal{U}$ is open with respect to the Hellinger distance, L(u) is a nondecreasing function of an arbitrary norm of u, and $\Sigma(P)$ is the covariance matrix of the canonical gradient of κ at P.

In this chapter we present a result for as. median unbiased estimators (Section 9.2) and a representation of limiting distributions as convolutions (Section 9.3). We abstain from reproducing the result of Koshevnik and Levit.

9.1. Comparison of concentrations

To compare the concentration of two estimators κ_i^n, $i = 1,2$, means to compare the concentration of the induced p-measures $P^n * \kappa_i^n$, $i = 1,2$, about $\kappa(P)$. Even with P fixed, we are left with the problem of comparing two p-measures. Since P varies over $\mathfrak{P}$, we have, in fact, to compare two *families* of p-measures. The outcome of such a comparison will be unequivocal only under special circumstances, for instance if the distributions $P^n * \kappa_i^n$ are approximately normal, which is usually the case if n is large. This kind of regularity makes a comparison feasible.

In other words, comparison of concentration has a certain regularity as a prerequisite. A fruitful discussion about methods for comparison of concentration requires, therefore, some knowledge about these regularities. Hence we take as a starting point of our discussion the following basic result on the concentration of estimators of real-valued functionals (see Section 9.2). With $\sigma^2(P) := P(\kappa^+(\cdot,P)^2)$ we have for any sufficiently regular estimator-sequence

(9.1.1) $$P^n\{\kappa^n \in I\} \leq N(\kappa(P), n^{-1}\sigma^2(P))(I) + o(n^0) ,$$

uniformly over all intervals I containing $\kappa(P)$, and even locally uniformly in P. The upper bound is sharp in the sense that it is attained (up to $o(n^0)$) for suitable estimator-sequences.

The interesting intervals I are, of course, those of order $O(n^{-1/2})$, so that (9.1.1) is usually written as

$$P^n\{n^{1/2}(\kappa^n - \kappa(P)) \in I\} \leq N(0,\sigma^2(P))(I) + o(n^0) ,$$

uniformly over all intervals I containing zero.

Let $L_P(u)$ denote the loss which occurs if P is the true p-measure, and u the estimate for $\kappa(P)$. It is natural to assume that $L_P(\kappa(P)) = 0$, and that $u \to L_P(u)$ is nondecreasing as u moves away from $\kappa(P)$ in either direction. Such loss functions are called *monotone* or *bowl shaped* (about $\kappa(P)$).

From (9.1.1) we obtain that the *distribution of losses*, $P^n*(L_P \circ \kappa^n)$, cannot be more concentrated about zero than the distribution $N(\kappa(P), n^{-1}\sigma^2(P))*L_P$. This holds true up to an error term $o(n^o)$, uniformly over all monotone loss functions.

(To see this observe that $(P^n*(L_P \circ \kappa^n))[0,r] = P^n\{L_P \circ \kappa^n \leq r\}$ $\leq N(\kappa(P), n^{-1}\sigma^2(P))\{L_P \leq r\} + o(n^o) = (N(\kappa(P), n^{-1}\sigma^2(P))*L_P)[0,r] + o(n^o)$, since for monotone loss functions, $\{u \in \mathbb{R}: L_P(u) \leq r\}$ is an interval containing $\kappa(P)$.)

Evaluated by their 'distributions of losses', two estimators will usually be incomparable. Comparing the distribution of losses leads only to a p a r t i a l order between the estimators.

This, perhaps, was the reason for introducing the 'expected loss', $P^n(L_P \circ \kappa^n)$ (for which Wald introduced the unfitting name 'risk'), thus obtaining a t o t a l order between the estimators. The merits of this invention seem to be open to questioning.

The partial order originating from the 'distribution of losses' is all we need if an estimator exists which is of maximal concentration in this partial order (i.e. which is comparable with any other estimator, and turns out to be at least as good). Then it is of no relevance that other - inferior - estimators may be incomparable with regard to their concentration. There is no need to enforce comparability by evaluating them by their risks.

To claim that there exists an estimator which is asymptotically of minimal risk with respect to monotone loss functions means to hold back the larger - and more interesting - part of the story, namely that estimators exist for which the distribution of losses itself is maximally concentrated.

Moreover, reducing the comparison of estimators to the comparison of risks brings about a certain technical difficulty. The relations between the distributions of losses hold up to an error term $o(n^o)$ only. Assume, for instance, that, with $Q_i^{(n)} := P^n * (L_P \circ \kappa_i^n)$, $i = 0,1$, we have uniformly for $r \geq 0$

(9.1.2) $$Q_1^{(n)}[0,r] \leq Q_o^{(n)}[0,r] + o(n^o) .$$

In general, (9.1.2) does not imply that

$$\int u Q_o^{(n)}(du) \leq \int u Q_1^{(n)}(du) + o(n^o) .$$

Even if such a relationship can be proved (for instance if L_P is bounded), it will turn out to be useless since $\int u Q_i^{(n)}(du)$ tends to zero for any reasonable estimator-sequence κ_i^n, $n \in \mathbb{N}$. To base the claim of superiority of κ_o^n over κ_1^n for large n on the comparison of risks, we need an assertion on the relative error of the risks for large n, say

$$\int u Q_o^{(n)}(du) / \int u Q_1^{(n)}(du) \leq 1 + o(n^o) .$$

Assertions of this kind, however, follow from (9.1.2) only for particular loss functions, those obeying a 'law of diminishing increment' (see Pfanzagl 1980b).

The usual way to avoid such difficulties is to let the loss function depend on n, i.e. to measure the loss connected with an estimate $\kappa^n(\underline{x})$ by $L_P(n^{1/2}(\kappa^n(\underline{x}) - \kappa(P)))$. We are still waiting to see a justification for the use of such loss functions.

Even if the notion of a loss function is appropriate for a certain problem, our knowledge about the loss function will usually be only vague. Hence it is important that the optimality results (expressed by concentration of the distribution of losses or by risk) hold true for all loss functions which come into question.

The considerations indicated so far refer to real-valued functionals. The situation is somewhat less favorable for multidimensional functionals. The multidimensional analogue of (9.1.1) is: For any sufficiently regular estimator-sequence κ^n, $n \in \mathbb{N}$,

(9.1.3) $$P^n\{\kappa^n \in C\} \leq N(\kappa(P), n^{-1}\Sigma(P))(C) + o(n^o)$$

holds uniformly over all convex sets C which are s y m m e t r i c about $\kappa(P)$.

Without symmetry, this assertion is not true any more. (See Pfanzagl, 1980a, pp. 19f., for a pertinent example.) Correspondingly, a bound for the concentration of the distribution of losses follows from (9.1.3) only for loss functions which are bowl shaped and symmetric about $\kappa(P)$. More precisely, we have to require that $L_P(\kappa(P)) = 0$, and that $\{u \in \mathbb{R}^k: L_P(u) \leq r\}$ is convex and symmetric about $\kappa(P)$ for every $r > 0$. Then (9.1.3) implies that - uniformly for all such loss functions - the distribution of losses, $P^n*(L_P \circ \kappa^n)$, cannot be more concentrated about zero than the distribution $N(\kappa(P), n^{-1}\Sigma(P))*L_P$, up to an error term $o(n^o)$.

This said, we restrict ourselves in the following to results concerning the concentration on appropriate sets.

The bounds given here are believed to be sharp in the sense that estimator-sequences attaining these bounds asymptotically do exist (under certain additional regularity conditions). This is certainly true for parametric families. General methods for constructing such estimator-sequences are indicated in Section 11.4.

9.2. Bounds for asymptotically median unbiased estimators

9.2.1. Definition. Let $\kappa: \mathfrak{P} \to \mathbb{R}$ be a functional. An estimator-sequence $\kappa^n: X^n \to \mathbb{R}$ is *as. median unbiased* for κ at P if for each $g \in T(P,\mathfrak{P})$ and each path $P_{t,g}$, $t \downarrow 0$, with derivative g the following relations hold uniformly for $t \geq 0$ in any bounded set:

$$P^n_{n^{-1/2},tg}\{\kappa^n \geq \kappa(P_{n^{-1/2},tg})\} \geq \frac{1}{2} + o(n^o),$$

$$P^n_{n^{-1/2},tg}\{\kappa^n \leq \kappa(P_{n^{-1/2},tg})\} \geq \frac{1}{2} + o(n^o).$$

In particular, any estimator-sequence κ^n, $n \in \mathbb{N}$, for which $P^n * n^{1/2}(\kappa^n - \kappa(P))$ approaches a normal limiting distribution with mean zero, locally uniformly in neighborhoods shrinking as $n^{-1/2}$, is as. median unbiased. Our definition requires, however, much less, since it forgoes uniformity in g.

In order to obtain bounds for the concentration of an as. median unbiased estimator-sequence we could use a locally uniform version of Theorem 8.4.1 on the power function of tests. The following direct proof gets along with somewhat weaker regularity conditions.

9.2.2. Theorem. *For $P \in \mathfrak{P}$ fixed, let $C(P,\mathfrak{P}) \subset T_w(P,\mathfrak{P})$ be a closed convex cone. Let $\kappa: \mathfrak{P} \to \mathbb{R}$ be differentiable at P, and let $\kappa^+(\cdot,P)$ denote the projection of any gradient into $C(P,\mathfrak{P})$, and $\sigma^2(P) := P(\kappa^+(\cdot,P)^2)$.*

If κ^n, $n \in \mathbb{N}$, is an estimator-sequence which is as. median unbiased for κ at P, then uniformly for $t',t'' \geq 0$ in any bounded set,

$$P^n\{\kappa(P) - n^{-1/2}t' < \kappa^n < \kappa(P) + n^{-1/2}t''\}$$
$$\leq N(0,\sigma^2(P))(-t',t'') + o(n^o).$$

Addendum. *If as. median unbiasedness holds uniformly for $t \in [0,t_n]$ with $t_n \uparrow \infty$, then the assertion holds uniformly for $t',t'' \geq 0$.*

9.2.3. Remark. The upper bound given by Theorem 9.2.2 is the better the larger $C(P,\mathfrak{P})$, with $C(P,\mathfrak{P}) = T(P,\mathfrak{P})$ yielding the lowest upper bound. The reason for considering s u b s e t s of $T(P,\mathfrak{P})$ is the following. Often we know for sure that the functions in a certain closed cone $C(P,\mathfrak{P})$ belong to $T(P,\mathfrak{P})$, whereas it needs more restrictive conditions to describe $T(P,\mathfrak{P})$ completely (see Sections 2.4 and 2.5). With this vague description of $T(P,\mathfrak{P})$ we obtain in general a bound which may be too large. However: If an estimator-sequence attaining this bound exists, then this bound is sharp, and the estimator-sequence is as. optimal. In such a situation a more complete description of the tangent set $T(P,\mathfrak{P})$ becomes unnecessary.

If this 'sure' subset of $T(P,\mathfrak{P})$ consists of a closed convex cone and its reflexion about 0, then we determine the projections into each of the two cones separately and take for $C(P,\mathfrak{P})$ the cone which yields the lower bound.

Proof of Theorem 9.2.2. Let $g \in C(P,\mathfrak{P})$ with $\|g\| = 1$ be fixed. By assumption, there exists a path $P_{t,-g}$, $t \downarrow 0$, in $\mathfrak{P}$ with derivative $-g$. For simplicity, write $P_{t,n} := P_{n^{-1/2}t,-g}$. Fix $t_o > 0$. The assumptions of Theorem 19.2.7 are fulfilled uniformly for $P_{t,-g}$, $t > 0$. The sequence of critical regions

$$C_{t,n} := \{\sum_{\nu=1}^{n} \log(p_{t,n}(x_\nu)/p(x_\nu)) < -t^2/2\}$$

is most powerful for testing $P^n_{t,n}$ against P^n. By Corollary 19.2.9 it is of level $\frac{1}{2} + o(n^o)$ uniformly for $t \in (0,t_o]$, and $P^n(C_{t,n}) = \Phi(t) + o(n^o)$ uniformly for $t \in (0,t_o]$. Since, by assumption, κ^n is as. median unbiased, we have uniformly for $t \in (0,t_o]$

$$P^n_{t,n}\{\kappa(P_{t,n}) < \kappa^n\} \leq \frac{1}{2} + o(n^o) .$$

The Neyman-Pearson Lemma 19.1.5 implies uniformly for $t \in (0,t_o]$

$$(9.2.4) \qquad P^n\{\kappa(P_{t,n}) < \kappa^n\} \leq \Phi(t) + o(n^o).$$

Since κ is differentiable at P, we have uniformly for $t \in (0,t_o]$

$$\kappa(P_{t,n}) = \kappa(P) - n^{-1/2}tP(\kappa^\cdot(\cdot,P)g) + o(n^{-1/2}).$$

Hence (9.2.4) implies uniformly for $t \in (0,t_o]$

$$P^n\{\kappa(P) - n^{-1/2}tP(\kappa^\cdot(\cdot,P)g) < \kappa^n\} \leq \Phi(t) + o(n^o),$$

where $\kappa^\cdot$ is an arbitrary gradient of κ. Presuming $P(\kappa^\cdot(\cdot,P)g) \neq 0$ we obtain uniformly for $t' \in (0,t_1]$

$$P^n\{\kappa(P) - n^{-1/2}t' < \kappa^n\} \leq \Phi(t'/P(\kappa^\cdot(\cdot,P)g)) + o(n^o).$$

Together with the corresponding inequality for the other tail this yields uniformly for $t', t'' \in (0, t_1]$

$$P^n\{\kappa(P) - n^{-1/2}t' < \kappa^n < \kappa(P) + n^{-1/2}t''\}$$
$$\leq N(0, P(\kappa^{\cdot}(\cdot,P)g)^2)(-t',t'') + o(n^0).$$

This relation holds for each fixed $g \in \mathcal{C}(P,\mathfrak{P})$ with $\|g\| = 1$ and $P(\kappa^{\cdot}(\cdot,P)g) \neq 0$, and therefore also with $P(\kappa^{\cdot}(\cdot,P)g)$ replaced by

$$\sigma(P) := \sup\{P(\kappa^{\cdot}(\cdot,P)g) : g \in \mathcal{C}(P,\mathfrak{P}), \|g\| = 1\}$$

(since the family of functions $\sigma \to N(0,\sigma^2)(-t',t'')$, $0 \leq t',t'' \leq t_1$, is equicontinuous).

By Proposition 4.3.5,

$$\sup\{P(\kappa^{\cdot}(\cdot,P)g) : g \in \mathcal{C}(P,\mathfrak{P}), \|g\| = 1\} = P(\kappa^{+}(\cdot,P)^2)^{1/2}.$$

Recall that this supremum is attained for $g = \kappa^{+}(\cdot,P)/P(\kappa^{+}(\cdot,P)^2)^{1/2}$. Hence the direction $\kappa^{+}(\cdot,P)$ is least favorable.

9.3. Multidimensional functionals

If the functional κ is k-dimensional and κ^n, $n \in \mathbb{N}$, a k-dimensional estimator-sequence for κ which is componentwise as. median unbiased $o(n^{-1/2})$, then a result like Theorem 9.2.2 holds for each component. In this section we present a result on bounds for the as. concentration of the k-dimensional distribution of the estimator.

The following result adapts a well-known result on parametric models to our more general framework. The parametric version of (9.3.3) is due to Kaufman (1966). The formulation (9.3.4), using convolutions, was introduced independently by Inagaki (1970) and Hâjek (1970). The proof given below uses a suggestion of Bickel (see Roussas, 1972, p. 136). So far, nonparametric versions of this theorem have been known only for certain special cases. Beran (1977a, p. 439, Theorem 6) gives an equivalent theorem for location functionals, Beran (1977b, p. 452, Theorem 5) for a particular functional extending the parameters from a parametric family to a nonparametric neighborhood, and Begun and Wellner (1981) for the proportionality factor in the proportional failure rate model (see also Section 18.6). A very general version of the convolution theorem is due to LeCam (1972, p. 257, Proposition 10). Moussatat (1976, p. 100, Theorem) obtains a convolution theorem for the parameter space being a separable Hilbert space.

9.3.1. Theorem. *For* $i = 1,\dots,k$ *let* $\kappa_i: \mathfrak{P} \to \mathbb{R}$ *be a functional which is differentiable at* P. *Assume that the canonical gradients* $\kappa_i^{\cdot}(\cdot,P) \in T_w(P,\mathfrak{P})$ *have a positive definite covariance matrix* $\Sigma(P) := (P(\kappa_i^{\cdot}(\cdot,P)\kappa_j^{\cdot}(\cdot,P)))_{i,j=1,\dots,k}$. *Let* κ^n, $n \in \mathbb{N}$, *be a regular estimator-sequence in the sense that there exists a p-measure* $M|\mathbb{B}^k$ *with the following property: For every* $a \in \mathbb{R}^k$ *there exists a path* $P_{t,a'\kappa^{\cdot}(\cdot,P)}$, $t \downarrow 0$, *with weak derivative* $a'\kappa^{\cdot}(\cdot,P)$, *such that*

$$P^n_{n^{-1/2},a'\kappa^{\cdot}(\cdot,P)} * n^{1/2}(\kappa^n - \kappa(P_{n^{-1/2},a'\kappa^{\cdot}(\cdot,P)})) \Rightarrow M . \tag{9.3.2}$$

Then there exists a p-measure $R|\mathbb{B}^k$ *such that*

$$P^n*(\tilde{\kappa}^{\cdot}(\cdot,P), n^{1/2}(\kappa^n - \kappa(P)) - \tilde{\kappa}^{\cdot}(\cdot,P)) \Rightarrow N(0,\Sigma(P)) \times R . \tag{9.3.3}$$

Relation (9.3.3) implies in particular

$$M = N(0,\Sigma(P)) \circledast R . \tag{9.3.4}$$

9.3.5. Remark. Assumption (9.3.2) is in particular fulfilled if $Q^n*n^{1/2}(\kappa^n-\kappa(Q)) \Rightarrow M_Q$ locally uniformly in a neighborhood of P, and if $M_Q \Rightarrow M_P$ for $Q \to P$ (in some appropriate sense). In fact, it is hard to understand the operational significance of a condition like (9.3.2), unless it is a consequence of local uniformity.

Without a uniformity condition for the convergence to the limiting distribution (like (9.3.2)), the theorem is not true any more.

This follows from the well-known examples of superefficient estimators (see LeCam, 1953, p. 280). The following nonparametric example of superefficiency can be found in Bickel and Lehmann (1975b, p. 1058, Note).

Let $\mathfrak{P}$ be the class of all sufficiently regular p-measures over $\mathbb{B}$, and $\kappa(P)$ the median of P. Let κ^n be the sample mean if the sample distribution $Q_n(\underline{x},\cdot)$ is approximately normal (more precisely: if $\inf_{\mu,\sigma}|Q_n(\underline{x},(-\infty,y])-\Phi(y-\mu)/\sigma| < n^{-1/4}$), and the sample median otherwise. Then the as. variance of $n^{1/2}(\kappa^n-\kappa(P))$ under P is $1/4p(\kappa(P))^2$ if P is not normal, but $\sigma^2 = 1/2\pi p(\kappa(P))$ if $P = N(\mu,\sigma^2)$. Observe that κ^n is even location and scale equivariant.

Local uniformity will be further investigated in Section 9.4.

9.3.6. Corollary. *Componentwise efficiency implies joint efficiency: Assume that an estimator-sequence κ^n, $n \in \mathbb{N}$, fulfills condition (9.3.2). If this sequence is componentwise as. efficient, i.e., if* $P^n * n^{1/2}(\kappa_i^n - \kappa_i(P))$, $n \in \mathbb{N}$, *converges weakly to* $N(0,P(\kappa_i^{\bullet}(\cdot,P)^2))$ *for* $i = 1,\ldots,k$, *then* $P^n * n^{1/2}(\kappa^n - \kappa(P))$, $n \in \mathbb{N}$, *converges weakly to* $N(0,\Sigma(P))$.

Proof. By Theorem 9.3.1 there exists a p-measure $R|\mathbb{B}^k$ such that

$$P^n * n^{1/2}(\kappa^n - \kappa(P)) \Rightarrow N(0,\Sigma(P)) \circledast R .$$

The covariance matrix of the convolution is the sum of the covariance matrices of $N(0,\Sigma(P))$ and R. Since by assumption

$$P^n * n^{1/2}(\kappa_i^n - \kappa_i(P)) \Rightarrow N(0,\Sigma_{ii}(P))$$

for $i = 1,\ldots,k$, it follows that the covariance matrix of R has zeros in its main diagonal and is, therefore, identically zero. Hence $R\{0\} = 1$.

Proof. of Theorem 9.3.1. Let $a \in \mathbb{R}^k$ be arbitrary. To simplify our notations, we write $\kappa^{\bullet}$ for $\kappa^{\bullet}(\cdot,P)$ and Σ for $\Sigma(P)$. Moreover, let $k := a'\kappa^{\bullet}$ and $Q_n := P_{n^{-1/2},k}$. Since κ is differentiable at P, we have

$$\kappa(Q_n) = \kappa(P) + n^{-1/2}\Sigma a + o(n^{-1/2}) .$$

Together with (9.3.2) this implies

(9.3.7) $$Q_n^n * n^{1/2}(\kappa^n - (\kappa(P) + n^{-1/2}\Sigma a)) \Rightarrow M .$$

We have $P^n * \tilde{\kappa}^{\cdot} \Rightarrow N(0,\Sigma)$ and (by (9.3.2)) $P^n * n^{1/2}(\kappa^n - \kappa(P)) \Rightarrow M$. In particular, both sequences are weakly sequentially compact. By Prohorov's theorem (Billingsley, 1968, p. 37, Theorems 6.1 and 6.2), a sequence of p-measures on a complete separable metric space is tight iff it is weakly sequentially compact. Since tightness of the marginals implies tightness of the joint distribution, the sequence $P^n * (\tilde{\kappa}^{\cdot}, n^{1/2}(\kappa^n - \kappa(P)))$, $n \in \mathbb{N}$, is weakly sequentially compact. Hence it contains a subsequence converging to some p-measure $K|\mathbb{B}^{2k}$. For convenience of notation we assume that the whole sequence converges to K. Hence

$$P^n * (\tilde{k} - \tfrac{1}{2}P(k^2), n^{1/2}(\kappa^n - \kappa(P))) \Rightarrow L,$$

where

$$L := K * ((u,v) \to (a'u - \tfrac{1}{2}a'\Sigma a, v)) .$$

By Theorem 19.2.7 and Slutzky's lemma, this implies

$$P^n * (\Lambda_n, n^{1/2}(\kappa^n - \kappa(P))) \Rightarrow L,$$

where

$$\Lambda_n(\underline{x}) := \sum_{\nu=1}^{n} \log(q_n(x_\nu)/p(x_\nu)) .$$

By Theorem 19.2.7 and Corollary 19.2.9,

$$P^n * \Lambda_n \Rightarrow N(-\tfrac{1}{2}P(k^2), P(k^2)) .$$

Furthermore, $\int \exp[u] N(-\frac{1}{2}P(k^2), P(k^2))(du) = 1$. Hence by a well-known theorem of LeCam (1960, p. 40, Theorem 2.1) the sequences P^n and Q_n^n, $n \in \mathbb{N}$, are contiguous, and $Q_n^n * (\Lambda_n, n^{1/2}(\kappa^n - \kappa(P)))$, $n \in \mathbb{N}$, converges weakly to the p-measure with L-density $(u,v) \to \exp[u]$. Together with (9.3.7) this implies that for any bounded and continuous function f,

(9.3.8) $$M(f) = \int f(v - \Sigma a)\exp[u] L(d(u,v))$$
$$= \int f(v - \Sigma a)\exp[a'u - \tfrac{1}{2}a'\Sigma a] K(d(u,v)) .$$

Applying (9.3.8) for $f(v) = \exp[it'v]$ we obtain

(9.3.9) $$\int \exp[it'v]M(dv) = \int \exp[a'u - \frac{1}{2}a'\Sigma a + it'(v-\Sigma a)]K(d(u,v)).$$

Since $a \in \mathbb{R}^k$ was arbitrary, (9.3.9) holds for all $a \in \mathbb{R}^k$. With the right side analytic in a (see Lehmann, 1959, p. 52, Theorem 9; recall that K does not depend on a) it holds for arbitrary $a \in \mathbb{C}^k$ and therefore, in particular, for $a = i(s-t)$, so that

(9.3.10) $$\int \exp[it'v]M(dv) = \exp[\frac{1}{2}s'\Sigma s - \frac{1}{2}t'\Sigma t] \int \exp[i(s'u + t'(v-u)]K(d(u,v)) .$$

With $K_o := K*((u,v) \to (u,v-u))$ this implies

(9.3.11) $$\int \exp[i(s'u+t'v)]K_o(d(u,v)) = \exp[-\frac{1}{2}s'\Sigma s] \exp[\frac{1}{2}t'\Sigma t]\int \exp[it'v]M(dv) .$$

Since the characteristic function of $K_o|\mathbb{B}^{2k}$ is a product, K_o is the independent product of its marginals on $\mathbb{B}^k$, the first marginal being $N(0,\Sigma)$ (with characteristic function $s \to \exp[-\frac{1}{2}s'\Sigma s]$).

From the definition of K as the weak limit of a subsequence of $P^n*(\tilde{\kappa}^{\cdot}, n^{1/2}(\kappa^n-\kappa(P)))$ it follows that K_o is the weak limit of this subsequence for $P^n*(\tilde{\kappa}^{\cdot}, n^{1/2}(\kappa^n-\kappa(P))-\tilde{\kappa}^{\cdot})$. Since K_o is uniquely determined by (9.3.11), it follows that the whole sequence converges to K_o, which proves (9.3.3).

9.3.12. Remark. Relation (9.3.3) suggests the interpretation that $N(0,\Sigma(P))$ is asymptotically the maximal concentration for any estimator-sequence with distributions converging continuously to a limiting distribution. The intuitive idea that $N(0,\Sigma(P)) \circledast R$ is less concentrated than $N(0,\Sigma(P))$ can be made more precise by means on Anderson's theorem (see Anderson, 1955, p. 172, Theorem 2) which asserts that $N(0,\Sigma(P))(C) \geq N(0,\Sigma(P)) \circledast R(C)$ for any convex set C which is symmetric about zero. Correspondingly, $N(0,\Sigma(P)) * L$ is stochastically smaller than $(N(0,\Sigma(P)) \circledast R) * L$ for any symmetric non-negative bowl-shaped loss function L.

Perhaps one could also think of another interpretation in terms of as. sufficiency. If R did not depend on P, such an interpretation would be straightforward: Knowing an estimator κ^n such that $n^{1/2}(\kappa^n - \kappa(P))$ is as. distributed as $N(0,\Sigma(P))$ one would be in the position to match any other estimator (attaining its limiting distribution uniformly in the sense of (9.3.2)) by adding to $\kappa^n(\underline{x})$ a term $n^{-1/2}z$, where z is an independent realization governed by R. The crux of this interpretation is that R depends on P, so that we need in addition to $\kappa^n(\underline{x})$ an (inefficient) estimate $P_n(\underline{x},\cdot)$ of P. (If R depends on P continuously, we may then determine z as an independent realization governed by $R(P_n(\underline{x},\cdot))$.)

The convolution theorem leads to a simple criterion for the as. efficiency of a given estimator-sequence.

__9.3.13. Corollary__. *Assume that, under the assumptions of Theorem 9.3.1, the limit p-measure* M *in (9.3.2) is normal* $N(0,\hat{\Sigma}(P))$. *Then* $\hat{\Sigma}(P) - \Sigma(P)$ *is nonnegative definite.*

If $\hat{\Sigma}(P) = P(gg')$, *where* $g = (g_1,\ldots,g_k)$ *with* $g_i \in T_w(P,\mathfrak{P})$, *and if (9.3.2) holds with* $\kappa^{\cdot}(\cdot,P)$ *replaced by* g_i, *then* $\hat{\Sigma}(P) = \Sigma(P)$ *(and* g *is the canonical gradient).*

__Proof__. As in the proof of Theorem 9.3.1 it can be shown (compare (9.3.10)) that for $h = g$ and $h = \kappa^{\cdot}(\cdot,P)$,

$$(9.3.14) \qquad \int \exp[it'v]M(dv)$$
$$= \exp[\tfrac{1}{2}P((t'(h-\kappa^{\cdot}(\cdot,P)))^2) - \tfrac{1}{2}P((t'\kappa^{\cdot}(\cdot,P))^2)]\psi(t)$$

with a certain characteristic function ψ. Since $M = N(0,P(gg'))$, we have

$$\int \exp[it'v]M(dv) = \exp[-\tfrac{1}{2}P((t'g)^2)].$$

Applying (9.3.14) for $h = g$ and $h = \kappa^\bullet(\cdot,P)$, we obtain for $t \in \mathbb{R}^k$

$$P((t'(g-\kappa^\bullet(\cdot,P)))^2) = 0.$$

Hence $g = \kappa^\bullet(\cdot,P)$ P-a.e. and therefore $P(gg') = \Sigma(P)$.

9.4. Locally uniform convergence

In this section we show that the limit distribution of an estimator-sequence under Q^n depends V-continuously on Q if the convergence is V-locally uniform. As a consequence we obtain that the distributions of as. efficient estimator-sequences necessarily fail to approach their limiting p-measure V-locally uniformly if the basic family of p-measures is too large.

9.4.1. Proposition. *Let $\kappa: \mathfrak{P} \to \mathbb{R}$ be continuous at some $P \in \mathfrak{P}$ with respect to a neighborhood system $\mathcal{U}$ of P which is at least as strong as the neighborhood system of the sup-metric. Let κ^n, $n \in \mathbb{N}$, be an estimator-sequence, and assume that there exists $U \in \mathcal{U}$ such that $Q^n * n^{1/2}(\kappa^n - \kappa(Q))$, $n \in \mathbb{N}$, converges weakly to a p-measure M_Q, uniformly for $Q \in U$. Then $M_Q \Rightarrow M_P$ if $Q \to P$ with respect to $\mathcal{U}$.*

9.4.2. Corollary. *If $M_Q = N(0,\Sigma(Q))$, then $Q \to \Sigma(Q)$ is continuous at P with respect to $\mathcal{U}$.*

Proof of Proposition 9.4.1. Let $f: \mathbb{R}^k \to [0,1]$ be an arbitrary uniformly continuous function. It suffices to show that $M_Q(f) \to M_P(f)$ if $Q \to P$ with respect to $\mathcal{U}$. With $Q_n := Q^n * n^{1/2}(\kappa^n - \kappa(Q))$ we have

$$\begin{aligned}
&|M_Q(f) - M_P(f)| \\
&\leq |M_Q(f) - Q_n(f)| \\
&\quad + |Q_n(f) - \int f(t-n^{1/2}(\kappa(Q)-\kappa(P)))P_n(dt)| \\
&\quad + |\int f(t-n^{1/2}(\kappa(Q)-\kappa(P)))P_n(dt) - \int f(t-n^{1/2}(\kappa(Q)-\kappa(P)))M_P(dt)| \\
&\quad + |\int f(t-n^{1/2}(\kappa(Q)-\kappa(P)))M_P(dt) - M_P(f)|.
\end{aligned}$$

By assumption, for every $\varepsilon > 0$ there exists n'_ε such that $n \geq n'_\varepsilon$ implies

$$|M_Q(f) - Q_n(f)| < \varepsilon/4 \qquad \text{for all } Q \in U.$$

Moreover, $P_n \Rightarrow M_P$ implies $P_n(f) \to M_P(f)$, uniformly for f in any class of uniformly bounded equicontinuous functions (see Parthasarathy, 1967, p. 51, Theorem 6.8). With f uniformly continuous, the family of functions $t \to f(t+a)$, $a \in \mathbb{R}^k$, is equicontinuous. Hence there exists $n_\varepsilon \geq n'_\varepsilon$ such that for all $Q \in U$

$$|\int f(t-n_\varepsilon^{1/2}(\kappa(Q)-\kappa(P)))P_{n_\varepsilon}(dt) - \int f(t - n_\varepsilon^{1/2}(\kappa(Q)-\kappa(P)))M_P(dt)| < \frac{\varepsilon}{4}.$$

Further, we obtain for $n = n_\varepsilon$

$$\begin{aligned}
&|Q_n(f) - \int f(t-n^{1/2}(\kappa(Q)-\kappa(P)))P_n(dt)| \\
&= |Q^n(f\circ(n^{1/2}(\kappa^n-\kappa(Q)))) - P^n(f\circ(n^{1/2}(\kappa^n-\kappa(Q))))| \\
&\leq V(Q^n,P^n) \leq nV(Q,P) < \frac{\varepsilon}{4}
\end{aligned}$$

if

$$(9.4.3) \qquad V(Q,P) \leq \frac{\varepsilon}{4}\, n_\varepsilon^{-1}.$$

Since f is uniformly continuous, there exists δ_ε such that

$$|\int f(t-n^{1/2}(\kappa(Q)-\kappa(P))M_P(dt) - M_P(f)| < \frac{\varepsilon}{4}$$

if

$$(9.4.4) \qquad |\kappa(Q) - \kappa(P)| < \delta_\varepsilon n_\varepsilon^{-1/2}.$$

Since $Q \to V(Q,P)$ and $Q \to \kappa(Q)$ are continuous at P, there exists $U_\varepsilon \in \mathcal{U}$ with $U_\varepsilon \subset U$ such that (9.4.3) and (9.4.4) are fulfilled for $Q \subset U_\varepsilon$. Hence $Q \in U_\varepsilon$ implies

$$|M_Q(f) - M_P(f)| < \varepsilon.$$

Corollary 9.4.2 can be used to prove for certain families $\mathfrak{P}$ that the normal approximation to as. efficient estimator-sequences cannot be locally uniform.

9.4.5. Corollary. *Let $\mathcal{U}$ be a neighborhood system of* P *which is at least as strong as the neighborhood system of the sup-metric, and for which* $Q \to \kappa(Q)$ *is continuous at* P.

If $Q \to Q(\kappa^{\cdot}(\cdot,Q)^2)$ *is discontinuous at* P, *then as. efficient estimator-sequences with a $\mathcal{U}$-locally uniform normal approximation cannot exist.*

In other words: The normal approximation can be locally uniform only with respect to a neighborhood system of P which is strong enough to render $Q \to Q(\kappa^{\cdot}(\cdot,Q)^2)$ continuous at P.

9.4.6. Remark. In order to apply Corollary 9.4.5 to any given neighborhood system $\mathcal{U}^*$ of P at least as strong as the sup-metric, it suffices to exhibit a sequence of p-measures Q_m, $m \in \mathbb{N}$, with the following properties:

(i) $Q_m \to P$ with respect to $\mathcal{U}^*$,

(ii) $\kappa(Q_m) \to \kappa(P)$,

(iii) $Q_m(\kappa^{\cdot}(\cdot,Q_m)^2)$ fails to converge to $P(\kappa^{\cdot}(\cdot,P)^2)$.

(Hint: Take $\mathfrak{P} = \{P\} \cup \{Q_m: m \in \mathbb{N}\}$, and take for $\mathcal{U}$ the neighborhood system generated by $\mathcal{U}^* \cap \{Q_m: m \in \mathbb{N}\}$ and $\{Q_m: |\kappa(Q_m)-\kappa(P)| < \varepsilon, m \in \mathbb{N}, \varepsilon > 0.)$

To obtain a strong nonexistence theorem, the neighborhood system should be as large as possible. One particular choice which suggests itself in our framework is the neighborhood system generated by the distance function $\Delta(Q;P) = P((\frac{q}{p} - 1)^2)^{1/2}$. (Recall that $V(Q,P) \leq \Delta(Q;P)$, so that the neighborhood system based on Δ is, in fact, stronger than that based on V.)

The following example illustrates the application of Corollary 9.4.5.

9.4.7. Example. Let $\mathfrak{P}$ be the family of all symmetric p-measures Q with $Q((q'/q)^2)$ finite which are equivalent to the Lebesgue measure. We shall show that an as. efficient estimator-sequence for the median necessarily fails to converge to its limiting distribution Δ-locally uniformly at $N(\mu,\sigma^2)$. (This remains true if we restrict $\mathfrak{P}$ to the subfamily with arbitrarily high moments and bounded Lebesgue density.)

To simplify our notations, we assume w.l.g. $\mu = 0$, $\sigma^2 = 1$. The as. efficient estimator-sequence for the median has as. variance $1/Q((q'/q)^2)$ (see (13.1.5)). According to Remark 9.4.6, it suffices to exhibit a sequence $Q_m \in \mathfrak{P}$, $m \in \mathbb{N}$, with $\kappa(Q_m) = 0$ for $m \in \mathbb{N}$ for which $N(0,1)((\frac{q_m}{\varphi} - 1)^2) \to 0$, whereas $Q_m((q_m'/q_m)^2)$ fails to converge to $N(0,1)((\varphi'/\varphi)^2)$. An example of such a sequence is

$$q_m = \varphi(1 + g_m)$$

with

$$g_m(x) = \exp[-\tfrac{m}{2}x^2] - (1+m)^{-1/2} .$$

Since $g_m > -1$ and $\int g_m(x)\varphi(x)dx = 0$, q_m is the density of a p-measure. Since g_m is symmetric about 0, we have $\kappa(Q_m) = 0$. An elementary computation shows that

$$\int\left(\frac{q_m(x)}{\varphi(x)} - 1\right)^2\varphi(x)dx \to 0 , \tag{9.4.8}$$

whereas

$$\int(q_m'(x)/q_m(x))^2 q_m(x)dx \to \infty . \tag{9.4.9}$$

(Hint: (9.4.8) is equivalent to $\int g_m(x)^2\varphi(x)dx \to 0$. For (9.4.9), the essential point is that both, $\int g_m'(x)\varphi'(x)dx \to \infty$ and $\int g_m'(x)^2\varphi(x)dx \to \infty$.)

9.4.10. Remark. Considering the fact that lack of uniformity is an inherent property, it is not advisable to consider an estimator-sequence as. efficient only if its distribution converges to the optimal limiting distribution locally uniformly, or if it is locally as. minimax.

Even though such a property may not be fulfilled for the full family, there may exist 'restricted' subsets (e.g. in the sense that the derivatives of the densities fulfill a uniform Lipschitz condition) on which the convergence is locally uniform, and which exhaust the whole family. Such a weakened requirement seems to be suggested by Ibragimov and Hasminskii (1981, pp. 229ff.) in their examples of as. efficient estimator-sequences, although in the definition of as. efficiency (p. 219, (1.9)) they require local uniformity in the full family.

A still weaker requirement is local uniformity over all (smooth) finite-dimensional subfamilies of $\mathfrak{P}$. A local as. minimax property in this spirit is considered by Fabian and Hannan (1982).

The phenomenon of nonuniformity in nonparametric procedures has first been exhibited by Bahadur and L.J.Savage (1956). A nonexistence result in the same spirit as above is contained in Klaassen (1979). He shows that an equivariant, antisymmetric estimator of the median cannot approach its limiting distribution locally uniformly if the family $\mathfrak{P}$ is large. More precisely, let $\mathfrak{P}$ denote the class of all symmetric p-measures $P|\mathbb{B}$ with Lebesgue density p and finite $\sigma(P) := (P((p'/p)^2))^{1/2}$. Then (see Klaassen, 1979, p. 253, Remark) for any sequence κ^n, $n \in \mathbb{N}$, of equivariant and antisymmetric estimators,

$$\lim_{n\to\infty} \inf_{P\in\mathfrak{P}_0} P\{n^{1/2} \frac{\kappa^n - \kappa(P)}{\sigma(P)} < t\} = \frac{1}{2} \qquad \text{for every } t > 0 \, ,$$

where $\mathfrak{P}_0$ is a subclass of $\mathfrak{P}$ with fixed $\kappa(P)$ (e.g., $\mathfrak{P}_0 = \{P \in \mathfrak{P}: \kappa(P) = 0\}$).

9.5. Restrictions of the basic family

In the foregoing sections we discussed the problem of estimating a functional κ defined on a basic family $\mathfrak{P}$. Now we consider the problem of how the as. variance bound can be reduced if the prior knowledge becomes more precise, i.e. if the basic family $\mathfrak{P}$ is replaced by a smaller family, say $\overline{\mathfrak{P}}$.

The restriction from $\mathfrak{P}$ to $\overline{\mathfrak{P}}$ brings about a restriction of the corresponding tangent spaces from $T(P,\mathfrak{P})$ to $T(P,\overline{\mathfrak{P}})$. Since we now require as. median unbiasedness (resp. condition (9.3.2)) to hold for directions in the smaller tangent space $T(P,\overline{\mathfrak{P}})$ only, we impose less restrictive conditions on our estimators, so that the as. variance bound becomes smaller in general. Whether such a decrease takes place or not is easy to decide: The as. variance bound connected with the basic family $\overline{\mathfrak{P}}$ is determined by the canonical gradient of κ in $T(P,\overline{\mathfrak{P}})$. This canonical gradient can be obtained from the gradient of κ in $T(P,\mathfrak{P})$, say $\kappa^{+}(\cdot,P)$, by projection into $T(P,\overline{\mathfrak{P}})$. The conclusion: *The restriction from $\mathfrak{P}$ to $\overline{\mathfrak{P}}$ will not reduce the as. variance bound for estimators of κ if the canonical gradient of κ in $T(P,\mathfrak{P})$ happens to be in* $T(P,\overline{\mathfrak{P}})$.

Notice that this is exactly the same condition which guarantees that the restriction from $\mathfrak{P}$ to $\overline{\mathfrak{P}}$ does not improve the as. envelope power function for tests of any hypothesis $\kappa(P) = c_o$ (see Remark 8.6.9).

<u>9.5.1. Example</u>. Let $\mathfrak{P}$ be an arbitrary family and $\kappa_i: \mathfrak{P} \to \mathbb{R}$ a differentiable functional with gradient $\kappa_i^{\bullet}(\cdot,P) \in T(P,\mathfrak{P})$, $i = 0,1$. The as. variance bound for estimators of $\kappa_1(P)$ will not be reduced by a

restriction to the family $\overline{\mathfrak{P}} = \{P \in \mathfrak{P}: \kappa_o(P) = 0\}$ iff $\kappa_1^{\bullet}(\cdot,P) \in T(P,\overline{\mathfrak{P}})$. Since $T(P,\overline{\mathfrak{P}}) = \{g \in T(P,\mathfrak{P}): P(g\kappa_o^{\bullet}(\cdot,P)) = 0\}$, this condition is equivalent to $P(\kappa_1^{\bullet}(\cdot,P)\kappa_o^{\bullet}(\cdot,P)) = 0$. If this is the case, the knowledge of the value of κ_o does not help to reduce the as. variance bound for estimators of $\kappa_1(P)$.

Numerous examples are of this type, e.g.: The knowledge of the location parameter does not help to increase the as. variance bound for the scale parameter, if the distribution is symmetric. Other examples can be modeled in analogy to Examples 8.6.8 and 8.6.11 referring to testing problems.

9.5.2. Example. Let $\mathfrak{P}$ denote the family of all p-measures $P|\mathbb{B}$ with positive Lebesgue density p. Let $\kappa(P) = \int p(\xi)^2 d\xi$. This functional, occurring in connection with nonparametric problems, was considered by a number of authors (Hodges and Lehmann, 1956, G.K.Bhattacharyya and Roussas, 1969, Dmitriev and Tarasenko, 1974, Schüler and Wolff, 1976). Bhattacharyya and Roussas suggest for $\kappa(P)$ the estimator

$$\frac{1}{n^2} \sum_{\nu=1}^{n} \sum_{\mu=1}^{n} \frac{1}{a_n} K\left(\frac{x_\nu - x_\mu}{a_n}\right)$$

with a bounded kernel K fulfilling $uK(u) \to 0$ as $|u| \to \infty$, and a sequence a_n, $n \in \mathbb{N}$, fulfilling $a_n \to 0$ and $na_n \to \infty$. Dmitriev and Tarasenko compute (p. 393, formula (20)) its as. variance,

$$4\left(\int p(\xi)^3 d\xi - \kappa(P)^2\right) .$$

It is straightforward to show that the canonical gradient of κ in $\mathscr{L}_*(P)$ is

$$\kappa^{\bullet}(x,P) = 2(p(x) - \kappa(P)) .$$

Since $\int \kappa^{\bullet}(\xi,P)^2 p(\xi) d\xi = 4\left(\int p(\xi)^3 d\xi - \kappa(P)^2\right)$, this proves that the estimator-sequence suggested by Bhattacharyya and Roussas is as. efficient.

Is it possible to obtain estimator-sequences which are as. superior if it is known that the true p-measure is symmetric? The answer is no, since then $\kappa^{\bullet}(\cdot,P)$ is symmetric about the same center of symmetry as P, and therefore an element of the tangent space of the family of all symmetric distributions (see Example 8.1.1).

The following example covers a number of special cases.

9.5.3. Example. Consider a family of p-measures $\mathfrak{P} = \{P_{\theta,\eta}: (\theta,\eta) \in \Theta \times H\}$, where Θ and H are arbitrary sets, and the subfamilies $\mathfrak{Q}_\theta := \{P_{\theta,\eta}: \eta \in H\}$ and $\mathfrak{P}_\eta := \{P_{\theta,\eta}: \theta \in \Theta\}$. Let $\kappa: \mathfrak{P} \to \mathbb{R}$ be a differentiable functional with canonical gradient $\kappa^{\bullet}(\cdot,P_{\theta,\eta}) \in T(P_{\theta,\eta},\mathfrak{P})$.

If $\kappa(P_{\theta,\eta})$ depends on θ only, i.e. $\kappa(P_{\theta,\eta'}) = \kappa(P_{\theta,\eta''})$, then

(9.5.4) $$\kappa^{\bullet}(\cdot,P_{\theta,\eta}) \perp T(P_{\theta,\eta},\mathfrak{Q}_\theta) .$$

To see this, let $g \in T(P_{\theta,\eta},\mathfrak{Q}_\theta)$ be arbitrary, and let P_{θ,η_t}, $t \downarrow 0$, be a path with derivative g. By Definition 4.1.1,

$$0 = \kappa(P_{\theta,\eta_t}) - \kappa(P_{\theta,\eta}) = t\int \kappa^{\bullet}(\xi,P_{\theta,\eta})g(\xi)P_{\theta,\eta}(d\xi) + o(t);$$

hence $\kappa^{\bullet}(\cdot,P_{\theta,\eta}) \perp g$. Since $g \in T(P_{\theta,\eta},\mathfrak{Q}_\theta)$ was arbitrary, this implies (9.5.4).

Let us now consider the question whether the knowledge of η leads to an increase in the as. variance bound for κ. In other words: Are as. better estimators available for $\kappa|\mathfrak{P}_\eta$ than the as. optimal estimators for $\kappa|\mathfrak{P}$? This will be so in general, unless $\kappa^{\bullet}(\cdot,P_{\theta,\eta}) \in T(P_{\theta,\eta},\mathfrak{P}_\eta)$.

In Section 2.6 it was indicated that

(9.5.5) $$T(P_{\theta,\eta},\mathfrak{P}_\eta) + T(P_{\theta,\eta},\mathfrak{Q}_\theta) \subset T(P_{\theta,\eta},\mathfrak{P}) .$$

Under stronger regularity conditions, equality will hold in (9.5.5). (This holds true in a number of cases which can be subsumed under this model. See, for instance, Sections 2.2 - 2.5.)

If equality holds in (9.5.5), then $T(P_{\theta,\eta},\mathfrak{P}_\eta) \perp T(P_{\theta,\eta},\mathfrak{Q}_\theta)$ implies $\kappa^{\bullet}(\cdot,P_{\theta,\eta}) \in T(P_{\theta,\eta},\mathfrak{P}_\eta)$ (as a consequence of (9.5.4)).

The conclusion: For a sufficiently regular family with two parameters, the orthogonality of the two tangent spaces $T(P_{\theta,\eta},\mathfrak{P}_\eta)$ and $T(P_{\theta,\eta},\mathfrak{Q}_\theta)$ implies that for a n y functional $\kappa(P_{\theta,\eta})$ depending on θ only estimator-sequences which are as. efficient on $\mathfrak{P}$ are necessarily also as. efficient on any of the subfamilies $\mathfrak{P}_\eta$ with η known. (In other words: Any such estimator-sequence is 'adaptive'.)

The orthogonality of $T(P_{\theta,\eta},\mathfrak{P}_\eta)$ and $T(P_{\theta,\eta},\mathfrak{Q}_\theta)$ seems to be a natural generalization of Stein's (1956) condition for $\Theta \subset \mathbb{R}^k$ which, in the more concise version of Bickel (1981, p. 42), requires orthogonality for any parametric subfamily of $\mathfrak{Q}_\theta$. Bickel (1981, p. 43) replaces Stein's condition by a condition S* which is perhaps unnecessarily restrictive in that it requires any P_{θ,η_o} to project e x a c t l y (in the sense of (7.2.2)) on P_{θ,η_1} in any of the subfamilies $\mathfrak{P}_{\eta_1}$.

Bickel (1982, Theorem 3.1) indicates for the case $\Theta \subset \mathbb{R}^k$ and $\kappa(P_{\theta,\eta}) = \theta$ how 'adaptive' estimators can be obtained from the Newton-Raphson improvement procedure (see 11.4.1), provided there exists an estimator $\ell^{(\cdot)}(\cdot,\theta,x_1,\ldots,x_n)$ for $\ell^{(\cdot)}(\cdot,\theta,\eta)$ fulfilling a certain condition H (see p. 15). One aspect of this condition, the assumption $\int \ell^{\cdot}(\xi,\theta,x_1,\ldots,x_n)P_{\theta,\eta}(d\xi) = 0$ for all $(x_1,\ldots,x_n) \in X^n$ and all $\eta \in H$ (in our notations) needs certainly to be relaxed to make this procedure more applicable.

Fabian and Hannan (1982, Theorem 7.10) come close to the idea that under Stein's orthogonality condition (certain) estimator-sequences which are as. efficient for the whole family are also as. efficient for any of the parametric subfamilies $\mathfrak{P}_\eta$, η known.

<u>9.5.6. Remark</u>. It is clear from the results of Sections 9.2 and 9.3 that we are able to use more accurate estimators if our prior knowledge about the possible p-measures is more accurate. Consider now a situation where we strongly believe that the true p-measure belongs to a certain family $\mathfrak{P}_o$, but that we cannot wholly exclude the possi-

bility that the true p-measure is only in the neighborhood of $\mathfrak{P}_o$, in a somewhat larger family $\mathfrak{P}$. If we want the estimator to be as. median unbiased also if the true p-measure is in $\mathfrak{P}_1$, then this will, in general, be possible only if we accept a certain reduction of the accuracy of the estimator in case the true p-measure belongs, in fact, to $\mathfrak{P}_o$. On the other hand, estimators which are as. optimal for estimating a functional κ on $\mathfrak{P}_o$, may be grossly wrong if the true p-measure deviates even slightly from $\mathfrak{P}_o$. In such a situation it may be advisable to resort to an estimator which is less efficient if the true p-measure is in $\mathfrak{P}_o$, but still useful if the true p-measure is not in $\mathfrak{P}_o$.

The decision whether one should sacrifice a certain amount of accuracy to obtain a useful estimate also in the - unlikely - case of the true p-measure being in $\mathfrak{P}-\mathfrak{P}_o$ depends strongly on subjective judgments and is not easy to formalize. There is one important exception: If the canonical gradient of κ in $T(P,\mathfrak{P})$ belongs to $T(P,\mathfrak{P}_o)$. Then the as. optimal estimators for κ on $\mathfrak{P}$ are as. optimal even for the restricted problem of estimating κ on $\mathfrak{P}_o$.

This suggests to proceed in the following way: Assume that the functional κ is defined on $\mathfrak{P}$, and that the true probability measure is known to belong to $\mathfrak{P}_o \subset \mathfrak{P}$. Then find a family $\mathfrak{P}_1$, $\mathfrak{P}_o \subset \mathfrak{P}_1 \subset \mathfrak{P}$, as large as possible, such that the canonical gradient of κ in $T(P,\mathfrak{P}_1)$ belongs to $T(P,\mathfrak{P}_o)$ (whenever $P \in \mathfrak{P}_o$). Use the estimator-sequence for κ which is as. optimal in $\mathfrak{P}_1$.

If we proceed in this way, we gain something (if - contrary to our expectation - the true p-measure belongs to $\mathfrak{P}_1-\mathfrak{P}_o$), without loosing anything (if - in accord with our expectation - the true p-measure belongs to $\mathfrak{P}_o$).

Warning: This is true only if we judge the performance of estimators by an approximation of first order. For small samples, estimators with identical first order approximations may behave quite different!

For the time being, we are unable to provide a systematic way of realizing this methodological principle, i.e. a method for constructing $\mathfrak{P}_1$. The following considerations may provide some intuitive clue. Let $\kappa^{\cdot}(\cdot,P)$ denote the gradient of κ in $T(P,\mathfrak{P})$, and $\kappa^{+}(\cdot,P)$ - for $P \in \mathfrak{P}_o$ - the projection of $\kappa^{\cdot}(\cdot,P)$ into $T(P,\mathfrak{P}_o)$. If $\mathfrak{P}_1$ is such that - for $P \in \mathfrak{P}_o$ - the tangent space $T(P,\mathfrak{P}_1)$ is orthogonal to $\kappa^{\cdot}(\cdot,P)-\kappa^{+}(\cdot,P)$, then the projection of $\kappa^{\cdot}(\cdot,P)$ into $T(P,\mathfrak{P}_1)$ coincides with $\kappa^{+}(\cdot,P)$. Hence as. optimal estimators in $\mathfrak{P}_1$ are even as. optimal in $\mathfrak{P}_o$.

9.6. Functionals of induced measures

Let $\mathfrak{P}|\mathscr{A}$ be a family of p-measures with tangent space $T_s(P,\mathfrak{P})$, and $T: (X,\mathscr{A}) \to (Y,\mathscr{B})$ a measurable map. Let $\kappa: \mathfrak{P} \to \mathbb{R}$ be a differentiable functional with canonical gradient $\kappa^{\cdot}(\cdot,P) \in T_s(P,\mathfrak{P})$. If $\kappa(P') \neq \kappa(P'')$ implies $P'*T \neq P''*T$, then we may define a functional $\kappa_*|\mathfrak{P}*T$ by

(9.6.1) $\qquad \kappa_*(P*T) = \kappa(P), \qquad P \in \mathfrak{P}.$

Assume that κ_* is differentiable, and let $\kappa_*^{\cdot}(\cdot,P*T) \in T_s(P*T,\mathfrak{P}*T)$ denote the canonical gradient. For estimators of κ based on observations x, we obtain (see Theorem 9.2.2) the as. variance bound $P(\kappa^{\cdot}(\cdot,P)^2)$; for estimators of κ_*, based on $T(x)$, the as. variance bound is $P*T(\kappa_*^{\cdot}(\cdot,P*T)^2)$. Since κ_* and κ are the same functionals by (9.6.1), we expect that optimal estimators based on $T(x)$ will, in general, be less accurate than optimal estimators based on x.

<u>9.6.2. Proposition</u>. $P(\kappa^{\cdot}(\cdot,P)^2) \leq P*T(\kappa_*^{\cdot}(\cdot,P*T)^2)$ *for all* $P \in \mathfrak{P}$.

For a parametric family of p-measures and for $\kappa(P_\theta) = \theta$, this relation reduces to

$$P_\theta(\ell'(\cdot,\theta)^2) \le P_\theta * T(\ell_T'(\cdot,\theta)^2) ,$$

where $\ell_T(\cdot,\theta)$ is the logarithm of the density of $P_\theta * T$. This is the well-known 'information inequality' of R.A.Fisher (see also Rao, 1973, p. 330).

<u>Proof</u>. By definition of the gradient, we have for all $g \in T_s(P,\mathfrak{P})$

$$t^{-1}(\kappa(P_{t,g})-\kappa(P)) = \int \kappa^\bullet(x,P)g(x)P(dx) + o(t^o) ,$$

$$t^{-1}(\kappa_*(P_{t,g}*T)-\kappa_*(P*T)) = \int \kappa_*^\bullet(y,P*T)(P^T g)(y)P*T(dy) + o(t^o),$$

since $dP_{t,g}/dP = 1 + t(g + r_t)$ implies $dP_{t,g}*T/dP*T = 1 + t(P^T g + P^T r_t)$ (with $P*T((P^T r_t)^2) \le P(r_t^2) = o(n^o))$. Hence

$$\int \kappa^\bullet(x,P)g(x)P(dx) = \int \kappa_*^\bullet(y,P*T)(P^T g)(y)P*T(dy) \quad \text{for all } g \in T_s(P,\mathfrak{P}).$$

By definition of the conditional expectation, we have

$$\int \kappa_*^\bullet(y,P*T)(P^T g)(y)P*T(dy) = \int \kappa_*^\bullet(T(x),P*T)g(x)P(dx).$$

Hence

$$\int \kappa^\bullet(x,P)g(x)P(dx) = \int \kappa_*^\bullet(T(x),P*T)g(x)P(dx) \qquad \text{for all } g \in T_s(P,\mathfrak{P}).$$

Since $\kappa^\bullet(\cdot,P) \in T_s(P,\mathfrak{P})$, this relation implies that $\kappa^\bullet(\cdot,P)$ is the projection of $x \to \kappa_*^\bullet(T(x),P*T)$ into $T_s(P,\mathfrak{P})$. Hence

$$P(\kappa^\bullet(\cdot,P)^2) \le \int \kappa_*^\bullet(T(x),P*T)^2 P(dx),$$

which proves the assertion.

<u>9.6.3. Remark</u>. If T is sufficient for P, then equality holds in Proposition 9.6.2. This follows immediately from Remark 1.5.3, according to which $T_s(P*T,\mathfrak{P}*T)\circ T = T_s(P,\mathfrak{P})$. This implies that $\kappa^\bullet(\cdot,P) \in T_s(P,\mathfrak{P})$ is a contraction of T, say $\kappa^\bullet(x,P) = \kappa_o^\bullet(T(x),P)$, so that $\kappa_o^\bullet(\cdot,P) \in T_s(P*T,\mathfrak{P}*T)$, considered as a function of P*T, is a canonical gradient of κ on $\mathfrak{P}*T$. Therefore,

$$P*T(\kappa_o^{\bullet}(\cdot,P)^2) = \int \kappa_o^{\bullet}(T(x),P)^2 P(dx) = P(\kappa^{\bullet}(\cdot,P)^2).$$

(See also Pitman, 1979, p. 19, Theorem.)

9.6.4. Remark. For $i = 1,\ldots,k$ let κ_i be a differentiable functional on $\mathfrak{P}$ with canonical gradient $\kappa_i^{\bullet}(\cdot,P)$, and let $\kappa_{i*}^{\bullet}(\cdot,P)$ denote the corresponding canonical gradient for κ_i, considered as a functional on $P*T$. Then

$$(P(\kappa_i^{\bullet}(\cdot,P)\kappa_j^{\bullet}(\cdot,P)))_{i,j=1,\ldots,k}$$

$$\leq (P*T(\kappa_{i*}^{\bullet}(\cdot,P*T)\kappa_{j*}^{\bullet}(\cdot,P*T)))_{i,j=1,\ldots,k}$$

(in the sense that the difference between these two matrices is positive semidefinite).

This result follows immediately by applying the one-dimensional version (see Proposition 9.6.1) to the functional $\kappa(P) := c_i\kappa_i(P)$.

9.6.5. Example. Let $\mathfrak{P}$ be a family of p-measures and $\kappa: \mathfrak{P} \to \mathbb{R}$ a differentiable functional with canonical gradient $\kappa^{\bullet}(\cdot,P)$. Assume that instead of the realization x, governed by P, one can only observe $S(x,y)$, where y is stochastically independent of x, and distributed according to an unknown p-measure Q, belonging to a certain family $\mathfrak{Q}$. If $\kappa(P') \neq \kappa(P'')$ implies $(P'\times Q)*S \neq (P''\times Q)*S$, we may define a functional κ_* on $\{(P\times Q)*S: P \in \mathfrak{P}, Q \in \mathfrak{Q}\}$ by

$$\kappa_*((P\times Q)*S) = \kappa(P) .$$

Since our observations are perturbed (instead of x we can only observe $S(x,y)$), we expect that optimal estimators based on observations $S(x_\nu,y_\nu)$, $\nu = 1,\ldots,n$, will be less accurate than optimal estimators based on x_ν, $\nu = 1,\ldots,n$. Under suitable regularity conditions, this is indeed the case. If κ_* admits a gradient, say $\kappa_*^{\bullet}(\cdot,(P\times Q)*S)$, we obtain from Proposition 9.6.2

$$P(\kappa^{\bullet}(\cdot,P)^2) \leq (P\times Q)*S(\kappa_*^{\bullet}(\cdot,(P\times Q)*S)^2) \quad \text{for all } P \in \mathfrak{P},\ Q \in \mathfrak{Q}.$$

Apparently, it has remained unnoticed that this relation is an immediate consequence of the information inequality in the parametric case, for Kale (1962) gives a direct proof for $S(x,y) = x + y$, $\mathfrak{P}|\mathbb{B}$ an exponential family, and $\mathfrak{Q} = \{Q_o\}$, where $Q_o|\mathbb{B}$ is a normal distribution with zero mean and known variance.

10. EXISTENCE OF ASYMPTOTICALLY EFFICIENT ESTIMATORS FOR PROBABILITY MEASURES

10.1. Asymptotic efficiency

For $\underline{x} \in X^n$, let $P_n(\underline{x},\cdot)$ denote a p-measure. Our problem is to evaluate the performance of P_n as an estimator for a p-measure known to belong to $\mathfrak{P}$. Throughout the following we assume that the estimator is strict, i.e. that $P_n(\underline{x},\cdot) \in \mathfrak{P}$ for every $\underline{x} \in X^n$. The basic problem is to define as. efficiency.

In parametric theory, this definition is straightforward, at least in regular cases. The problem of estimating P_θ is identical with the problem of estimating the parameter θ, and it is hard to think of any purpose for which an estimator of P_θ different from P_{θ^n} may be as. preferable, if θ^n is as. efficient for θ. The situation seems to be different in nonparametric theory. If our final goal is to estimate the value of a functional κ on $\mathfrak{P}$, we are out for an estimator $\underline{x} \to P_n(\underline{x},\cdot)$ such that $\underline{x} \to \kappa(P_n(\underline{x},\cdot))$ is as. efficient for κ, i.e. that it has minimal as. variance $\sigma^2(P) := P(\kappa^\cdot(\cdot,P)^2)$ for every $P \in \mathfrak{P}$. An estimator P_n well suited for this purpose may be unfit for estimating $\sigma(P)$, i.e. $\underline{x} \to \sigma(P_n(\underline{x},\cdot))$ may be useless as an estimator for $\sigma(P)$. (Think of cases where $\kappa^\cdot(\cdot,P)$ involves the derivative of the density!)

In the following we confine ourselves to defining as. efficiency for a particular purpose, namely for estimating P-integrals. (Estimators which are as. efficient in this sense may be rather poor for other purposes. So, for instance, the sup-distance $V(P_n(\underline{x},\cdot),P)$, $n \in \mathbb{N}$, may even fail to converge to zero.)

10.1.1. Definition. An estimator-sequence $\underline{x} \to P_n(\underline{x},\cdot) \in \mathfrak{P}$, $n \in \mathbb{N}$, is *as. efficient at* P *in* $\mathfrak{P}$ if $P_n(\underline{x},\cdot) \in \mathfrak{P}$ for every $\underline{x} \in X^n$, and if

$$n^{1/2}\int f(\xi)P_n(\cdot,d\xi) = \tilde{f} + o_p(n^0) \qquad \text{for every } f \in T(P,\mathfrak{P}).$$

Notice the importance of the phrase 'at P in $\mathfrak{P}$' in this definition. If $P \in \overline{\mathfrak{P}} \subset \mathfrak{P}$, an estimator-sequence which is as. efficient 'at P in $\mathfrak{P}$' will, in general, fail to be as. efficient 'at P in $\overline{\mathfrak{P}}$' unless $P_n(\underline{x},\cdot) \in \overline{\mathfrak{P}}$ for every $\underline{x} \in X^n$.

If an estimator-sequence is as. efficient at P in $\mathfrak{P}$ for every $P \in \mathfrak{P}$, we call it *as. efficient on* $\mathfrak{P}$.

10.1.2. Remark. If an estimator-sequence is as. efficient at P in $\mathfrak{P}$, then for every $f \in T(P,\mathfrak{P})$ the sequence of induced p-measures $P^n * n^{1/2}\int f(\xi)P_n(\cdot,d\xi)$, $n \in \mathbb{N}$, converges weakly to $N(0,P(f^2))$.

10.1.3. Example. For $i = 1,\dots,m$ let $(X_i,\mathscr{A}_i)$ be a measurable space and $\mathfrak{P}_i|\mathscr{A}_i$ a family of p-measures with tangent space $T(P_i,\mathfrak{P}_i)$. Assume that the estimator-sequence $\underline{x}_i \to P_{in}(\underline{x}_i,\cdot)|\mathscr{A}_i$, $n \in \mathbb{N}$, (with $\underline{x}_i = (x_{i\nu})_{\nu=1,\dots,n}$) is as. efficient on $\mathfrak{P}_i$. Then the estimator-sequence

$$(\underline{x}_1,\dots,\underline{x}_m) \to \mathop{\times}_{i=1}^{m} P_{in}(\underline{x}_i,\cdot), \qquad n \in \mathbb{N},$$

is as. efficient on

$$\mathfrak{P} = \{\mathop{\times}_{i=1}^{m} P_i : \; P_i \in \mathfrak{P}_i,\; i = 1,\dots,m\}.$$

This follows immediately from Proposition 2.4.1, stating that $T(\times_{i=1}^{m} P_i,\mathfrak{P}) = \{(x_1,\dots,x_m) \to \sum_{i=1}^{m} f_i(x_i): f_i \in T(P_i,\mathfrak{P}_i)\}$.

Examples of as. efficient estimator-sequences for particular models will be given in Sections 10.2 and 10.3.

10.2. Density estimators

Let $\mathfrak{P}$ be dominated by a σ-finite measure $\mu|\mathscr{A}$. Let $F_n: X \times X \to \mathbb{R}$, $n \in \mathbb{N}$, be a sequence of functions such that for every $f \in T(P,\mathfrak{P})$ the function

(10.2.1) $$r_n(x) := \int f(\xi) F_n(x,\xi)\mu(d\xi) - f(x)$$

fulfills

(10.2.2) $$n^{-1/2} \sum_{\nu=1}^{n} r_n(x_\nu) = o_P(n^o) .$$

For $n \in \mathbb{N}$ let $P_n(\underline{x},\cdot)$ denote the measure with μ-density

$$\xi \to \frac{1}{n} \sum_{\nu=1}^{n} F_n(x_\nu,\xi) .$$

(Note that this is a p-measure only if $F_n(x,\xi) \geq 0$ for all $x,\xi \in X$ and $\int F_n(x,\xi)\mu(d\xi) = 1$ for all $x \in X$.) We have

$$n^{1/2}\int f(\xi) P_n(\underline{x},d\xi) = n^{-1/2} \sum_{\nu=1}^{n} \int f(\xi) F_n(x_\nu,\xi)\mu(d\xi)$$
$$= \tilde{f}(\underline{x}) + o_P(n^o) .$$

Hence the estimator-sequence P_n, $n \in \mathbb{N}$, is as. efficient at P in $\mathfrak{P}$ in the sense of Definition 10.1.1, provided $P_n(\underline{x},\cdot) \in \mathfrak{P}$ for all $\underline{x} \in X^n$.

By the degenerate convergence criterion (see, e.g., Loève, 1977, p. 329), the following conditions are sufficient for (10.2.3):

(10.2.3) $$P(r_n 1_{\{|r_n| \leq n^{1/2}\}}) = o(n^{-1/2}) ,$$

(10.2.4) $P\{|r_n| > n^{1/2}\} = o(n^{-1})$,

(10.2.5) $P(r_n^2 \, 1_{\{|r_n| \le n^{1/2}\}}) = o(n^{o})$.

In order to establish condition (10.2.2) for a certain function f and a p-measure P we may use different sets of regularity conditions, involving f, P, and the sequence F_n. Since also discontinuous functions f are of interest (see Proposition 5.4.2), it is advisable to place only minimal conditions upon f, and use instead assumptions on F_n and, if necessary, on the density of P. The following example presents such a set of conditions.

<u>10.2.6. Example</u>. *Kernel estimators*: Let $X = \mathbb{R}$ and $\mu = \lambda$. For $n \in \mathbb{N}$ let $k_n: \mathbb{R} \to \mathbb{R}$ be measurable and symmetric about 0. Assume that there exist $c_n \downarrow 0$ such that $k_n(x) = 0$ for $x \notin [-c_n, c_n]$. Moreover, assume that $\int k_n(\xi)d\xi = 1$ and $\int \xi^2 |k_n(\xi)| d\xi = o(n^{-1/2})$. An important particular case is $k_n(\xi) = b_n^{-1} k(b_n^{-1}\xi)$ with $\int k(\xi)d\xi = 1$ and $k(x) = 0$ for $x \notin [-1,1]$, say, provided $b_n = o(n^{-1/4})$.

Then $F_n(x,\xi) := k_n(\xi - x)$ has properties (10.2.3) - (10.2.5) for f and P if

(i) f is smooth in the sense that there exist $\varepsilon > 0$ and a function $g \in \mathscr{L}_2(P)$ such that $|f(\xi)| \le g(x)$ for all $\xi, x \in \mathbb{R}$ with $|\xi - x| < \varepsilon$,

(ii) P admits a Lebesgue density p which is smooth in the sense that there exist $\varepsilon > 0$ and a function $g \in \mathscr{L}_2(P)$ such that $p''(\xi) \le g(x)p(x)$ for all $\xi, x \in \mathbb{R}$ with $|\xi - x| < \varepsilon$.

In the following we prove that $P(r_n) = o(n^{-1/2})$ and $P(r_n^2) = o(n^{o})$. From this, (10.2.3) - (10.2.5) follow immediately.

By (10.2.1),

$$r_n(x) = \int f(\xi)k_n(\xi - x)d\xi - f(x) = \int (f(x+\xi) - f(x))k_n(\xi)d\xi.$$

(i) $P(r_n) = \int f(\eta)(p(\eta - \xi) - p(\eta))k_n(\xi)d\xi d\eta$. Using

$$\int (p(x-\xi) - p(x))k_n(\xi)d\xi = \int \xi^2 \int_o^1 p''(x - u\xi)du k_n(\xi)d\xi$$

we obtain for n large enough so that $c_n < \varepsilon$

$$|\int(p(x-\xi)-p(x))k_n(\xi)d\xi| \leq g(x)p(x)\int\xi^2|k_n(\xi)|d\xi .$$

Hence

$$|P(r_n)| \leq \int|f(\eta)|g(\eta)p(\eta)d\eta\int\xi^2|k_n(\xi)|d\xi = o(n^{-1/2}) .$$

(ii) $P(r_n^2) \leq \int\int(f(\eta+\xi)-f(\eta))^2 p(\eta)|k_n(\xi)|d\xi.$

Since k_n is concentrated on $[-c_n, c_n]$ and $c_n \downarrow 0$, it remains to show that

$$\lim_{x\to 0}\int(f(\eta+x)-f(\eta))^2 p(\eta)d\eta = 0 .$$

This follows from Lemma 19.1.4.

<u>10.2.7. Example</u>. *Orthogonal series estimators*. Let $(X,\mathscr{A})$ be a measurable space, $\mu|\mathscr{A}$ a p-measure, and h_k, $k\in\mathbb{N}$, an orthonormal base of $\mathscr{L}_2(\mu)$. Let

$$F_K(x,\xi) := \sum_{k=1}^{K} h_k(x)h_k(\xi) .$$

In the following, we indicate the existence of a sequence $K(n)\uparrow\infty$, $n\in\mathbb{N}$, such that $F_{K(n)}$, $n\in\mathbb{N}$, fulfills conditions (10.2.3) - (10.2.5) for all p-measures $P|\mathscr{A}$ with μ-square integrable density p, and a sufficiently large class of functions $f\in\mathscr{L}_*(P)\cap\mathscr{L}_2(\mu)$.

Let $a_k := \mu(fh_k)$ and

$$r_K(x) := \sum_{k=1}^{K} a_k h_k(x) - f(x) = \int f(\xi)F_K(x,\xi)\mu(d\xi) - f(x) .$$

We have $P(|r_K|) = \mu(p|r_K|) \leq \mu(p^2)^{1/2}\mu(r_K^2)^{1/2}$ and

$$P(r_K^2 1_{\{|r_K|\leq n^{1/2}\}}) \leq n^{1/2}P(|r_K|) \leq n^{1/2}\mu(p^2)^{1/2}\mu(r_K^2)^{1/2} .$$

Hence relations (10.2.3) - (10.2.5) hold true if K(n) can be chosen such that $\mu(r_{K(n)}^2)^{1/2} = o(n^{-1/2})$. Since r_K depends on f, we need for this purpose that $\mu(r_K^2)$, $K\in\mathbb{N}$, converges to zero at a uniform rate. This can be ascertained by further regularity conditions, holding uniformly in f. As an example, take $\mu = N(0,1)$, and let $h_k(x) = (-1)^k\exp[x^2](\partial^k/\partial x^k)\exp[-x^2]$, $k\in\mathbb{N}_0$, be the Hermite polynomials.

In this case we have $N(0,1)(r_K^2) \leq K^{-1} c(K)$, where $c(K)$, $K \in \mathbb{N}$, is a null-sequence depending on f, provided $(x \to xf(x)) \in \mathscr{L}_2(N(0,1))$ and $(x \to f'(x)) \in \mathscr{L}_2(N(0,1))$. (This can be proven similarly as Theorem 1 in Walter, 1977, p. 1261.)

The reader interested in the problem of density estimation in general is referred to Walter and Blum (1979) and to the survey papers by Wegman (1972a,b) and Wertz (1978).

10.3. Parametric families

In this section we consider a parametric family $\mathfrak{P} = \{P_\theta : \theta \in \Theta\}$ with $\Theta \subset \mathbb{R}^k$. Let θ^n, $n \in \mathbb{N}$, be a sequence of estimators for θ. It is tempting to say that P_{θ^n}, $n \in \mathbb{N}$, considered as a sequence of estimators for P_θ, is as. efficient on $\mathfrak{P}$ iff the estimator-sequence θ^n, $n \in \mathbb{N}$, is as. efficient for θ. The following considerations show that this is almost true.

We call θ^n, $n \in \mathbb{N}$, an *as. maximum likelihood estimator-sequence* if it admits for every $\theta \in \Theta$ a stochastic expansion

$$(10.3.1) \qquad n^{1/2}(\theta^n - \theta) = \Lambda(\theta)\tilde{\ell}^{(\cdot)}(\cdot,\theta) + r_n(\cdot,\theta)$$

with $r_n(\cdot,\theta) = o_\theta(n^0)$. (For typographical reasons we write o_θ for o_{P_θ}.)

10.3.2. Proposition. *Let θ^n, $n \in \mathbb{N}$, be an as. maximum likelihood estimator-sequence. Under the assumptions of Proposition 2.2.1, the estimator-sequence $\underline{x} \to P_{\theta^n(\underline{x})}$, $n \in \mathbb{N}$, is as. efficient on $\mathfrak{P}$.*

Proof. For $|\tau-\theta| < \varepsilon$, say, the P_θ-density of P_τ can be approximated by

$$1 + (\tau-\theta)'\ell^{(\cdot)}(\xi,\theta) + r(\xi,\theta,\tau)$$

with

(10.3.3) $\quad |r(\xi,\theta,\tau)| \leq |\tau-\theta|^2 M(\xi,\theta)$

and P_θ-square integrable $M(\cdot,\theta)$. Hence the P_θ-density of $P_{\theta^n(\underline{x})}$ admits for $|\theta^n(\underline{x}) - \theta| < \varepsilon$ the approximation

(10.3.4) $\quad 1 + (\theta^n(\underline{x}) - \theta)'\ell^{(\cdot)}(\xi,\theta) + r(\xi,\theta,\theta^n(\underline{x}))$.

Let $f \in \mathscr{L}_*(P_\theta)$ be arbitrary. Then

(10.3.5) $\quad n^{1/2}\int f(\xi)P_{\theta^n(\underline{x})}(d\xi) = n^{1/2}(\theta^n(\underline{x}) - \theta)'c + \hat{r}_n(\underline{x},\theta)$

with

$$c := \int f(\xi)\ell^{(\cdot)}(\xi,\theta)P_\theta(d\xi),$$

$$\hat{r}_n(\underline{x},\theta) := n^{1/2}\int f(\xi)r(\xi,\theta,\theta^n(\underline{x}))P_\theta(d\xi).$$

Using (10.3.1) we obtain from (10.3.5)

(10.3.6) $\quad n^{1/2}\int f(\xi)P_{\theta^n(\underline{x})}(d\xi) = c'\Lambda(\theta)\tilde{\ell}^{(\cdot)}(\underline{x},\theta) + c'r_n(\underline{x},\theta) + \hat{r}_n(\underline{x},\theta)$.

By Proposition 2.2.1 we have

$$T(P_\theta,\mathfrak{P}) = [\ell^{(1)}(\cdot,\theta),\dots,\ell^{(k)}(\cdot,\theta)] .$$

Since $c'\Lambda(\theta)\ell^{(\cdot)}(\cdot,\theta)$ is the projection of f into $T(P_\theta,\mathfrak{P})$, relation (10.3.6), applied for $f \in T(P_\theta,\mathfrak{P})$, is a representation of the kind required by Definition 10.1.1. In order to show that the remainder term is $o_\theta(n^\circ)$, apply (10.3.3) and the Schwarz inequality to $\hat{r}_n$.

<u>10.3.7. Remark</u>. The estimator-sequence P_{θ^n}, $n \in \mathbb{N}$, is n o t as. efficient on $\mathfrak{P}$ any more if $P_\theta^n * n^{1/2}(\theta^n-\theta)$, $n \in \mathbb{N}$, converges to a normal distribution $N(0,\Sigma(\theta))$ with $\Sigma(\theta) > \Lambda(\theta)$. Relation (10.3.5) implies that $P_\theta^n * n^{1/2}\int f(\xi)P_{\theta^n}(d\xi)$ converges weakly to $N(0,c'\Sigma(\theta)c)$. For $f \in T(P_\theta,\mathfrak{P})$ the representation required by Definition 10.1.1 implies that $P_\theta^n * n^{1/2}\int f(\xi)P_{\theta^n}(d\xi)$ converges weakly to $N(0,P_\theta(f^2))$. Since $P_\theta(f^2) = c'\Lambda(\theta)c$, this contradicts $\Sigma(\theta) > \Lambda(\theta)$.

10.4. Projections of estimators

In this section we investigate estimator-sequences P_n, $n \in \mathbb{N}$, as. efficient at P in $\mathfrak{P}$, the densities of which have, in addition, a representation

(10.4.1) $$p_n(\underline{x},\xi) = p(\xi)(1 + g_n(\underline{x},\xi) + r_n(\underline{x},\xi))$$

with $g_n(\underline{x},\cdot) \in T(P,\mathfrak{P})$, and with a remainder term fulfilling

(10.4.2) $$\int f(\xi) r_n(\cdot,\xi) P(d\xi) = o_P(n^{-1/2}) \qquad \text{for all } f \in \mathscr{L}_*(P).$$

Condition (10.4.2) is in particular fulfilled if

(10.4.3) $$\int r_n(\cdot,\xi)^2 P(d\xi) = o_P(n^{-1}).$$

The latter condition is fulfilled if $\mathfrak{P}$ is at P approximable by $T(P,\mathfrak{P})$ (see Definition 1.1.8), and $\Delta(P_n(\underline{x},\cdot);P) = O_P(n^{-1/2})$. Hence (10.4.3) is in particular fulfilled for parametric families $\mathfrak{P} = \{P_\theta: \theta \in \Theta\}$, $\Theta \subset \mathbb{R}^k$, and $P_n(\underline{x},\cdot) := P_{\theta^n(\underline{x})}$ under the assumptions of Proposition 2.2.1 if $\theta^n - \theta = O_\theta(n^{-1/2})$. For as. maximum likelihood estimator-sequences, representation (10.4.1) is given in the following proposition.

10.4.4. Proposition. *Let θ^n, $n \in \mathbb{N}$, be an as. maximum likelihood estimator-sequence (see (10.3.1)). Under the assumptions of Proposition 2.2.1, the estimator-sequence $\underline{x} \to P_{\theta^n(\underline{x})}$, $n \in \mathbb{N}$, admits a representation*

$$p(\xi,\theta^n(\underline{x})) = p(\xi,\theta)(1 + n^{-1/2}\tilde{\ell}^{(\cdot)}(\underline{x},\theta)'\Lambda(\theta)\ell^{(\cdot)}(\xi,\theta) + r_n(\underline{x},\xi))$$

with a remainder term fulfilling $\int r_n(\cdot,\xi)^2 P_\theta(d\xi) = o_\theta(n^{-1})$.

Proof. By (10.3.4) and (10.3.1), $P_{\theta^n(\underline{x})}$ has for $|\theta^n(\underline{x}) - \theta| < \varepsilon$ a P_θ-density

$$1 + n^{-1/2}\tilde{\ell}^{(\cdot)}(\underline{x},\theta)'\Lambda(\theta)\ell^{(\cdot)}(\cdot,\theta) + r_{n,1}(\underline{x},\cdot) + r_{n,2}(\underline{x},\cdot)$$

with

$$r_{n,1}(\underline{x},\xi) = n^{-1/2}r_n(\underline{x},\theta)'\ell^{(\cdot)}(\xi,\theta),$$

$$r_{n,2}(\underline{x},\xi) = r(\xi,\theta,\theta^n(\underline{x})).$$

We have

$$\int r_{n,1}(\cdot,\xi)^2 P_\theta(d\xi) = n^{-1}r_n(\underline{x},\theta)'L(\theta)r_n(\underline{x},\theta) = o_\theta(n^{-1}).$$

Since $\theta^n-\theta = O_\theta(n^{-1/2})$, $c_n \uparrow \infty$ implies

$$P_\theta^n\{|\theta^n-\theta| > n^{-1/2}c_n\} = o(n^o).$$

For $|\theta^n(\underline{x}) - \theta| \leq n^{-1/2}c_n$ we obtain from (10.3.3)

$$r_{n,2}(\underline{x},\xi) \leq |\theta^n(\underline{x}) - \theta|^2 M(\xi,\theta) \leq n^{-1}c_n^2 M(\xi,\theta).$$

Choosing $c_n \uparrow \infty$ such that $n^{-1/2}c_n = o(n^{-1/4})$, we obtain

$$\int r_{n,2}(\cdot,\xi)^2 P_\theta(d\xi) = o_\theta(n^{-1}).$$

10.4.5. Proposition. *An estimator-sequence* $\underline{x} \to P_n(\underline{x},\cdot) \in \mathfrak{P}$, $n \in \mathbb{N}$, *admitting a representation (10.4.1) with a remainder term fulfilling (10.4.2) is as. efficient at* P *in* $\mathfrak{P}$ *iff for all* $f \in T(P,\mathfrak{P})$

(10.4.6) $$n^{1/2}\int f(\xi)g_n(\cdot,\xi)P(d\xi) = \tilde{f} + o_P(n^o).$$

Proof. From (10.4.1) and (10.4.2) we obtain for $f \in T(P,\mathfrak{P})$

$$n^{1/2}\int f(\xi)P_n(\cdot,d\xi) = n^{1/2}\int f(\xi)(1 + g_n(\cdot,\xi) + r_n(\cdot,\xi))P(d\xi)$$
$$= n^{1/2}\int f(\xi)g_n(\cdot,\xi)P(d\xi) + o_P(n^o)$$

By (10.4.6) this implies the assertion.

10.4.7. Proposition. *If the estimator-sequence* P_n, $n \in \mathbb{N}$, *is as. efficient at* P *in* $\mathfrak{P}$ *and admits a representation (10.4.1) with a remainder term fulfilling (10.4.2), then for all* $f \in \mathscr{L}_*(P)$,

$$n^{1/2}\int f(\xi)P_n(\cdot,d\xi) = \tilde{f}_o + o_P(n^o),$$

where f_o *denotes the projection of* f *into* $T(P,\mathfrak{P})$.

Proof. From (10.4.1) and (10.4.2) we obtain for $f \in \mathscr{L}_*(P)$

$$\int f(\xi)P_n(\cdot,d\xi) = \int f(\xi)g_n(\cdot,\xi)P(d\xi) + o_P(n^{-1/2}).$$

Since $g_n(\underline{x},\cdot) \in T(P,\mathfrak{P})$, we have

$$\int f(\xi)g_n(\underline{x},\xi)P(d\xi) = \int f_o(\xi)g_n(\underline{x},\xi)P(d\xi).$$

Since P_n, $n \in \mathbb{N}$, is as. efficient at P in $\mathfrak{P}$, we have

$$n^{1/2}\int f_o(\xi)P_n(\cdot,d\xi) = \tilde{f}_o + o_P(n^o).$$

This implies the assertion.

Let $\underline{x} \to P_n(\underline{x},\cdot) \in \mathfrak{P}$, $n \in \mathbb{N}$, be an estimator-sequence which is as. efficient on $\mathfrak{P}$. If it is known that the true p-measure belongs, in fact, to a certain subfamily $\overline{\mathfrak{P}}$, then the estimators P_n are not wholly satisfactory any more under this more stringent condition. First of all, $P_n(\underline{x},\cdot)$ will, in general, not belong to $\overline{\mathfrak{P}}$, i.e., the estimators are not strict any more. More important, they will not be efficient any more on the more stringent model $\overline{\mathfrak{P}}$. How can the more precise knowledge (that $P \in \overline{\mathfrak{P}}$) be used to obtain better estimators? The following theorem suggests that this can be done by projection of P_n into $\overline{\mathfrak{P}}$.

Remember that the projection $\overline{P}_n(\underline{x},\cdot)$ of $P_n(\underline{x},\cdot)$ into $\overline{\mathfrak{P}}$ is defined (see 7.2.1) by

$$p_n(\underline{x},\cdot)/\overline{p}_n(\underline{x},\cdot) - 1 \quad \text{orthogonal to} \quad T(\overline{P}_n(\underline{x},\cdot),\overline{\mathfrak{P}}),$$

or equivalently by

$$\int f(\xi)P_n(\underline{x},d\xi) = 0 \qquad \text{for all } f \in T(\overline{P}_n(\underline{x},\cdot),\overline{\mathfrak{P}}).$$

<u>10.4.8. Theorem</u>. *Let* $T_s(\cdot,\overline{\mathfrak{P}})$ *be continuous at* P. *Assume that the estimator-sequence* P_n, $n \in \mathbb{N}$, *is as. efficient at* P *in* $\mathfrak{P}$, *provided* $P \in \overline{\mathfrak{P}}$, *and that* $\Delta(P_n(\underline{x},\cdot);P) = o_P(n^{-1/4})$. *If* $\Delta(\overline{P}_n(\underline{x},\cdot);P) = O_P(\Delta(P_n(\underline{x},\cdot);P))$, *then the estimator-sequence* $\overline{P}_n$, $n \in \mathbb{N}$, *is as. efficient at* P *in* $\mathfrak{P}$.

<u>Proof</u>. Since P_n, $n \in \mathbb{N}$, is as. efficient at P in $\mathfrak{P}$, we have for $f \in T_s(P,\overline{\mathfrak{P}}) \subset T_s(P,\mathfrak{P})$

$$n^{1/2}\int f(\xi)P_n(\cdot,d\xi) = \tilde{f} + o_P(n^o).$$

Define $k_n(\underline{x},\xi)$ by $p_n = p(1+k_n)$ and $\overline{k}_n(\underline{x},\xi)$ by $\overline{p}_n = p(1+\overline{k}_n)$. From Lemma 7.4.1 we obtain for $f \in T_s(P,\overline{\mathfrak{P}})$

$$|\int f(\xi)\overline{P}_n(\underline{x},d\xi) - \int f(\xi)P_n(\underline{x},d\xi)| = |\int f(\xi)(\overline{k}_n(\underline{x},\xi) - k_n(\underline{x},\xi))P(d\xi)|$$

$$= \|f\|_P o(\Delta(P_n(\underline{x},\cdot),P)) = o_P(n^{-1/2}).$$

Hence $\overline{P}_n$, $n \in \mathbb{N}$, is as. efficient at P in $\overline{\mathfrak{P}}$.

<u>10.4.9. Remark</u>. The projections $\overline{P}_n(\underline{x},\cdot)$ are assumed to be chosen such that $\Delta(\overline{P}_n;P) = O_P(\Delta(P_n;P))$. Such a restriction is perhaps somewhat artificial, since it depends on the (unknown) true p-measure P. Under appropriate restrictions on $\mathfrak{P}$ it follows from $\Delta(P_n;\overline{P}_n) = O_P(\Delta(P_n;\overline{\mathfrak{P}}))$ (see Remark 7.4.3).

<u>10.4.10. Remark</u>. Consider the case of an estimator-sequence P_n, $n \in \mathbb{N}$, which is as. efficient on $\mathfrak{P}$. By Theorem 10.4.8 the projection of P_n into $\mathfrak{P}_1 \subset \mathfrak{P}$, say $P_{n,1}$, is as. efficient on $\mathfrak{P}_1$. If the true p-measure P is known to belong to a subfamily $\mathfrak{P}_o \subset \mathfrak{P}_1$, we obtain from Theorem 10.4.8 two estimator-sequences which are as. efficient on $\mathfrak{P}_o$, namely the projection $P_{n,o}$ of P_n into $\mathfrak{P}_o$, and the projection $P_{n,1,o}$ of $P_{n,1}$ into $\mathfrak{P}_o$. Theorem 7.5.1 gives a reasonable explanation for this: The two estimator-sequences agree closely. Under the assumptions of Theorem 10.4.8, Theorem 7.5.1 yields $\Delta(P_{n,1,o};P_{n,o}) = o_P(\Delta(P_n;P_{n,o}))$. If $\Delta(P_n;P_{n,o}) = O_P(\Delta(P_n;P))$, then $\Delta(P_n;P) = O_P(n^{-1/2})$ guarantees that $n^{1/2}\Delta(P_{n,1,o};P_{n,o}) = o_P(n^o)$.

<u>10.4.11. Remark</u>. Theorem 10.4.8 and Remark 10.4.10 presume $\Delta(P_n(\underline{x},\cdot);P) = O_P(n^{-1/2})$. This assumption is unrealistic for large families $\mathfrak{P}$. It can, however, be weakened to $\Delta(P_n(\underline{x},\cdot);P) = o_P(n^{-1/4})$ for sufficiently regular $\mathfrak{P}$. Assume that $T_s(\cdot,\overline{\mathfrak{P}})$ is continuous in the stronger sense that Definition 1.1.12 holds with $o(\Delta(Q;P)^o)$ replaced by $O(\Delta(Q;P))$. It is easy to see that Lemma 7.4.1 then holds with an error term $\|f\| O(\Delta(Q;P)^2)$, and hence $\Delta(P_n(\underline{x},\cdot);P) = o_P(n^{-1/4})$ is sufficient for

$\overline{P}_n$, $n \in \mathbb{N}$, to be as. efficient at P in $\overline{\mathfrak{P}}$.

Remark 10.4.10 can be modified analogously.

<u>10.4.12. Remark</u>. In Section 7.3 we discussed the possibility of defining projections by minimization of distances. It was mentioned that the projections obtained from Definition 7.2.1 are approximately equivalent to projections obtained by minimization of distances which approximate Δ in the sense of Definition 6.2.1. Distances based on distribution functions (such as Cramér-von Mises distances or Kolmogorov-Smirnov distances) lead to essentially different results, and we cannot expect in general that estimators for p-measures obtained by minimization of such distances will be as. efficient in general. This follows immmediately from the fact that estimators for functionals derived from such estimators of p-measures are inefficient (see Remark 10.5.7).

10.5. Projections into a parametric family

For parametric families, as. efficient estimator-sequences for p-measures can be obtained by the maximum likelihood method (see Proposition 10.3.2). For the purpose of illustration, we show how they can also be obtained by projection of an estimator-sequence which is as. efficient on a larger family.

<u>10.5.1. Proposition</u>. *Let* $\overline{\mathfrak{P}} = \{P_\theta : \theta \in \Theta\}$, $\Theta \subset \mathbb{R}^k$, *fulfill the assumptions of Proposition 2.2.1. Assume, furthermore, that the second order logarithmic derivatives of the density fulfill a local Lipschitz condition, i.e.,*

$$|\ell^{(ij)}(x,\tau) - \ell^{(ij)}(x,\theta)| \leq |\tau-\theta| M(x,\theta)$$

for $i,j = 1,\dots,k$ *and* $|\tau-\theta| < \varepsilon$, *say, with* P_θ*-integrable* $M(\cdot,\theta)$.

Let P_n, $n \in \mathbb{N}$, *be an estimator-sequence which is as. efficient at* P_θ *in* $\mathfrak{P}$ *and has a representation (10.4.1) at* P_θ *with a remainder term fulfilling (10.4.2).*

Then any projection $P_{\underline{\theta}^n(\underline{x})}$ *of* $P_n(\underline{x},\cdot)$ *into* $\overline{\mathfrak{P}}$ *with* $\Delta(P_{\underline{\theta}^n(\underline{x})};P_\theta) = o_\theta(n^{-1/4})$ *defines an estimator-sequence* $\underline{\theta}^n$, $n \in \mathbb{N}$, *which is as. maximum likelihood (in the sense of (10.3.1)).*

<u>Proof</u>. By Taylor expansion we obtain

$$\ell^{(\cdot)}(\xi,\underline{\theta}^n(\underline{x}))' = \ell^{(\cdot)}(\xi,\theta)' + (\underline{\theta}^n(\underline{x})-\theta)' \int_0^1 \ell^{(\cdot\cdot)}(\theta,(1-u)\theta + u\underline{\theta}^n(\underline{x}))du .$$

By (7.2.2), $\underline{\theta}^n$ fulfills

$$\int \ell^{(\cdot)}(\xi,\underline{\theta}^n(\underline{x}))P_n(\underline{x},d\xi) = 0 .$$

Hence

$$(10.5.2) \qquad 0 = \int \ell^{(\cdot)}(\xi,\theta)'P_n(\cdot,d\xi) + (\underline{\theta}^n-\theta)'\int \ell^{(\cdot\cdot)}(\xi,\theta)P_n(\cdot,d\xi) + (\underline{\theta}^n-\theta)'\int(\int_0^1 \ell^{(\cdot\cdot)}(\xi,(1-u)\theta+u\underline{\theta}^n)du - \ell^{(\cdot\cdot)}(\xi,\theta))P_n(\cdot,d\xi).$$

Since $\ell^{(i)}(\cdot,\theta) \in T(P_\theta,\overline{\mathfrak{P}}) \subset T(P_\theta,\mathfrak{P})$, as. efficiency of P_n, $n \in \mathbb{N}$, at P_θ in $\mathfrak{P}$ implies

$$(10.5.3) \qquad \int \ell^{(\cdot)}(\xi,\theta)P_n(\cdot,d\xi) = n^{-1/2}\tilde{\ell}^{(\cdot)}(\cdot,\theta) + o_\theta(n^{-1/2}).$$

Using $L(\theta) = -P_\theta(\ell^{(\cdot\cdot)}(\cdot,\theta))$, we obtain from Proposition 10.4.7

$$(10.5.4) \qquad \int \ell^{(\cdot\cdot)}(\xi,\theta)P_n(\cdot,d\xi) = -L(\theta) + n^{-1/2}\ell_0^{(\cdot\cdot)}(\cdot,\theta) + o_\theta(n^{-1/2}),$$

where $\ell_0^{(ij)}(\cdot,\theta)$ denotes the projection of $\ell^{(ij)}(\cdot,\theta)$ into $T(P_\theta,\overline{\mathfrak{P}})$. From the Lipschitz condition on $\ell^{(ij)}(\cdot,\theta)$ we obtain for $|\underline{\theta}^n(\underline{x})-\theta| < \varepsilon$

$$(10.5.5) \qquad |\int(\int_0^1 \ell^{(\cdot\cdot)}(\xi,(1-u)\theta+u\underline{\theta}^n(\underline{x}))du - \ell^{(\cdot\cdot)}(\xi,\theta))P_n(\underline{x},d\xi)| \leq |\underline{\theta}^n(\underline{x})-\theta| \int M(\xi,\theta)P_n(\underline{x},d\xi).$$

As in (10.5.4) we infer from Proposition 10.4.7

$$(10.5.6) \qquad \int M(\xi,\theta)P_n(\cdot,d\xi) = P_\theta(M(\cdot,\theta)) + o_\theta(n^{-1/2}) = O_\theta(n^0).$$

By Proposition 6.2.7 we have

$$|\underline{\theta}^n(\underline{x}) - \theta| = O(\Delta(P_{\underline{\theta}^n(\underline{x})}; P_\theta)) = o_\theta(n^{-1/4}).$$

Since $L(\theta)$ is nonsingular with inverse $\Lambda(\theta)$, we obtain from (10.5.2) - (10.5.6) that

$$n^{1/2}(\underline{\theta}^n - \theta) = \Lambda(\theta)\tilde{\ell}^{(\cdot)}(\cdot,\theta) + o_\theta(n^0).$$

Hence (see (10.3.1)) the estimator-sequence $\underline{\theta}^n$, $n \in \mathbb{N}$, is as. maximum likelihood.

<u>10.5.7. Remark</u>. Let $\overline{P}_n$ denote the projection of P_n into $\overline{\mathfrak{P}}$ as defined by 7.2.1, and let $\tilde{P}_n$ denote the projection defined by minimization of $\delta(P_n, Q)$ for $Q \in \overline{\mathfrak{P}}$. If δ approximates Δ in the sense of Definition 6.2.1, then by Proposition 7.3.5, $\delta(P_n, P_\theta) = O_\theta(n^{-1/2})$ is sufficient for $\delta(\overline{P}_n, \tilde{P}_n) = o(\delta(P_n, P_\theta)) = o_\theta(n^{-1/2})$.

It is easy to see that under stronger assumptions Proposition 7.3.5 holds with $o(\delta(Q,\overline{Q}))$ replaced by $O(\delta(Q,\overline{Q})^2)$. Then even $\delta(P_n, P_\theta) = o_\theta(n^{-1/4})$ is sufficient for $\delta(\overline{P}_n, \tilde{P}_n) = o_\theta(n^{-1/2})$.

Define $\underline{\theta}^n$ and $\underset{\sim}{\theta}^n$ by $\overline{P}_n(\underline{x}, \cdot) = P_{\underline{\theta}^n(\underline{x})}$ and $\tilde{P}_n(\underline{x}, \cdot) = P_{\underset{\sim}{\theta}^n(\underline{x})}$, respectively. Then $\underline{\theta}^n - \underset{\sim}{\theta}^n = o(\delta(\overline{P}_n, \tilde{P}_n)) = o_\theta(n^{-1/2})$. Hence $P_\theta^n * n^{1/2}(\underline{\theta}^n - \theta)$ is as. equivalent to $P_\theta^n * n^{1/2}(\underset{\sim}{\theta}^n - \theta)$.

This shows that *minimum distance methods* (determining the estimate $\theta^n(\underline{x})$ by minimizing a distance $\delta(P_n(\underline{x}, \cdot), P_\theta)$ for $\theta \in \Theta$) lead to as. efficient estimator-sequences if we use distance functions δ which approximate Δ. For δ the Hellinger distance and P_n a kernel density estimator, this is shown by Beran (1977b, p. 450, Theorem 4).

It follows from Remark 7.6.6 that minimum distance methods lead, in general, to as. inefficient estimator-sequences for θ if distance functions based on the distribution function are used. Parr and De Wet (1981) use the distance function $\int (F_P(t) - F_Q(t))^2 w(t) dt$.

With a weight function w not depending on θ, the minimum distance estimator-sequence for θ is, in general, as. inefficient (see also Remark 15.2.8). An exception are, for instance, location parameter families $\overline{\mathfrak{P}}$, where as. efficient estimator-sequences can be obtained with a weight function which depends, of course, on $\overline{\mathfrak{P}}$, but not on θ; see also Boos (1981).

10.6. Projections into a family of product measures

Let $(X_i, \mathscr{A}_i)$, $i = 1,\dots,m$, be measurable spaces and $\mathfrak{P}$ a family of p-measures on $\times\mathscr{A}_i$. (Products and sums over i always run from 1 to m.) For $Q|\times\mathscr{A}_i$ let $\pi_i Q$ denote the i-th marginal distribution. Let $\mathfrak{P}_i$ be a family of p-measures on $\mathscr{A}_i$ such that $Q \in \mathfrak{P}$ implies $\pi_i Q \in \mathfrak{P}_i$, $i = 1,\dots,m$. Assume, finally, that $\overline{\mathfrak{P}} = \{\times P_i : P_i \in \mathfrak{P}_i\} \subset \mathfrak{P}$.

Trivial examples for $\mathfrak{P}$ are (i) the family of all p-measures on $\mathbb{B}^m$ which are equivalent to the Lebesgue measure (with $\mathfrak{P}_i$ the family of all p-measures on $\mathbb{B}$ equivalent to the Lebesgue measure), or (ii) the family of all m-variate normal distributions (with $\mathfrak{P}_i$ the family of all univariate normal distributions). Recall that (see Proposition 2.4.1)

$$T(\times P_i, \overline{\mathfrak{P}}) = \{(x_1,\dots,x_m) \to \Sigma g_i(x_i) : g_i \in T(P_i, \mathfrak{P}_i)\}.$$

Moreover, our assumptions on $\mathfrak{P}_i$, $i = 1,\dots,m$, imply that for every $P \in \mathfrak{P}$, every $g \in T(P,\mathfrak{P})$, and $k = 1,\dots,m$, we have $g_k \in T(\pi_k P, \mathfrak{P}_k)$, where g_k is defined by

$$g_k(x_k) := \int g(x_1,\dots,x_m) \underset{j \neq k}{\times} \pi_j P(dx_j).$$

This implies in particular that $(x_1,\dots,x_m) \to \Sigma g_i(x_i)$ is the projection of $g \in T(P,\mathfrak{P})$ into $T(\times\pi_i P, \overline{\mathfrak{P}})$.

Assume we are given an estimator-sequence $P_n(\underline{x},\cdot)|\times\mathscr{A}_i$, based on the sample $\underline{x} = (x_{1\nu},\dots,x_{m\nu})_{\nu=1,\dots,n}$, which is as. efficient on $\mathfrak{P}$.

If we know that the true p-measure belongs, in fact, to $\overline{\mathfrak{P}}$, is it then possible to obtain from $P_n(\underline{x},\cdot)$ an estimator-sequence which is as. efficient on $\overline{\mathfrak{P}}$? According to Theorem 10.4.8 this is possible by projection of $P_n(\underline{x},\cdot)$ into $\overline{\mathfrak{P}}$. By Section 7.7, this projection is $\times\pi_i P_n(\underline{x},\cdot)$.

We present this result mainly to illustrate the projection method. For practical purposes, it is obviously preferable to proceed in a different way. For, if P is a product measure, then our sample $(x_{1\nu},\ldots,x_{m\nu})_{\nu=1,\ldots,n}$ breaks up into m independent samples, $\underline{x}_i = (x_{i\nu})_{\nu=1,\ldots,n}$, $i = 1,\ldots,m$. If the family $\mathfrak{P}$ is such that we are able to obtain as. efficient estimator-sequences for the m-variate p-measure P based on $(x_{1\nu},\ldots,x_{m\nu})_{\nu=1,\ldots,n}$, then there is every chance that $\mathfrak{P}_i$ is such that we are able to obtain as. efficient estimator-sequences for the univariate p-measures P_i, say $\underline{x}_i \to P_{in}(\underline{x}_i,\cdot)$, $n \in \mathbb{N}$. Then the estimator-sequence $(\underline{x}_1,\ldots,\underline{x}_m) \to \times P_{in}(\underline{x}_i,\cdot)$, $n \in \mathbb{N}$, is as. efficient on $\overline{\mathfrak{P}}$ according to Remark 10.1.3.

The estimator $\times\pi_i P_n(\underline{x},\cdot)$, obtained by the projection method, shows the somewhat disturbing feature that $\pi_i P_n(\underline{x},\cdot)$ depends on the whole sample $(x_{1\nu},\ldots,x_{m\nu})_{\nu=1,\ldots,n}$, and not on $(x_{i\nu})_{\nu=1,\ldots,n}$ only. However, this dependence of $\pi_1 P_n(\underline{x},\cdot)$, say, on $(x_{2\nu})_{\nu=1,\ldots,n},\ldots,(x_{m\nu})_{\nu=1,\ldots,n}$ will necessarily be negligible. Let $P = \times P_i$ denote the true p-measure. Since $\times\pi_i P_n(\underline{x},\cdot)$ is as. efficient on $\overline{\mathfrak{P}}$, we have by Definition 10.1.1 for all $g_i \in T(P_i,\mathfrak{P}_i)$, $i = 1,\ldots,m$,

$$n^{1/2}\int \Sigma g_i(\xi_i)\times\pi_i P_n(\underline{x},d\xi_i) = n^{-1/2}\sum_{\nu=1}^{n}(\Sigma g_i(x_{i\nu})) + o_p(n^{o}) .$$

Hence for all $g \in T(P_i,\mathfrak{P}_i)$, $i = 1,\ldots,m$,

$$n^{1/2}\int g(\xi)\pi_i P_n(\underline{x},d\xi) = n^{-1/2}\sum_{\nu=1}^{n} g(x_{i\nu}) + o_p(n^{o}) ,$$

i.e., $n^{1/2}\int g(\xi)\pi_i P_n(\underline{x},d\xi)$ depends on $(x_{j\nu})_{\nu=1,\ldots,n}$ for $j \neq i$ only through a term of order $o_p(n^{o})$.

In particular cases, $\pi_i P_n(\underline{x},\cdot)$ depends on $(x_{i\nu})_{\nu=1,\dots,n}$ only, so, for instance, if $P_n(\underline{x},\cdot)$ is a kernel estimator with Lebesgue density

$$(\xi_1,\dots,\xi_m) \to \frac{1}{n}\sum_{\nu=1}^{n} k_n(\xi_1-x_{1\nu},\dots,\xi_m-x_{m\nu}) .$$

In this case, the marginal distribution $\pi_i P_n(\underline{x},\cdot)$ has Lebesgue density

$$\xi_i \to \frac{1}{n}\sum_{\nu=1}^{n} k_{ni}(\xi_i-x_{i\nu}) ,$$

with $k_{ni}(\xi_i) := \int k_n(\xi_1,\dots,\xi_m) \underset{j\neq i}{\times} d\xi_j$.

The same holds true if $\mathfrak{P}$ is the family of all m-variate normal distributions and $P_n(\underline{x}) = N(\bar{x},s)$, where $\bar{x}$ is the sample mean vector and s the sample covariance matrix: The i-th marginal of $N(\bar{x},s)$ is $N(\bar{x}_i,s_{ii})$, which depends on $(x_{i\nu})_{\nu=1,\dots,n}$ only.

The following is an illustration of Corollary 7.4.2. For the sake of simplicity we assume that $T(P,\mathfrak{P}) = \mathscr{L}_*(P)$ (and, correspondingly, $T(P_i,\mathfrak{P}_i) = \mathscr{L}_*(P_i)$ for $i = 1,\dots,m$). Let

$$k_n(\underline{x},\xi_1,\dots,\xi_m) := 1 - dP_n(\underline{x},\cdot)/d{\times}P_i .$$

With

$$k_{ni}(\underline{x},\xi_i) := \int k_n(\underline{x},\xi_1,\dots,\xi_m) \underset{j\neq i}{\times} P_j(d\xi_j)$$

we obtain

$$d\pi_i P_n(\underline{x},\cdot)/dP_i = 1 + k_{ni}(\underline{x},\xi_i),\ i = 1,\dots,m,$$

hence

$$d{\times}\pi_i P_n(\underline{x},\cdot)/d{\times}P_i = \Pi(1 + k_{ni}(\underline{x},\xi_i)) =: 1 + \bar{k}_n(\underline{x},\xi_1,\dots,\xi_m), \text{ say.}$$

Since

$$\int \bar{k}_n(\underline{x},\xi_1,\dots,\xi_m) \underset{j\neq i}{\times} P_j(d\xi_j) = k_{ni}(\underline{x},\xi_i) ,$$

the functions $k_n(\underline{x},\cdot)$ and $\bar{k}_n(\underline{x},\cdot)$ have the same projection into $T({\times}P_i,\bar{\mathfrak{P}})$, namely $(\xi_1,\dots,\xi_m) \to \Sigma k_{ni}(\underline{x},\xi_i)$. (According to Corollary 7.4.2, the difference between these projections is, in general, of order $o(\|k_n(\underline{x},\cdot)\|)$.)

Finally we consider the following problem: Let $Y = X^m$, and let $\mathfrak{Q}$ be a family of p-measures on $\mathscr{A}^m$. Let $\mathfrak{P} = \{\times Q_i : Q_i \in \mathfrak{Q}\}$, and $\overline{\mathfrak{P}} = \{Q^m : Q \in \mathfrak{Q}\}$. Assume we are given a sequence of estimators $\underline{x} \to Q_n(\underline{x},\cdot)$. If the true p-measure belongs to $\overline{\mathfrak{P}}$, i.e. if it is the product of i d e n t i c a l components Q, the projection of $\times Q_n(\underline{x}_i,\cdot)$ into $\overline{\mathfrak{P}}$ leads to the following estimator for Q,

$$(10.6.1) \qquad \overline{Q}_n(\underline{x}_1,\ldots,\underline{x}_m,\cdot) := \frac{1}{m}\Sigma\, Q_n(\underline{x}_i,\cdot) .$$

Since $Q_n(\underline{x}_i,\cdot)$ is the i-th marginal of $\times Q_n(\underline{x}_i,\cdot)$, this follows immediately from the result at the end of Section 7.7.

If the estimator-sequence Q_n, $n \in \mathbb{N}$, is as. efficient on $\mathfrak{Q}$, the estimator-sequence $\overline{Q}_n$, $n \in \mathbb{N}$, is as. efficient on $\overline{\mathfrak{P}}$.

Notice that (10.6.1) is obvious if Q_n is a kernel type estimator, for in this case the summation extends over a sample of $m\cdot n$ independent observations. (There is, however, a difference compared to the kernel estimator for the sample size $m\cdot n$, consisting in the use of k_n rather than $k_{m\cdot n}$.)

Since the product of as. efficient estimator-sequences is as. efficient (see Example 10.1.3), $(\underline{x}_1,\ldots,\underline{x}_m) \to \times Q_n(\underline{x}_i,\cdot)$, $n \in \mathbb{N}$, is as. efficient on $\mathfrak{P}$. Since the projection of as. efficient estimator-sequences is as. efficient (see Theorem 10.4.7), it follows that $(\underline{x}_1,\ldots,\underline{x}_m) \to Q_n(\underline{x}_1,\ldots,\underline{x}_m,\cdot)^m$, $n \in \mathbb{N}$, is as. efficient on $\overline{\mathfrak{P}}$, and this implies that $(\underline{x}_1,\ldots,\underline{x}_m) \to Q_n(\underline{x}_1,\ldots,\underline{x}_m,\cdot)$ is as. efficient on $\mathfrak{Q}$. This can, however, also be verified directly. If we consider $(x_{1\nu},\ldots,x_{m\nu})_{\nu=1,\ldots,n}$ as n realizations from Q^m, and $Q_n(\underline{x}_1,\ldots,\underline{x}_m,\cdot)$ as an estimate based on the sample $(x_{1\nu},\ldots,x_{m\nu})_{\nu=1,\ldots,n}$, of size n, the condition for as. efficiency reads (see Definition 10.1.1)

$$n^{1/2}\int \Sigma f(\xi_i) \times Q_n(\underline{x}_1,\ldots,\underline{x}_m,d\xi_i) = n^{-1/2}\sum_{\nu=1}^{n}\left(\Sigma f(x_{i\nu})\right) + o_P(n^{\circ}).$$

The alternative is to consider $x_{i\nu}$, $i = 1,\ldots,m$, $\nu = 1,\ldots,n$ as $m\cdot n$ realizations from Q, and $Q_n(\underline{x}_1,\ldots,\underline{x}_m,\cdot)$ as an estimate based on the sample $(x_{i\nu})_{i=1,\ldots,m,\ =1,\ldots,n}$ of size m·n. The condition for as. efficiency reads

$$(m\cdot n)^{1/2}\int f(\xi)Q_n(\underline{x}_1,\ldots,\underline{x}_m,d\xi) = (m\cdot n)^{-1/2}\sum_{\nu=1}^{n}\Sigma f(x_{i\nu}) + o_P(n^{o}).$$

It is straightforward to see that both conditions are equivalent, and follow from (10.6.1) if

$$n^{1/2}\int f(\xi)Q_n(\underline{x}_i,d\xi) = n^{-1/2}\sum_{\nu=1}^{n} f(x_{i\nu}) + o_P(n^{o}) .$$

11. EXISTENCE OF ASYMPTOTICALLY EFFICIENT ESTIMATORS FOR FUNCTIONALS

11.1. Introduction

Our problem is to estimate a real-valued functional $\kappa: \mathfrak{P} \to \mathbb{R}$. Theorem 9.2.2 contains as. bounds for the concentration of estimators of κ. Let κ be differentiable and $\kappa^{\bullet}(\cdot,P) \in T(P,\mathfrak{P})$ its canonical gradient. If an estimator-sequence κ^n, $n \in \mathbb{N}$, for κ is as. median unbiased, then, for $n \to \infty$, $P^n * n^{1/2}(\kappa^n - \kappa(P))$ cannot be more concentrated about 0 than $N(0,P(\kappa^{\bullet}(\cdot,P)^2))$. Hence it is justified to call an estimator-sequence which is as. median unbiased for κ on $\mathfrak{P}$, *as. efficient* for κ on $\mathfrak{P}$ if it is as. normal with variance $P(\kappa^{\bullet}(\cdot,P)^2)$.

In this chapter we suggest heuristic procedures for the construction of as. efficient estimator-sequences.

<u>11.1.1. Remark</u>. For k-dimensional functionals an appropriate definition of as. efficiency can be based on Theorem 9.3.1. This theorem suggests to call a k-dimensional estimator-sequence as. efficient if it converges to its limiting distribution continuously in the sense of (9.3.2), and if this limit distribution is $N(0,\Sigma(P))$. Since k-dimensional estimator-sequences which are as. efficient in this sense consist of components which are as. efficient, and since - conversely -

a k-tuple of as. efficient estimator-sequences is as. efficient provided the k-dimensional joint distribution converges to its limit continuously (see Corollary 9.3.6), we restrict ourselves to $k = 1$.

11.2. Asymptotically efficient estimators for functionals from asymptotically efficient estimators for probability measures

11.2.1. Theorem. *Let* $\kappa: \mathfrak{P} \to \mathbb{R}$ *be strongly differentiable at* P *in the sense that*

$$\kappa(Q) - \kappa(P) = Q(\kappa^{\bullet}(\cdot,P)) + o(\delta(Q,P)^{1+\varepsilon})$$

for some $\varepsilon > 0$. *Let* P_n, $n \in \mathbb{N}$, *be as. efficient at* P *in* $\mathfrak{P}$ *with* $\delta(P_n(\underline{x},\cdot),P) = o_P(n^{-1/2(1+\varepsilon)})$. *Then the estimator-sequence* $\underline{x} \to \kappa(P_n(\underline{x},\cdot))$, $n \in \mathbb{N}$, *is as. efficient for* κ *at* P.

Proof. Since κ is strongly differentiable, we have

$$n^{1/2}(\kappa(P_n(\underline{x},\cdot))-\kappa(P)) = n^{1/2}\int \kappa^{\bullet}(\xi,P)P_n(\underline{x},d\xi)+n^{1/2}o(\delta(P_n(\underline{x},\cdot),P)^{1+\varepsilon}) .$$

By Remark 10.1.2, the distribution of $\underline{x} \to n^{1/2}\int \kappa^{\bullet}(\xi,P)P_n(\underline{x},d\xi)$ under P^n converges weakly to $N(0,P(\kappa^{\bullet}(\cdot,P)^2))$. Hence the assertion follows from $\delta(P_n(\underline{x},\cdot),P)^{1+\varepsilon} = o_P(n^{-1/2})$.

11.2.2. Remark. Let $\mathfrak{P} = \{P_\theta: \theta \in \Theta\}$, $\Theta \subset \mathbb{R}^k$, and $\kappa(P_\theta) = \theta$. In Proposition 10.3.2 it was shown that $\underline{x} \to P_{\theta^n(\underline{x})}$ is as. efficient on $\mathfrak{P}$ if θ^n is as. efficient for θ. Theorem 11.2.1 implies the converse: If $\underline{x} \to P_{\theta^n(\underline{x})}$ is as. efficient on $\mathfrak{P}$, then $\underline{x} \to \theta^n(\underline{x}) = \kappa(P_{\theta^n(\underline{x})})$ is as. efficient for θ.

11.2.3. Remark. If κ is defined on $\mathfrak{P}$, but it is known that P belongs, in fact, to a subfamily $\overline{\mathfrak{P}} \subset \mathfrak{P}$, then any estimator-sequence which is as. efficient for κ on $\mathfrak{P}$ will, in general, cease to be as. efficient under the more stringent assumption that the true p-measure belongs to the subfamily $\overline{\mathfrak{P}}$. Theorem 11.2.1 tells us in connection with Theorem 10.4.8 how an as. efficient estimator-sequence for κ on $\overline{\mathfrak{P}}$ can be obtained: It suffices to determine the projection of P_n into $\overline{\mathfrak{P}}$, say $\overline{P}_n$, and to choose the estimator $\underline{x} \to \kappa(\overline{P}_n(\underline{x},\cdot))$. For $n \to \infty$, it attains the minimal as. variance, $P(\overline{\kappa}^{\bullet}(\cdot,P)^2)$ (where $\overline{\kappa}^{\bullet}(\cdot,P)$ is the canonical gradient of κ in $T(P,\overline{\mathfrak{P}})$, obtainable as the projection of $\kappa^{\bullet}(\cdot,P)$ into $T(P,\overline{\mathfrak{P}})$).

11.2.4. Remark. Often, the functional κ is defined for a class of p-measures including the discrete p-measures. In such cases it is tempting to try $\kappa(Q_n(\underline{x},\cdot))$ as an estimator, with $Q_n(\underline{x},\cdot)$ the empirical p-measure. In certain cases, this estimator turns out to be as. efficient. This is true in particular for von Mises functionals (see Section 5.1). If

$$\kappa(P) = \int k(\xi_1,\dots,\xi_m)P(d\xi_1)\dots P(d\xi_m),$$

we obtain

$$\kappa(P_n(\underline{x},\cdot)) = \frac{1}{n^m}\sum_{\nu_1=1}^{n}\dots\sum_{\nu_m=1}^{n} k(x_{\nu_1},\dots,x_{\nu_m}) + o_p(n^{-1/2})$$

if $P_n(\underline{x},\cdot)$, $n \in \mathbb{N}$, is as. efficient in the sense of Definition 10.1.1. Since

$$\begin{aligned}\frac{1}{n^m}\sum_{\nu_1=1}^{n}\dots\sum_{\nu_m=1}^{n} k(x_{\nu_1},\dots,x_{\nu_m}) &= \int k(\xi_1,\dots,\xi_m)Q_n(\underline{x},d\xi_1)\dots Q_n(\underline{x},d\xi_m)\\ &= \kappa(Q_n(\underline{x},\cdot)),\end{aligned}$$

we have $\kappa(P_n(\underline{x},\cdot)) = \kappa(Q_n(\underline{x},\cdot)) + o_p(n^{-1/2})$. Since $\kappa(P_n(\underline{x},\cdot))$ is as. efficient by Theorem 11.2.1, this shows that $\kappa(Q_n(\underline{x},\cdot))$ is as. efficient, too. A direct derivation of the as. distribution of $\kappa(Q_n(\underline{x},\cdot))$ for von Mises functionals κ was first given by Hoeffding (1948, p. 309, Theorem 7.4). It also follows from a more general result of von Mises (1936; see also 1947, p. 327, Theorem I) on the distribution of $\kappa(Q_n(\underline{x},\cdot))$ for general differentiable functionals.

11.3. Functions of asymptotically efficient estimators are asymptotically efficient

Let $\kappa_i: \mathfrak{P} \to \mathbb{R}$, $i = 1,\dots,k$, be differentiable functionals with canonical gradients $\kappa_i^{\bullet}(\cdot,P) \in T_w(P,\mathfrak{P})$. Let $\kappa := (\kappa_1,\dots,\kappa_k)'$. Assume that the covariance matrix $\Sigma(P) := P(\kappa^{\bullet}(\cdot,P)\kappa^{\bullet}(\cdot,P)')$ is positive definite. Let $K: \mathbb{R}^k \to \mathbb{R}$ be a function with continuous partial derivatives. Our problem is to estimate the functional $K\circ\kappa$.

11.3.1. Proposition. *For* $i = 1,\dots,k$ *let* κ_i^n, $n \in \mathbb{N}$, *be as. efficient estimator-sequences for* κ, *the joint distribution of which converges to a limiting distribution continuously in the sense of (9.3.2). Then* $K\circ\kappa^n$, $n \in \mathbb{N}$, *is as. efficient for* $K\circ\kappa$.

Proof. By Proposition 4.4.2 the canonical gradient of $K\circ\kappa$ is $K^{(i)}(\kappa(P))\kappa_i^{\bullet}(\cdot,P)$, the as. variance bound therefore

(11.3.2) $\quad K(\kappa(P))'\Sigma(P)K(\kappa(P))$.

It remains to be shown that this as. variance bound is attained by $K\circ\kappa^n$, $n \in \mathbb{N}$. We have

(11.3.3) $\quad K(\kappa^n(\underline{x})) = K(\kappa(P)) + K^{(i)}(\kappa(P))(\kappa_i^n(\underline{x}) - \kappa_i(P)) + R_n(\underline{x},P)$

with $\quad R_n(\underline{x},P) = (\kappa_i^n(\underline{x}) - \kappa_i(P))\int_0^1 (K^{(i)}((1-u)\kappa(P) + u\kappa^n(\underline{x})) - K^{(i)}(\kappa(P)))du$.

Since $R_n(\cdot,P) = o_P(n^{-1/2})$, we obtain from (11.3.3) that

(11.3.4) $\quad n^{1/2}(K(\kappa^n) - K(\kappa(P))) = K^{(i)}(\kappa(P))n^{1/2}(\kappa_i^n - \kappa_i(P)) + o_P(n^0)$.

By Corollary 9.3.6, the as. efficiency of κ_i^n for $i = 1,\dots,k$ implies that $P^n * n^{1/2}(\kappa^n - \kappa(P))$ converges weakly to $N(0,\Sigma(P))$. This implies that $P^n * n^{1/2}(K(\kappa^n) - K(\kappa(P)))$ converges weakly to the normal distribution with variance (11.3.2).

11.4. Improvement of asymptotically inefficient estimators

In Section 11.2 we have suggested that under certain conditions the estimator-sequence $\kappa(P_n)$, $n \in \mathbb{N}$, will be as. efficient for κ on $\mathfrak{P}$ if the estimator-sequence P_n, $n \in \mathbb{N}$, is as. efficient on $\mathfrak{P}$. If P_n, $n \in \mathbb{N}$, is as. inefficient, then $\kappa(P_n)$, $n \in \mathbb{N}$, will, in general, be inefficient, too. (Under special circumstances, an as. inefficient estimator-sequence P_n, $n \in \mathbb{N}$, may lead to an as. efficient estimator-sequence $\kappa(P_n)$, $n \in \mathbb{N}$. See Remark 4.4.3 for an example.)

In this section we suggest a heuristic procedure for the improvement of as. inefficient estimator-sequences $\kappa(P_n)$, $n \in \mathbb{N}$.

<u>11.4.1. Improvement procedure</u>. Assume that $\kappa: \mathfrak{P} \to \mathbb{R}$ is differentiable. For $P \in \mathfrak{P}$ let $\kappa^{\bullet}(\cdot,P) \in T(P,\mathfrak{P})$ denote its canonical gradient. Let P_n, $n \in \mathbb{N}$, be an estimator-sequence attaining its values in $\mathfrak{P}$. Then the *improved estimator-sequence* κ^n, $n \in \mathbb{N}$, is defined by

$$(11.4.2)\qquad \kappa^n(\underline{x}) := \kappa(P_n(\underline{x},\cdot)) + n^{-1} \sum_{\nu=1}^{n} \kappa^{\bullet}(x_\nu, P_n(\underline{x},\cdot)), \qquad \underline{x} \in X^n,\ n \in \mathbb{N}.$$

The reader familiar with parametric statistical theory will easily recognize that (11.4.2) generalizes the improvement procedure based on the Newton-Raphson approximation to the solution of the likelihood equation (see, e.g., LeCam, 1956, p. 139).

Let $\mathfrak{P} = \{P_\theta : \theta \in \Theta\}$, $\Theta \subset \mathbb{R}^k$, and $\kappa(P_\theta) = \theta_1$. Then (see Proposition 5.3.1) $\kappa^{\bullet}(x,P_\theta) = \Lambda_{1j}(\theta)\ell^{(j)}(x,\theta)$. If $\hat{\theta}^n$ is a preliminary estimator of θ, then (11.4.2) specializes to

$$\theta_1^n(\underline{x}) = \hat{\theta}_1^n(\underline{x}) + \frac{1}{n} \sum_{\nu=1}^{n} \Lambda_{1j}(\hat{\theta}^n(\underline{x}))\ell^{(j)}(x_\nu, \hat{\theta}^n(\underline{x})) .$$

That this improvement procedure yields as. efficient estimator-sequences in the parametric case is well known. It would now be in order to present a general theorem, asserting that the heuristic procedure, given by (11.4.2), leads to as. efficient estimator-sequences under rather general circumstances. Such a theorem is not available by now, and it seems unlikely that there will ever be one, general enough to cover all special cases under nearly optimal regularity conditions. The reason is that the improvement procedure requires an estimate $\kappa^{\bullet}(\cdot,P_n(\underline{x},\cdot))$ for $\kappa^{\bullet}(\cdot,P)$. In some cases, this is an easy task, so for instance in the case of a parametric family or, more generally, if $\kappa^{\bullet}(\cdot,P)$ depends on P through a finite-dimensional functional only. (Pertinent examples are minimum contrast functionals; see Proposition 5.2.3.) In other cases, $\kappa^{\bullet}(\cdot,P)$ depends on P in a more complex way, e.g. through the density or its derivatives. (This is the case for quantiles; see Proposition 5.4.2.)

Whatever the particular case might be, the essential point is that, when choosing the estimator P_n, one has to bear in mind that not only should $\kappa(P_n)$ be a good estimator for $\kappa(P)$, but also $\kappa^{\bullet}(\cdot,P_n)$ a reasonably good estimator for $\kappa^{\bullet}(\cdot,P)$.

Warning: The improvement procedure will not work, in general, if different estimators P_n are used in $\kappa(P_n)$ and $\kappa^{\bullet}(\cdot,P_n)$.

In view of the difficulties connected with the estimation of $\kappa^{\bullet}(\cdot,P)$, we confine ourselves to give a general outline of a proof which can be used as a model for proofs tailored to the particular cases.

In order to prove that $P^n * n^{1/2}(\kappa^n - \kappa(P))$, $n \in \mathbb{N}$, converges weakly to $N(0,P(\kappa^{\bullet}(\cdot,P)^2))$, we have to show that

$$(11.4.3) \qquad n^{1/2}(\kappa^n - \kappa(P)) = \tilde{\kappa}^{\bullet}(\cdot,P) + o_P(n^0) .$$

From this, the assertion follows by the Cramér-Slutzky lemma.

To prove (11.4.3), assume that

$$(11.4.4)\qquad \kappa(P_n(\underline{x},\cdot)) = \kappa(P) + \int \kappa^{\cdot}(\xi,P)P_n(\underline{x},d\xi) + o_P(n^{-1/2}) .$$

This holds, for instance, if κ is strongly differentiable in the sense that

$$\kappa(Q) = \kappa(P) + \int \kappa^{\cdot}(\xi,P)Q(d\xi) + O(\delta(Q,P)^2)$$

uniformly for $Q \in \mathfrak{P}$, and if P_n fulfills $\delta(P_n(\underline{x},\cdot),P) = o_P(n^{-1/4})$. Notice, however, that (11.4.4) is, in fact, much less restrictive than strong differentiability, because it requires the error term to be of order $o(n^{-1/2})$ stochastically. It avoids the uniformity in Q required by the strong differentiability condition. The difference becomes obvious if we consider, e.g., von Mises functionals (see Remark 11.2.4).

From (11.4.4) we obtain for the improved estimator κ^n defined in (11.4.2),

$$\begin{aligned} n^{1/2}(\kappa^n(\underline{x})-\kappa(P)) &= n^{1/2}(\kappa(P_n(\underline{x},\cdot))-\kappa(P) \\ &\quad + n^{-1/2}\sum_{\nu=1}^{n} \kappa^{\cdot}(x_\nu,P_n(\underline{x},\cdot)) + o_P(n^{\circ}) \\ &= n^{-1/2}\sum_{\nu=1}^{n}[\kappa^{\cdot}(x_\nu,P_n(\underline{x},\cdot)) + \int\kappa^{\cdot}(\xi,P)P_n(\underline{x},d\xi)] + o_P(n^{\circ}) . \end{aligned}$$

The heuristic considerations in Section 11.5 suggest that $P_n \to P$ stochastically implies that the distribution of

$$n^{-1/2}\sum_{\nu=1}^{n}[\kappa^{\cdot}(x_\nu,P_n(\underline{x},\cdot)) + \int\kappa^{\cdot}(\xi,P)P_n(\underline{x},d\xi)]$$

behaves under P^n asymptotically like the distribution of $\tilde{\kappa}^{\cdot}(\cdot,P)$. If this is the case, (11.4.3) follows.

<u>11.4.5. Remark</u>. Hasminskii and Ibragimov (1979, pp. 44-46, Theorems 2-4) establish the as. efficiency of improved estimator-sequences for non-parametric models. For technical reasons, they use a slightly modified version of the improvement procedure, in which the estimator $P_n(\underline{x},\cdot)$ is based on $x_1,\ldots,x_{k(n)}$, whereas the summation in

$\Sigma\kappa^{\bullet}(x_\nu, P_n(\underline{x},\cdot))$ extends over $\nu = k(n)+1,\dots,n$. By this trick they achieve that x_ν for $\nu = k(n)+1,\dots,n$ are stochastically independent of $P_n(\underline{x},\cdot)$. Letting $k(n)$, $n \in \mathbb{N}$, tend to infinity sufficiently f a s t , one obtains estimators $P_n(\underline{x},\cdot)$ which are sufficiently close to P. If $k(n)$, $n \in \mathbb{N}$, tends to infinity sufficiently s l o w l y , the omission of $x_1,\dots,x_{k(n)}$ in the summation $\Sigma\kappa^{\bullet}(x_\nu, P_n(\underline{x},\cdot))$ is of negligible influence. $k(n) \sim n/\log n$ turns out to be an appropriate rate. (This device was also used by Hájek (1962, p. 1146, Theorem 9.1) to construct an adaptive sequence of rank tests in the linear model, by van Eeden (1970, p. 175, Theorem 2.1) resp. Takeuchi (1971) to construct an as. efficient estimator-sequence of the location parameter in the two- resp. one-sample problem, and by Bickel (1982, Theorem 3.1) to construct adaptive estimator-sequences.) The obvious objection is that this trick serves technical purposes only, and we have some doubts whether this variety of the improvement procedures will meet wide approval among applied statisticians. What we want to know is how estimators resulting from a practically reasonable improvement procedure behave.

The regularity conditions imposed by Hasminskii and Ibragimov (1979, p. 44) on the functional are rather severe. The general framework is that of a family of p-measures on $\mathbb{B}$ with Lebesgue densities. One condition on the functional is that $\int (q(\xi)-p(\xi))^2 d\xi \to 0$ implies $\int (\kappa^{\bullet}(\xi,Q)-\kappa^{\bullet}(\xi,P))^2 p(\xi)d\xi \to 0$, the other one requires that $\int (\kappa^{\bullet}(\xi,Q)-\kappa^{\bullet}(\xi,P))^2 d\xi \le c(\int (q(\xi)-p(\xi))^2 d\xi)^{\delta}$ with $\delta > 0$ sufficiently large. It is clear that such conditions exclude many interesting cases. As to the first condition, think, for instance, of the family of all sufficiently regular p-measures on $\mathbb{B}$ with symmetric density, and $\kappa(P)$ the median of P, in which case $\kappa^{\bullet}(x,P) = -\ell'(x,P)/P(\ell'(\cdot,P)^2)$ (see (15.1.2)). For the second condition think of $\mathfrak{P} = \{N(0,\sigma^2): \sigma > 0\}$, and $\kappa(N(\mu,\sigma^2)) = \sigma$, in which case $\kappa^{\bullet}(x,N(0,\sigma^2))$ is a quadratic function of x.

<u>11.4.6. Remark.</u> If

$$(11.4.7)\qquad \sum_{\nu=1}^{n} \kappa^{\bullet}(x_\nu, P_n(\underline{x},\cdot)) = 0 \qquad \text{for all } \underline{x} \in X^n,\ n \in \mathbb{N},$$

then the improvement procedure leaves $\kappa(P_n)$ unchanged. If regularity conditions are fulfilled which guarantee that the improved estimators are as. efficient, the additional fulfillment of (11.4.7) guarantees that the estimator-sequence $\kappa(P_n)$, $n \in \mathbb{N}$, itself is as. efficient. Hence in special cases it may be useful to construct the estimators P_n such that they fulfill the 'estimating equation'

$$\sum_{\nu=1}^{n} \kappa^{\bullet}(x_\nu, P) = 0 .$$

This is a straightforward generalization of the likelihood equation. Whether it can be useful under more general circumstances remains to be explored.

<u>11.4.8. Example.</u> Let f be a contrast function and κ the pertaining minimum contrast functional. Then the canonical gradient (see Proposition 5.2.3) is $\kappa^{\bullet}(\cdot,P) = -F(P)^{-1} f^{(\cdot)}(\cdot,\kappa(P))$. Hence (11.4.7) is, in this case, equivalent to

$$(11.4.9)\qquad \sum_{\nu=1}^{n} f^{(\cdot)}(x_\nu, \kappa(P_n(\underline{x},\cdot))) = 0, \qquad \underline{x} \in X^n,\ n \in \mathbb{N},$$

i.e., it leads to minimum contrast estimators. These estimators are written as $\kappa(P_n(\underline{x},\cdot))$ (instead of $\kappa^n(\underline{x})$) only for reasons of consistency. Relation (11.4.9) leads immediately to the function $\underline{x} \to \kappa(P_n(\underline{x},\cdot))$ (without determining $P_n(\underline{x},\cdot)$ itself). Hence we obtain from our general results that minimum contrast estimators are as. efficient if the family of p-measures is full. This result has first been established by Levit (1975, p. 732, Theorem 3.1).

11.5. A heuristic justification of the improvement procedure

Let $k(\cdot,Q) \in \mathscr{L}_*(Q)$ for $Q \in \mathfrak{P}$. In this section we suggest a way of proving

$$(11.5.1) \qquad n^{-1/2} \sum_{\nu=1}^{n} [k(x_\nu, P_n(\underline{x},\cdot)) + \int k(\xi,P)P_n(\underline{x},d\xi)] = \tilde{k}(\underline{x},P) + o_P(n^0).$$

In Section 11.4 this relation was applied for $k = \kappa^\cdot$ to prove that the estimator-sequence defined by (11.4.2) is as. efficient.

Relation (11.5.1) is relatively easy to establish if the function $Q \to k(\cdot,Q)$ is differentiable. The concept of differentiability which we need here is not identical with the differentiability of $Q \to k(x,Q)$ for every $x \in X$, because a remainder term converging pointwise to zero is not sufficient.

$Q \to k(\cdot,Q)$ is *differentiable* at P with derivative $k^\cdot(\cdot,\cdot,P) \in \mathscr{L}_*(P^2)$ if $k^\cdot(x,\cdot,P) \in T_s(P,\mathfrak{P})$ for every $x \in X$, and if for each $g \in T_s(P,\mathfrak{P})$ and each path $P_{t,g}$, $t \downarrow 0$, with strong derivative g,

$$(11.5.2) \qquad k(x,P_{t,g}) = k(x,P) + t\int k^\cdot(x,\eta,P)g(\eta)P(d\eta) + R_t(x,P), \qquad x \in X,$$

where $R_t(\cdot,P) \to 0$ if $t \downarrow 0$ in some appropriate sense.

<u>11.5.3. Lemma</u>. *If* $Q \to k(\cdot,Q)$ *is differentiable in the sense of (11.5.2) with* $\|R_t(\cdot,P)\| = o(t)$, *then* $y \to \int k^\cdot(\xi,y,P)P(d\xi)$ *is the projection of* $-k(\cdot,P)$ *into* $T_s(P,\mathfrak{P})$.

In particular, if $k(\cdot,P)$ *is orthogonal to* $T_s(P,\mathfrak{P})$, *then* $\int k^\cdot(\xi,\cdot,P)P(d\xi) \equiv 0$.

<u>Proof</u>. Let $g \in T_s(P,\mathfrak{P})$ and $P_{t,g}$, $t \downarrow 0$, a path with P-density $1+t(g+r_t)$. Integrating (11.5.2) over x with respect to $P_{t,g}$, we obtain for all $t > 0$

$$-\int k(\xi,P)P_{t,g}(d\xi) = \int\int k^{\bullet}(\xi,\eta,P)P_{t,g}(d\eta)P_{t,g}(d\xi) + \int R_t(\xi,P)P_{t,g}(d\xi) .$$

Since $\int k^{\bullet}(\cdot,\eta,P)P(d\eta) \equiv 0$, this implies for all $g \in T_s(P,\mathfrak{P})$

$$-\int k(\xi,P)g(\xi)P(d\xi) = \int\int k^{\bullet}(\xi,\eta,P)P(d\xi)g(\eta)P(d\eta) .$$

Since $k^{\bullet}(x,\cdot,P) \in T_s(P,\mathfrak{P})$ for every $x \in X$, we have $\int k^{\bullet}(\xi,\cdot,P)P(d\xi) \in T_s(P,\mathfrak{P})$, and the assertion follows.

If $Q \to k(x,Q)$ is differentiable, it is not unreasonable to expect that

(11.5.4) $\quad k(x,P_n(\underline{x},\cdot)) = k(x,P) + \int k^{\bullet}(x,\eta,P)P_n(\underline{x},d\eta) + R_n(x,P_n(\underline{x},\cdot))$

with $\int R_n(\xi,P_n(\underline{x},\cdot))P(d\xi) = o_P(n^{-1/2})$.

Assume that P_n, $n \in \mathbb{N}$, is an estimator-sequence with a representation (10.4.1), i.e.,

$$p_n(\underline{x},\xi) = p(\xi)(1 + g_n(\underline{x},\xi) + r_n(\underline{x},\xi)) ,$$

where $g_n(\underline{x},\cdot) \in T_s(P,\mathfrak{P})$ and $\int r_n(\underline{x},\xi)^2P(d\xi) = o_P(n^{-1})$. Then

$$k(x,P_n(\underline{x},\cdot)) = k(x,P) + \int k^{\bullet}(x,\eta,P)g_n(\underline{x},\eta)P(d\eta) + \int k^{\bullet}(x,\eta,P)r_n(\underline{x},\eta)P(d\eta).$$

By Lemma 11.5.3,

(11.5.5) $\quad \int\int k^{\bullet}(\xi,\eta,P)P(d\xi)g_n(\underline{x},\eta)P(d\eta) = -\int k(\xi,P)g_n(\underline{x},\xi)P(d\xi).$

(This is the point where we use the fact that only one estimator-sequence appears in (11.5.1). It can be seen from this relation that the improvement procedure (11.4.2) will not work, in general, if two different estimator-sequences are used.)

From (11.5.4) and (11.5.5) we obtain

$$n^{-1/2}\sum_{\nu=1}^{n}[k(x_\nu,P_n(\underline{x},\cdot)) + \int k(\xi,P)P_n(\underline{x},d\xi)]$$
$$= \tilde{k}(\underline{x},P) + n^{-1/2}\sum_{\nu=1}^{n}[\int k^{\bullet}(x_\nu,\eta,P)g_n(\underline{x},\eta)P(d\eta) - \int\int k^{\bullet}(\xi,\eta,P)g_n(\underline{x},\eta)P(d\eta)P(d\xi)]$$

$$+ n^{-1/2} \sum_{\nu=1}^{n} \int k^{\bullet}(x_\nu,\eta,P) r_n(\underline{x},\eta) P(d\eta) + n^{-1/2} \sum_{\nu=1}^{n} R_n(x_\nu, P_n(\underline{x},\cdot))$$
$$+ n^{1/2} \int k(\xi,P) r_n(\underline{x},\xi) P(d\xi) .$$

These considerations suggest that one can prove (11.5.1) under reasonable regularity conditions, using the peculiarities of the family $\mathfrak{P}$, and of the estimator-sequence P_n, $n \in \mathbb{N}$. Notice that

$$x \to \int k^{\bullet}(x,\eta,P) g_n(\underline{x},\eta) P(d\eta) - \int\int k^{\bullet}(\xi,\eta,P) g_n(\underline{x},\eta) P(d\xi) P(d\eta)$$

has expectation zero and variance $o_P(n^o)$, so that the only problem in proving

$$n^{-1/2} \sum_{\nu=1}^{n} [\int k^{\bullet}(x_\nu,\eta,P) g_n(\underline{x},\eta) P(d\eta) - \int\int k^{\bullet}(\xi,\eta,P) g_n(\underline{x},\eta) P(d\xi) P(d\eta)]$$
$$= o_P(n^o)$$

is the stochastic dependence between x_ν and $g_n(\underline{x},\cdot)$. Another delicate point is whether

$$n^{-1/2} \sum_{\nu=1}^{n} R_n(x_\nu, P_n(\underline{x},\cdot)) = o_P(n^o) .$$

This is essentially equivalent to $\int R_n(\xi, P_n(\underline{x},\cdot)) P(d\xi) = o_P(n^{-1/2})$.

<u>11.5.6. Remark</u>. Let $\mathfrak{P}_o \subset \mathfrak{P}$ and $k(\cdot,P) \perp T_s(P,\mathfrak{P}_o)$. Assume that the estimator-sequence P_n, $n \in \mathbb{N}$, maps into $\mathfrak{P}_o$, and has a representation

$$p_n(\underline{x},\xi) = p(\xi)(1 + g_n(\underline{x},\xi) + r_n(\underline{x},\xi)) ,$$

where $g_n(\underline{x},\cdot) \in T_s(P,\mathfrak{P}_o)$ and $\int r_n(\underline{x},\xi)^2 P(d\xi) = o_P(n^{-1})$. Then

$$\int k(\xi,P) P_n(\underline{x},d\xi) = \int k(\xi,P) r_n(\underline{x},\xi) P(d\xi) = o_P(n^{-1/2}) .$$

In this case, (11.5.1) reduces to

$$n^{-1/2} \sum_{\nu=1}^{n} k(x_\nu, P_n(\underline{x},\cdot)) = \tilde{k}(\underline{x},P) + o_P(n^o) .$$

<u>11.5.7. Remark</u>. Another possibility to prove (11.5.1) is to focus attention on how $k(\cdot,P_n(\underline{x},\cdot))$ depends on $x_1,\ldots,x_n$. Under appropriate regularity conditions we obtain from (11.5.4) by integration over x

$$\int k(\xi,P) P_n(\underline{x},d\xi) = -\int k(\xi,P_n(\underline{x},\cdot)) P(d\xi) + o_P(n^{-1/2}) .$$

Hence proving (11.5.1) is equivalent to proving

$$(11.5.8) \qquad n^{-1/2} \sum_{\nu=1}^{n} h_n(x_\nu, \underline{x}) = o_P(n^o) ,$$

where

$$h_n(\xi, \underline{x}) = k(\xi, P_n(\underline{x}, \cdot)) - \int k(\xi, P_n(\underline{x}, \cdot)) P(d\xi) - k(\xi, P).$$

(11.5.8) follows from convergence to zero in quadratic mean. Since the order statistic is sufficient for families of product measures with identical components, we may assume w.l.g. that $h_n(\xi, \underline{x})$ is symmetric in $x_1, \dots, x_n$. We have

$$(n^{-1/2} \sum_{\nu=1}^{n} h_n(x_\nu, \underline{x}))^2 = n^{-1} \sum_{\nu=1}^{n} h_n(x_\nu, \underline{x})^2 + n^{-1} \sum_{\nu \neq \mu} h_n(x_\nu, \underline{x}) h_n(x_\mu, \underline{x}).$$

Hence (11.5.8) follows from

$$(11.5.9) \qquad \int h_n(x_1, \underline{x})^2 P^n(d\underline{x}) = o(n^o) ,$$

$$(11.5.10) \qquad \int h_n(x_1, \underline{x}) h_n(x_2, \underline{x}) P^n(d\underline{x}) = o(n^{-1}).$$

There is a fair chance that (11.5.9) and (11.5.10) are fulfilled if $k(\xi, P_n(\underline{x}, \cdot))$ admits a stochastic expansion

$$(11.5.11) \qquad n^{1/2}(k(\xi, P_n(\underline{x}, \cdot)) - k(\xi, P)) \doteq n^{-1/2} \sum_{\nu=1}^{n} g(\xi, x_\nu).$$

Then

$$n^{1/2} \int k(\xi, P_n(\underline{x}, \cdot)) P(d\xi) \doteq n^{-1/2} \sum_{\nu=1}^{n} \int g(\xi, x_\nu) P(d\xi),$$

and, with $g_o(\xi, x) := g(\xi, x) - \int g(\xi, x) P(d\xi)$,

$$n^{1/2} h_n(\xi, \underline{x}) \doteq n^{-1/2} \sum_{\nu=1}^{n} g_o(\xi, x_\nu) .$$

Hence

$$h_n(x_1, \underline{x}) \doteq n^{-1} \sum_{\nu=1}^{n} g_o(x_1, x_\nu) = n^{-1} \sum_{\nu=2}^{n} g_o(x_1, x_\nu) + n^{-1} g_o(x_1, x_1),$$

so that (11.5.9) - (11.5.10) can be expected to be fulfilled.

11.6. Estimators with stochastic expansion

An estimator-sequence κ^n, $n \in \mathbb{N}$, has a *stochastic expansion* of order $o(n^o)$ if for every $Q \in \mathfrak{P}$ there exists a function $k(\cdot,Q) \in \mathscr{L}_*(Q)$ such that

$$n^{1/2}(\kappa^n - \kappa(P)) = \tilde{k}(\cdot,P) + o_P(n^o) .$$

Without a certain amount of local uniformity in P, this condition is technically useless. The following condition is sufficient for our purposes and less restrictive than local uniformity in P:

For each $g \in T_s(P,\mathfrak{P})$ and each path $P_{t,g}$, $t \downarrow 0$, with strong derivative g,

$$n^{1/2}(\kappa^n - \kappa(P_{n^{-1/2},g})) = \tilde{k}(\cdot,P_{n^{-1/2},g}) + o_P(n^o) . \tag{11.6.1}$$

Most estimators obtained by one of the standard methods fulfill such a condition, so for instance maximum likelihood estimators, minimum contrast estimators, maximum probability estimators, and Bayes estimators.

11.6.2. Proposition. *If the function* $Q \to k(\cdot,Q)$ *occuring in the stochastic expansion of the functional* κ *(see (11.6.1)) is differentiable in the sense of (11.5.2) with* $\|R_t(\cdot,P)\| = o(t)$, *then* $k(\cdot,P)$ *is a gradient of* κ *at* P.

Proof. Let $P_{t,g}$, $t \downarrow 0$, be a path with strong derivative $g \in T_s(P,\mathfrak{P})$. Then, for all $x \in X$,

$$k(x,P_{n^{-1/2},g}) = k(x,P) + n^{-1/2}\int k^{\cdot}(x,\eta,P)g(\eta)P(d\eta) + R_{n^{-1/2}}(x,P).$$

Hence by Lemma 11.5.3,

$$(11.6.3)\quad n^{1/2}(\kappa^n(\underline{x}) - \kappa(P_{n^{-1/2},g}))$$
$$= \tilde{k}(\underline{x},P) + n^{-1}\sum_{\nu=1}^{n}\int k^{\cdot}(x_\nu,\eta,P)g(\eta)P(d\eta) + o_P(n^o)$$
$$= \tilde{k}(\underline{x},P) + \int\int k^{\cdot}(\xi,\eta,P)g(\eta)P(d\eta)P(d\xi) + o_P(n^o)$$
$$= \tilde{k}(\underline{x},P) - \int k(\xi,P)g(\xi)P(d\xi) + o_P(n^o)\ .$$

On the other hand,

$$(11.6.4)\quad \kappa(P_{n^{-1/2},g}) = \kappa(P) + n^{-1/2}\int\kappa^{\cdot}(\xi,P)g(\xi)P(d\xi) + o(n^{-1/2}).$$

From (11.6.3) and (11.6.4) we obtain for all $g \in T_s(P,\mathfrak{P})$,

$$P(\kappa^{\cdot}(\cdot,P)g) = P(k(\cdot,P)g).$$

Hence $k(\cdot,P)$ is a gradient.

By the Cramér-Slutzky lemma, for any estimator-sequence κ^n, $n \in \mathbb{N}$, with stochastic expansion (11.6.1), the sequence $P^n * n^{1/2}(\kappa^n - \kappa(P))$, $n \in \mathbb{N}$, converges weakly to $N(0,P(k(\cdot,P)^2))$. Since $k(\cdot,P)$ is a gradient, we have (see (4.3.4)) $P(k(\cdot,P)^2) \geq P(\kappa^{\cdot}(\cdot,P)^2)$. Hence $N(0,P(\kappa^{\cdot}(\cdot,P)^2))$ is a lower as. bound for the concentration of any estimator-sequence with a (sufficiently regular) as. expansion.

From Theorem 9.3.1 we know already that $N(0,P(\kappa^{\cdot}(\cdot,P)^2))$ is a lower as. bound for the concentration of any estimator-sequence which converges to its limiting distribution uniformly in a sense which corresponds to the uniformity in (11.6.1). We just could not resist the temptation to point out that this is almost trivial in the special case of an estimator-sequence with stochastic expansion.

12. EXISTENCE OF ASYMPTOTICALLY EFFICIENT TESTS

12.1. Introduction

Given the problem of testing the hypothesis $\mathfrak{P}_o \subset \mathfrak{P}$, we obtained in Section 8.4 the as. envelope power function $\Phi(N_\alpha + n^{1/2}\delta(Q,\mathfrak{P}_o))$. According to Theorem 8.5.3, it is impossible that a test-sequence attains this as. envelope power function for alternatives Q deviating from the hypothesis in different directions. Hence the only situation in which no further problems arise is that of a hypothesis with one-dimensional co-space (see Section 8.2 for a more detailed discussion of this point).

In the following Section 12.2 we suggest a method for constructing sequences of critical regions which are as. efficient against alternatives in a given direction; in Section 12.3 we consider the particular case of a hypothesis specified in terms of a functional, and critical regions based on estimators of this functional.

12.2. An asymptotically efficient critical region

Let P_n, $n \in \mathbb{N}$, be an estimator-sequence for $P \in \mathfrak{P}_o$. By Remark 11.5.6 it is plausible that

$$n^{-1/2} \sum_{\nu=1}^{n} k(x_\nu, P_n(\underline{x},\cdot)) = \tilde{k}(\underline{x},P) + o_P(n^o)$$

under suitable regularity conditions. Moreover, we may assume that

$$(\int k(\xi,P)^2 P(d\xi))^{1/2} = (\int k(\xi,P_n(\underline{x},\cdot))^2 P_n(\underline{x},d\xi))^{1/2} + o_P(n^o).$$

The critical region for testing against alternatives in direction $k(\cdot,P) \in T^{\perp}(P;\mathfrak{P}_o,\mathfrak{P})$ is

$$(12.2.1) \qquad C_n := \{\underline{x}: n^{-1/2} \sum_{\nu=1}^{n} k(x_\nu, P_n(\underline{x},\cdot)) > -N_\alpha (\int k(\xi,P_n(\underline{x},\cdot))^2 P_n(\underline{x},d\xi))^{1/2}\}.$$

It is straightforward to show that Corollary 19.2.9 implies under appropriate regularity conditions that

$$P^n_{n^{-1/2}t,k(\cdot,P)}(C_n) = \Phi(N_\alpha + t\|k(\cdot,P)\|) + o(n^o),$$

where $P_{t,k(\cdot,P)}$, $t \downarrow 0$, is any path with strong derivative $k(\cdot,P)$.

Since $n^{1/2}\delta(P_{n^{-1/2}t,k(\cdot,P)},\mathfrak{P}_o) = t\|k(\cdot,P)\| + o(n^o)$ by Proposition 6.2.18, the sequence of critical regions C_n, $n \in \mathbb{N}$, is as. most powerful against alternatives $P_{n^{-1/2}t,k(\cdot,P)}$.

12.3. Hypotheses on functionals

If a hypothesis is specified in terms of functionals, then the co-space is one-dimensional iff it is specified by o n e functional, i.e. iff

$$(12.3.1) \qquad \mathfrak{P}_o = \{P \in \mathfrak{P}: \kappa(P) = c\}.$$

If κ is differentiable with canonical gradient $\kappa^{\bullet}(\cdot,P) \in T(P,\mathfrak{P})$, we have (see Example 8.1.2)

$$T^{\perp}(P;\mathfrak{P}_o,\mathfrak{P}) = [\kappa^{\bullet}(\cdot,P)].$$

If an estimator for $\kappa^{\bullet}(\cdot,P)$ is easily accessible, then the critical region (12.2.1) with $k(\cdot,P) = \kappa^{\bullet}(\cdot,P)$ can be used for testing the hypothesis $\mathfrak{P}_o$.

If an estimator κ^n for κ is available, it may be easier to base the test for the hypothesis $\mathfrak{P}_0$ on this estimator. Assume that for every $Q \in \mathfrak{P}$, $Q^n * n^{1/2}(\kappa^n - \kappa(Q)) \Rightarrow N(0,\sigma(Q)^2)$. If σ^n, $n \in \mathbb{N}$, is a sequence of estimators for $\sigma(Q)$, it suggests itself to use the critical region

(12.3.2) $\quad C_n := \{\underline{x} \in X^n : n^{1/2}(\kappa^n(\underline{x}) - c) > -N_\alpha \sigma^n(\underline{x})\}$

for testing the hypothesis $\mathfrak{P}_0$ against alternatives $\kappa(Q) > c$.

To determine the as. power of C_n, $n \in \mathbb{N}$, we proceed as follows: For $Q \in \mathfrak{P}$,

$$\begin{aligned} Q^n(C_n) &= Q^n\{\underline{x} \in X^n : n^{1/2}(\kappa^n(\underline{x}) - \kappa(Q)) \\ &\qquad > -N_\alpha \sigma^n(\underline{x}) - n^{1/2}(\kappa(Q) - c)\} \\ &= Q^n\{\underline{x} \in X^n : n^{1/2}(\kappa^n(\underline{x}) - \kappa(Q))/\sigma(Q) \\ &\qquad > -N_\alpha - n^{1/2}(\kappa(Q)-c)/\sigma(Q) + N_\alpha(1-\sigma^n(\underline{x})/\sigma(Q))\}. \end{aligned}$$

If σ^n, $n \in \mathbb{N}$, is consistent, $N_\alpha(1-\sigma^n(\underline{x})/\sigma(Q))$ converges stochastically to zero, and we obtain from the Cramér-Slutzky lemma that

$$Q^n(C_n) = \Phi(N_\alpha + n^{1/2}(\kappa(Q)-c)/\sigma(Q)) + o(n^0) .$$

Comparing this relation with Remark 8.4.6, we see that the sequence C_n, $n \in \mathbb{N}$, is as. efficient iff $\sigma(Q)^2 = P(\kappa^\bullet(\cdot,Q)^2)$, i.e. iff the estimator-sequence κ^n, $n \in \mathbb{N}$, is as. efficient.

In contrast to this, the as. efficiency of the estimator-sequence for $\sigma(Q)$ is of secondary importance. Not more than its consistency is required to obtain as. efficiency of the test-sequence. Of course, for small samples, more accurate estimators for $\sigma(Q)$ are likely to lead to better tests, and only the unlimited increase of n makes this effect disappear.

<u>12.3.3. Remark</u>. It appears natural to try to improve the finite-sample performance of tests by using for $\sigma(Q)$ an estimator-sequence which is as. efficient for the smaller family $\mathfrak{P}_0$ (and therefore for $P \in \mathfrak{P}_0$ of

higher accuracy than an estimator-sequence which is as. efficient for the whole family $\mathfrak{P}$). Since it is the behavior of $Q^n * \sigma^n$ for $Q \in \mathfrak{P}$ with $\delta(Q, \mathfrak{P}_o) \leq cn^{-1/2}$ (and not only for $Q \in \mathfrak{P}_o$), the test-power depends on 'robustness'-properties of σ^n. Hence it is not easy to say whether the use of such estimators is recommendable or not. The example of the parametric family shows that the use of estimators efficient for $\mathfrak{P}_o$ need not necessarily lead to an improvement of test-power.

If $\mathfrak{P} = \{P_{\delta,\tau}: \delta \in \mathbb{R}, \tau \in T\}$ and $\mathfrak{P}_o = \{P_{\delta_o,\tau}: \tau \in T\}$, one may use for $\sigma(P_{\delta,\tau})^2 = \Lambda_{oo}(\delta,\tau)$ the estimator $\Lambda_{oo}(\delta_o, \tau^n(\cdot,\delta_o))$, where $\tau^n(\cdot,\delta_o)$, $n \in \mathbb{N}$, is an as. efficient estimator-sequence for $\mathfrak{P}_o$. Or one may use $\Lambda_{oo}(\delta^n, \tau^n)$, where (δ^n, τ^n) is as. efficient for $\mathfrak{P}$, or $\Lambda_{oo}(\delta_o, \tau^n)$. By Remark 8 in Pfanzagl and Wefelmeyer (1978, p. 60), all these estimators lead to critical regions which have up to $o(n^{-1})$ the same power and therefore not too different small-sample properties.

It needs a deeper investigation to find out how the different estimators for $\sigma(P)$ influence the power of the critical regions in general.

13. INFERENCE FOR PARAMETRIC FAMILIES

13.1. Estimating a functional

Let $\mathfrak{P}$ be a family of p-measures, and $\kappa: \mathfrak{P} \to \mathbb{R}$ a differentiable functional. Let $\kappa^{\bullet}(\cdot,P) \in T(P,\mathfrak{P})$ denote the canonical gradient of κ at P.

Let $\mathfrak{P}_o = \{P_\theta: \theta \in \Theta\}$, $\Theta \subset \mathbb{R}^k$, be a parametric subfamily of $\mathfrak{P}$. Our problem is to find an estimator for κ which is as. efficient for $\mathfrak{P}_o$.

From the results obtained so far it is straightforward how such an estimator can be obtained: If θ^n is an as. efficient estimator for θ, then P_{θ^n} is an as. efficient estimator for P_θ (see Proposition 10.3.2). Since P_{θ^n} is as. efficient for P_θ, $\kappa(P_{\theta^n})$ is as. efficient for $\kappa(P_\theta)$ (see Theorem 11.2.1). The following is a direct proof for the particular case in question.

To obtain the as. variance bound, we have to determine the projection of $\kappa^{\bullet}(\cdot,P_\theta)$ into $T(P_\theta,\mathfrak{P}_o)$. By Proposition 2.2.1 we have

$$T(P_\theta,\mathfrak{P}_o) = [\ell^{(i)}(\cdot,\theta): i = 1,\dots,k].$$

The projection of $\kappa^{\bullet}(\cdot,P_\theta)$ into $T(P_\theta,\mathfrak{P}_o)$, say $\bar{a}'(\theta)\ell^{(\cdot)}(\cdot,\theta)$, is uniquely determined by (see Proposition 4.3.2)

$$\int (\kappa^{\bullet}(\xi,P_\theta) - \bar{a}'(\theta)\ell^{(\cdot)}(\xi,\theta))a'\ell^{(\cdot)}(\xi,\theta)P_\theta(d\xi) = 0 \text{ for all } a \in \mathbb{R}^k.$$

This yields

$$\bar{a}(\theta) = \Lambda(\theta)P_\theta(\kappa^{\bullet}(\cdot,P_\theta)\ell^{(\cdot)}(\cdot,\theta)) ,$$

i.e., the c a n o n i c a l gradient for $\mathfrak{P}_o$ is

(13.1.1) $\kappa^*(\cdot,P_\theta) = P_\theta(\kappa^\cdot(\cdot,P_\theta)\ell^{(\cdot)}(\cdot,\theta))'\Lambda(\theta)\ell^{(\cdot)}(\cdot,\theta)$.

From this we obtain the as. variance bound for κ on $\mathfrak{P}_o$

(13.1.2) $P_\theta(\kappa^*(\cdot,P_\theta)^2) = P_\theta(\kappa^\cdot(\cdot,P_\theta)\ell^{(\cdot)}(\cdot,\theta))'\Lambda(\theta)P_\theta(\kappa^\cdot(\cdot,P_\theta)\ell^{(\cdot)}(\cdot,\theta))$.

It is easy to check that this variance bound is smaller than the as. variance bound for κ on $\mathfrak{P}$, $P_\theta(\kappa^\cdot(\cdot,P_\theta)^2)$ (unless $\kappa^\cdot(\cdot,P_\theta) \in T(P_\theta,\mathfrak{P}_o)$, in which case the two bounds are equal).

It remains to be shown that the estimator-sequence $\kappa(P_{\theta^n})$ attains this as. variance bound if θ^n is as. efficient. If the functional κ is differentiable on $\mathfrak{P}_o$, the function $\theta \to \kappa(P_\theta)$ is differentiable, since

$$\begin{aligned}\kappa(P_{\theta+ta}) - \kappa(P_\theta) &= P_{\theta+ta}(\kappa^\cdot(\cdot,P_\theta)) + o(t)\\ &= ta'P_\theta(\kappa^\cdot(\cdot,P_\theta)\ell^{(\cdot)}(\cdot,\theta)) + o(t),\end{aligned}$$

hence

$$\lim_{t\to 0} t^{-1}(\kappa(P_{\theta+ta}) - \kappa(P_\theta)) = a'P_\theta(\kappa^\cdot(\cdot,P_\theta)\ell^{(\cdot)}(\cdot,\theta)).$$

Hence the function $\theta \to \kappa(P_\theta)$ has partial derivatives $P_\theta(\kappa^\cdot(\cdot,P_\theta)\ell^{(i)}(\cdot,\theta))$, $i = 1,\ldots,k$. Assuming that these derivatives are continuous, we obtain immediately (see (11.3.4))

(13.1.3) $n^{1/2}(\kappa(P_{\theta^n})-\kappa(P_\theta)) = P_\theta(\kappa^\cdot(\cdot,P_\theta)\ell^{(\cdot)}(\cdot,\theta))'n^{1/2}(\theta^n-\theta) + o_\theta(n^o)$.

Since $P_\theta^n * n^{1/2}(\theta^n-\theta) \Rightarrow N(0,\Lambda(\theta))$, this implies that $P_\theta^n * n^{1/2}(\kappa(P_{\theta^n})-\kappa(P_\theta))$ is as. normal with variance given by (13.1.2). Hence $\kappa(P_{\theta^n})$ is as. efficient for κ on $\mathfrak{P}_o$.

<u>13.1.4. Example</u>. Let $\mathfrak{P}$ be the family of all p-measures on $\mathbb{B}$ with $\int x^2 P(dx) < \infty$, and let $\kappa(P)$ be the variance of P, i.e.

(13.1.5) $\kappa(P) = \frac{1}{2}\int (x_1-x_2)^2 P(dx_1)P(dx_2)$.

The gradient of κ in $\mathscr{L}_*(P)$ is (see Proposition 5.1.5)

$$\kappa^\cdot(x,P) = \int (x-x_2)^2 P(dx_2) - \kappa(P) = (x-\mu(P))^2 - \kappa(P)$$

with $\mu(P) = \int xP(dx)$, the as. variance bound

(13.1.6) $$P(\kappa^\bullet(\cdot,P)^2) = \int (x-\mu(P))^4 P(dx) - \kappa(P)^2.$$

This as. variance bound is attained by the estimator-sequence

(13.1.7) $$\underline{x} \to n^{-1} \sum_{\nu=1}^{n} (x_\nu - \bar{x}_n)^2, \qquad \bar{x}_n := n^{-1} \sum_{\nu=1}^{n} x_\nu .$$

If the true p-measure is known to belong to a certain parametric family, and if an as. efficient estimator-sequence θ^n is available, then $\frac{1}{2}\int (x_1-x_2)^2 P_{\theta^n}(dx_1)P_{\theta^n}(dx_2)$ will, in general, be a better estimator-sequence. This will not be the case if this parametric family is the family $\mathfrak{P}_o = \{N(\mu,\sigma^2): \mu \in \mathbb{R}, \sigma^2 > 0\}$, for in this case $\kappa^\bullet(\cdot,N(\mu,\sigma^2)) = \sigma^{-3} \frac{\partial}{\partial\sigma}\ell(\cdot,\mu,\sigma) \in T(N(\mu,\sigma^2),\mathfrak{P}_o)$, so that the as. variance bound of κ for $\mathfrak{P}_o$ is the same as that for $\mathfrak{P}$.

If the available estimator-sequence θ^n, $n \in \mathbb{N}$, fails to be as. efficient, the improvement procedure suggested in 11.4.1 can be used to obtain an as. efficient estimator for κ. Using the canonical gradient given by (13.1.1), this leads to the estimator-sequence

(13.1.8) $$\kappa^n(\underline{x}) := \kappa(P_{\theta^n(\underline{x})}) + \bar{a}(\theta^n(\underline{x}))' n^{-1} \sum_{\nu=1}^{n} \ell^{(\cdot)}(x_\nu, \theta^n(\underline{x})),$$

for which

(13.1.9) $$n^{1/2}(\kappa^n(\underline{x}) - \kappa(P)) = n^{1/2}(\kappa(P_{\theta^n(\underline{x})}) - \kappa(P)) + \bar{a}(\theta^n(\underline{x}))' n^{-1/2} \sum_{\nu=1}^{n} \ell^{(\cdot)}(x_\nu, \theta^n(\underline{x})).$$

We have (see (13.1.3))

$$n^{1/2}(\kappa(P_{\theta^n(\underline{x})}) - \kappa(P_\theta)) = P_\theta(\kappa^\bullet(\cdot,P_\theta)\ell^{(\cdot)}(\cdot,\theta))' n^{1/2}(\theta^n(\underline{x})-\theta) + o_\theta(n^0)$$

and

$$n^{-1/2} \sum_{\nu=1}^{n} \ell^{(\cdot)}(x_\nu, \theta^n(\underline{x})) = n^{-1/2} \sum_{\nu=1}^{n} \ell^{(\cdot)}(x_\nu,\theta) + P_\theta(\ell^{(\cdot\cdot)}(\cdot,\theta)) n^{1/2}(\theta^n(\underline{x})-\theta) + o_\theta(n^0).$$

Since $\bar{a}(\theta)P_\theta(\ell^{(\cdot\cdot)}(\cdot,\theta)) = -P_\theta(\kappa^\bullet(\cdot,P_\theta)\ell^{(\cdot)}(\cdot,\theta))$, (13.1.9) implies

$$n^{1/2}(\kappa^n(\underline{x}) - \kappa(P)) = \bar{a}(\theta)' n^{-1/2} \sum_{\nu=1}^{n} \ell^{(\cdot)}(x_\nu,\theta) + o_\theta(n^0).$$

Therefore, $P_\theta^n * n^{1/2}(\kappa^n - \kappa(P))$ is as. normal with variance

$\bar{a}(\theta)'P_\theta(\ell^{(\cdot)}(\cdot,\theta)\ell^{(\cdot)}(\cdot,\theta)')\bar{a}(\theta)$, which equals the as. variance bound given in (13.1.2). Hence the estimator-sequence defined by (13.1.8) is as. efficient for κ on $\mathfrak{P}_o$.

13.2. Variance bounds for parametric subfamilies

Let now $\mathfrak{P} = \{P_\theta: \theta \in \Theta\}$, $\Theta \subset \mathbb{R}^k$, and consider a subfamily $\mathfrak{P}_o$. Such a restriction can, for instance, be due to side conditions, i.e. $\mathfrak{P}_o = \{P_\theta: \theta \in \Theta,\ F(\theta) = 0\}$ for some $F: \Theta \to \mathbb{R}^m$; or $\mathfrak{P}_o$ can be a curved subfamily, say $\mathfrak{P}_o = \{P_{c(\tau)}: \tau \in T\}$ for some $T \subset \mathbb{R}^q$ and $c: T \to \Theta$.

For as. theory, the essential point is the restriction of the tangent space from $T(P_\theta,\mathfrak{P}) = [\ell^{(i)}(\cdot,\theta): i = 1,\dots,k]$ to a certain subspace, say $T(P_\theta,\mathfrak{P}_o) = [f_\alpha(\cdot,\theta): \alpha = 1,\dots,m]$ with $f_\alpha(\cdot,\theta) = c_{\alpha i}\ell^{(i)}(\cdot,\theta)$, $\alpha = 1,\dots,m$; i.e. $T(P_\theta,\mathfrak{P}_o)$ is the m-dimensional subspace spanned by $f_\alpha(\cdot,\theta)$, $\alpha = 1,\dots,m$ (see Remark 2.2.7). If θ^n is as. efficient for $\mathfrak{P}$, then $\kappa(P_{\theta^n})$ will be as. efficient for κ on $\mathfrak{P}$, but not necessarily for κ on $\mathfrak{P}_o$.

The canonical gradient of κ in $T(P_\theta,\mathfrak{P}_o)$ is now

(13.2.1) $$\kappa^+(\cdot,P_\theta) = P_\theta(\kappa^\cdot(\cdot,P_\theta)f(\cdot,\theta))'B(\theta)f(\cdot,\theta)$$

with the $m \times m$ matrix $B(\theta)$ being the inverse of

$$(P_\theta(f_\alpha(\cdot,\theta)f_\beta(\cdot,\theta)))_{\alpha,\beta=1,\dots,m}$$
$$= (c_{\alpha i}(\theta)c_{\beta j}(\theta)L_{i,j}(\theta))_{\alpha,\beta=1,\dots,m}$$

The corresponding variance bound is

(13.2.2) $$P_\theta(\kappa^+(\cdot,P_\theta)^2) = P_\theta(\kappa^\cdot(\cdot,P_\theta)f(\cdot,\theta))'B(\theta)P_\theta(\kappa^\cdot(\cdot,P_\theta)f(\cdot,\theta)).$$

The problem of how to obtain as. efficient estimators for κ will be discussed in Section 13.3. We conclude the present section with two examples illustrating that the restriction to a parametric subfamily does not necessarily lead to lower as. variance bounds.

<u>13.2.3. Example</u>. Let $\mathfrak{P} = \{P_{(\theta_1,\theta_2)}: (\theta_1,\theta_2) \in \Theta\}$, and let $\mathfrak{P}_o = \{P_{(\theta_1,\theta_2^o)}: (\theta_1,\theta_2^o) \in \Theta\}$, where θ_2^o is fixed. Our problem is to estimate θ_1, or, more formally, the functional $\kappa(P_{(\theta_1,\theta_2)}) = \theta_1$. We have $\kappa^\cdot(\cdot,P_{(\theta_1,\theta_2)}) = \Lambda_{1i}(\theta_1,\theta_2)\ell^{(i)}(\cdot,\theta_1,\theta_2)$. If the parameters θ_1,θ_2 are unrelated, i.e. $L_{1,2}(\theta_1,\theta_2) = 0$, then $\Lambda_{12}(\theta_1,\theta_2) = 0$ and $\kappa^\cdot(\cdot,P_{(\theta_1,\theta_2)}) = \Lambda_{11}(\theta_1,\theta_2)\cdot\ell^{(1)}(\cdot,\theta_1,\theta_2)$. Since this gradient belongs to $T(P_{(\theta_1,\theta_2^o)},\mathfrak{P}_o) = [\ell^{(1)}(\cdot,\theta_1,\theta_2^o)]$, the knowledge of the value θ_2^o does not help to find an as. better estimator for θ_1.

<u>13.2.4. Example</u>. Let $\mathfrak{P} = \{N(\mu_1,\mu_2,\sigma_1^2,\sigma_2^2,\rho): \mu_i \in \mathbb{R},\ \sigma_i^2 > 0,\ \rho\in(-1,1)\}$, and $\mathfrak{P}_o = \{N(\mu_1,\mu_2,\sigma^2,\sigma^2,\rho): \mu_i \in \mathbb{R},\ \sigma^2 > 0,\ \rho \in (-1,1)\}$. Consider the problem of estimating ρ, i.e. the functional $\kappa(N(\mu_1,\mu_2,\sigma_1^2,\sigma_2^2,\rho)) = \rho$. We have

$$T(N(\mu_1,\mu_2,\sigma_1^2,\sigma_2^2,\rho),\mathfrak{P}) = \{(x_1,x_2) \to a_1 \frac{x_1-\mu_1}{\sigma_1} + a_2 \frac{x_2-\mu_2}{\sigma_2} + b_1\Big(\frac{(x_1-\mu_1)^2}{\sigma_1^2} - 1\Big) + b_2\Big(\frac{(x_2-\mu_2)^2}{\sigma_2^2} - 1\Big) + c\Big(\frac{x_1-\mu_1}{\sigma_1}\frac{x_2-\mu_2}{\sigma_2} - \rho\Big): (a_1,a_2,b_1,b_2,c) \in \mathbb{R}^5\}$$

and

$$T(N(\mu_1,\mu_2,\sigma^2,\sigma^2,\rho),\mathfrak{P}_o) = \{(x_1,x_2) \to a_1 \frac{x_1-\mu_1}{\sigma} + a_2 \frac{x_2-\mu_2}{\sigma} + b\Big(\frac{(x_1-\mu_1)^2}{\sigma^2} + \frac{(x_2-\mu_2)^2}{\sigma^2} - 2\Big) + c\Big(\frac{(x_1-\mu_1)(x_2-\mu_2)}{\sigma^2} - \rho\Big): (a_1,a_2,b,c)\in\mathbb{R}^4\}.$$

The gradient of κ in $T(N(\mu_1,\mu_2,\sigma_1^2,\sigma_2^2,\rho),\mathfrak{P})$ is

$$\kappa^\cdot(x_1,x_2;\mu_1,\mu_2,\sigma_1^2,\sigma_2^2,\rho) = -\frac{\rho}{2}\Big(\frac{(x_1-\mu_1)^2}{\sigma_1^2} + \frac{(x_2-\mu_2)^2}{\sigma_2^2} - 2\Big) + \frac{x_1-\mu_1}{\sigma_1}\frac{x_2-\mu_2}{\sigma_2} - \rho.$$

An as. efficient estimator for ρ is the sample correlation coefficient. Since $\kappa^\cdot(\cdot;\mu_1,\mu_2,\sigma^2,\sigma^2,\rho)$ belongs not only to $T(N(\mu_1,\mu_2,\sigma^2,\sigma^2,\rho),\mathfrak{P})$, but even to the subspace $T(N(\mu_1,\mu_2,\sigma^2,\sigma^2,\rho),\mathfrak{P}_o)$, the knowledge that

$\sigma_1^2 = \sigma_2^2$ does not allow to obtain estimators for ρ which are as. better than the sample correlation coefficient. (There exist estimators which are superior, of course, but the improvement is at the deficiency level only.) An analogous result holds for tests on ρ (see Example 8.6.8).

13.3. Asymptotically efficient estimators for parametric subfamilies

Section 13.2 contains bounds for the as. variance of estimators in parametric subfamilies. In this section we discuss the possibility of obtaining estimator-sequences attaining these bounds.

If an estimator-sequence θ^n for $\mathfrak{P}$ is available, it is tempting to use P_{θ^n} as an initial estimator and to apply the improvement procedure suggested in 11.4.1 to obtain an as. efficient estimator for κ on $\mathfrak{P}_o$, say

$$(13.3.1)\qquad \kappa^n(\underline{x}) = \kappa(P_{\theta^n(\underline{x})}) + n^{-1} \sum_{\nu=1}^{n} \kappa^+(x_\nu, P_{\theta^n(\underline{x})}) ,$$

where κ^+ is the canonical gradient of κ in $\mathfrak{P}_o$, given by (13.2.1). Regrettably, this idea fails because $P_{\theta^n(\underline{x})}$ is not an element of the subfamily $\mathfrak{P}_o$.

In the following we discuss the possibility of obtaining as. efficient estimators for $\mathfrak{P}_o$ by the projection method. Instead of applying Theorem 10.4.8, it seems preferable to use the idea of projection as a heuristic principle, to allow slight modifications if technically convenient, and to check afterwards whether the resulting estimator-sequence is as. efficient. From the as. efficient estimators for $\mathfrak{P}_o$, as. efficient estimators for κ on $\mathfrak{P}_o$ are obtained immediately (see Theorem 11.2.1).

At first we focus our attention to the case of a *curved subfamily*, i.e. $\mathfrak{P}_o = \{P_{c(\tau)}: \tau \in T\}$, $T \subset \mathbb{R}^q$.

As. efficient estimators can in this case be obtained by the maximum likelihood method, applied for the parametrization $\tau \to P_{c(\tau)}$. But in certain cases it might be preferable to make use of estimators θ^n already available for the larger family $\mathfrak{P}$, so for instance in the case dealt with in Remark 13.3.4. This can be done by the projection method.

According to Remark 2.2.7, the tangent space of $P_{c(\tau)}$ in $\mathfrak{P}_o$ is spanned by $c_{\alpha j}(\tau)\ell^{(j)}(\cdot,c(\tau))$, $\alpha = 1,\dots,q$, where $c_{\alpha j}(\tau) = (\partial/\partial\tau_\alpha)c_j(\tau)$. By Definition 7.2.1, the projection of $P_{\theta^n(\underline{x})}$ into $\mathfrak{P}_o$ is determined as the solution in τ of

(13.3.2) $$P_{\theta^n(\underline{x})}(c_{\alpha j}(\tau)\ell^{(j)}(\cdot,c(\tau))) = 0 \qquad \text{for } \alpha = 1,\dots,q.$$

Denote this solution by $\tau^n(\underline{x})$. Expanding $\theta \to p(\xi,\theta)$ about $c(\tau^n(\underline{x}))$ we obtain

$$\frac{p(\xi,\theta^n(\underline{x}))}{p(\xi,c(\tau^n(\underline{x})))} = 1 + (\theta_i^n(\underline{x})-c_i(\tau^n(\underline{x})))\ell^{(i)}(\xi,c(\tau^n(\underline{x}))) + o_{c(\tau)}(n^{-1/2}).$$

Hence condition (13.3.2) for the determination of $\tau^n(\underline{x})$ becomes

$$c_{\alpha i}(\tau^n(\underline{x}))L_{i,j}(c(\tau^n(\underline{x})))(\theta_j^n(\underline{x})-c_j(\tau^n(\underline{x})))$$
$$= o_{c(\tau)}(n^{-1/2}) \qquad \text{for } \alpha = 1,\dots,q.$$

A system of equations for τ as. equivalent to (13.3.2) is, therefore

$$c_{\alpha i}(\tau)L_{i,j}(c(\tau))(\theta_j^n(\underline{x})-c_j(\tau)) = 0 \quad \text{for } \alpha = 1,\dots,q,$$

or, technically more convenient,

(13.3.3) $$c_{\alpha i}(\tau)L_{i,j}(\theta^n(\underline{x}))(\theta_j^n(\underline{x})-c_j(\tau)) = 0 \text{ for } \alpha = 1,\dots,q.$$

From this we easily obtain

(13.3.4) $$\tau_\alpha^n-\tau_\alpha = B_{\alpha\beta}(\tau)c_{\beta j}(\tau)L_{j,i}(c(\tau))(\theta_i^n-c_i(\tau)) + o_{c(\tau)}(n^{-1/2}),$$
$$\alpha = 1,\dots,q,$$

where the $q\times q$-matrix B is the inverse of $(c_{\alpha i}c_{\beta j}L_{i,j})_{\alpha,\beta=1,\dots,q}$. Since $P^n_{c(\tau)}*n^{1/2}(\theta^n-c(\tau)) \Rightarrow N(0,\Lambda(c(\tau)))$, this implies that

$P^n_{c(\tau)} * n^{1/2}(\tau^n - \tau) \Rightarrow N(0, B(\tau))$. (Hint: $BCL\Lambda(BCL)' = BCLC'B = B$.)

It remains to be shown that the covariance matrix B is minimal. This, however, follows immediately from

$$P_{c(\tau)}\left(\frac{\partial}{\partial\tau_\alpha}\ell(\cdot, c(\tau))\frac{\partial}{\partial\tau_\beta}\ell(\cdot, c(\tau))\right) = c_{\alpha i}(\tau)c_{\beta j}(\tau)L_{i,j}(c(\tau)),$$

which is the inverse of B.

13.3.5. Remark. The model of a curved subfamily comprises in particular the case that some parameters become known: Let $\theta = (\theta_1, \ldots, \theta_q, \theta_{q+1}, \ldots, \theta_k)$. If $\mathfrak{P}_0$ is the subfamily of all P_θ with $\theta_i = \theta_i^o$ for $i = q+1, \ldots, k$, let T be the section of Θ at $(\theta^o_{q+1}, \ldots, \theta^o_k)$, and define

$$c_i(\theta_1, \ldots, \theta_q) = \begin{cases} \theta_i & \text{for } i = 1, \ldots, q \\ \theta_i^o & \text{for } i = q+1, \ldots, k. \end{cases}$$

In this case it is tempting to use the estimator $\underline{\theta}^n(\underline{x}) = (\theta^n_1(\underline{x}), \ldots, \theta^n_q(\underline{x}), \theta^o_{q+1}, \ldots, \theta^o_k)$ (if $\theta^n(\underline{x}) = (\theta^n_1(\underline{x}), \ldots, \theta^n_q(\underline{x}), \theta^n_{q+1}(\underline{x}), \ldots, \theta^n_k(\underline{x}))$). This, however, does not solve our problem, since $\underline{\theta}^n$ fails to be as. efficient for $\mathfrak{P}_0$ in the general case, even if θ^n was as. efficient for $\mathfrak{P}$.

(Hint: $P^n_\theta * n^{1/2}(\theta^n_i(\underline{x}) - \theta_i)_{i=1,\ldots,q} \Rightarrow N(0, \underline{\Lambda}(\theta))$, where $\underline{\Lambda} := (\Lambda_{\alpha\beta})_{\alpha,\beta=1,\ldots,q}$, whereas the minimal covariance matrix is in this case the inverse of $(L_{\alpha,\beta})_{\alpha,\beta=1,\ldots,q}$, say Λ^*.)

The obvious reason for this is that $P_{(\theta_1, \ldots, \theta_q, \theta^o_{q+1}, \ldots, \theta^o_k)}$ is n o t the projection of $P_{(\theta_1, \ldots, \theta_q, \theta_{q+1}, \ldots, \theta_k)}$ into $\mathfrak{P}_0$.

As. efficient estimators $(\underline{\theta}^n_1, \ldots, \underline{\theta}^n_q)$ can be obtained from (13.3.3) which leads in this case to

(13.3.6) $\underline{\theta}^n_\alpha(\underline{x}) = \theta^n_\alpha(\underline{x}) + \Lambda^*_{\alpha\beta}(\hat{\theta}^n(\underline{x}))L_{\beta,a}(\hat{\theta}^n(\underline{x}))(\theta^n_a(\underline{x}) - \theta^o_a)$, $\alpha = 1, \ldots, q$,

where the summation over β extends from 1 to q, the summation over a from q+1 to k. For $\hat{\theta}^n(\underline{x})$ we may either choose $\theta^n(\underline{x})$ or $(\theta^n_1(\underline{x}), \ldots, \theta^n_q(\underline{x}), \theta^o_{q+1}, \ldots, \theta^o_k)$.

Since $\underline{\theta}^n_\alpha(\underline{x})-\theta_\alpha = \Lambda^*_{\alpha\beta}(\theta)L_{\beta,j}(\theta)(\theta^n_j(\underline{x})-\theta_j) + o_\theta(n^{-1/2})$, $\alpha = 1,\dots,q$ (with $\theta_j = \theta^o_j$ for $j = q+1,\dots,k$), it follows immediately that

$$P^n_\theta * n^{1/2}(\underline{\theta}^n_\alpha - \theta_\alpha)_{\alpha=1,\dots,q} \Rightarrow N(0,\Lambda^*(\theta)).$$

Now we consider briefly the case of a subfamily $\mathfrak{P}_o$ determined by *side conditions*, i.e. $\mathfrak{P}_o = \{P_\theta: \theta \in \Theta,\ F(\theta) = 0\}$, where $F: \Theta \to \mathbb{R}^m$ is sufficiently regular.

According to Remark 2.2.7, the tangent space of P_θ in $\mathfrak{P}_o$ is

$$T(P_\theta,\mathfrak{P}_o) = \{a'\ell^{(\cdot)}(\cdot,\theta):\ a \in \mathbb{R}^k,\ D(\theta)a = 0\}$$

with $D(\theta) := ((\partial/\partial\theta_j)F_\alpha(\theta))_{\alpha=1,\dots,m;j=1,\dots,k}$. By Definition 7.2.1, the projection of $P_{\theta^n(\underline{x})}$ into $\mathfrak{P}_o$ is determined as the solution in θ of

(13.3.7) $\quad P_{\theta^n(\underline{x})}(a'\ell^{(\cdot)}(\cdot,\theta)) = 0 \qquad$ for all $a \in \mathbb{R}^k$ fulfilling $D(\theta)a = 0$.

Denote this solution by $\underline{\theta}^n(\underline{x})$. Expanding $\theta \to \ell^{(\cdot)}(\cdot,\theta)$ about θ^n we obtain

$$\ell^{(\cdot)}(\xi,\underline{\theta}^n) = \ell^{(\cdot)}(\xi,\theta^n) + \ell^{(\cdot\cdot)}(\xi,\theta^n)(\underline{\theta}^n-\theta^n) + o_\theta(n^{-1/2}).$$

Hence condition (13.3.7) becomes

$$a'L(\theta)(\underline{\theta}^n-\theta^n) = o_\theta(n^{-1/2}) \quad \text{for all } a \in \mathbb{R}^k \text{ fulfilling } D(\underline{\theta}^n)a = 0.$$

This condition is as. equivalent to

(13.3.8) $\quad a'L(\theta^n)(\underline{\theta}^n-\theta^n) = 0 \qquad$ for all $a \in \mathbb{R}^k$ fulfilling $D(\theta^n)a = 0$,

i.e., if a is orthogonal to the rows of $D(\theta^n)$, then a is orthogonal to $L(\theta^n)(\underline{\theta}^n-\theta^n)$, so that (13.3.8) is equivalent to

$$L(\theta^n)(\underline{\theta}^n-\theta^n) = D(\theta^n)'c,$$

in other words,

$$\underline{\theta}^n - \theta^n = \Lambda(\theta^n)D(\theta^n)'c,$$

where $c \in \mathbb{R}^m$ is determined from the side condition $F(\underline{\theta}^n) = 0$. This implies that asymptotically

$$0 = F(\underline{\theta}^n) = F(\theta^n) + D(\theta^n)\Lambda(\theta^n)D(\theta^n)'c .$$

If $D(\theta)\Lambda(\theta)D(\theta)'$ is nonsingular with inverse $B(\theta)$, then

$$c = -B(\theta^n)F(\theta^n),$$

so that asymptotically

$$\underline{\theta}^n = \theta^n - \Lambda(\theta^n)D(\theta^n)'B(\theta^n)F(\theta^n).$$

This relation can be used to determine the as. distribution of $n^{1/2}(\underline{\theta}^n-\theta)$. Since $F(\theta) = 0$, we have $F(\theta^n) = D(\theta)(\theta^n-\theta) + o_\theta(n^{-1/2})$, hence

$$\underline{\theta}^n-\theta = (I - \Lambda(\theta)D'(\theta)B(\theta)D(\theta))(\theta^n-\theta).$$

If θ^n is as. efficient for $\mathfrak{P}$, i.e. $P_\theta^n * n^{1/2}(\theta^n-\theta) \Rightarrow N(0,\Lambda(\theta))$, then $n^{1/2}(\underline{\theta}^n-\theta)$ is as. normal with covariance matrix

(13.3.9) $\quad (I - \Lambda D'BD)\Lambda(I - \Lambda D'BD)' = \Lambda - \Lambda D'BD\Lambda.$

To see that this is the minimal as. covariance matrix, we proceed as follows. $\kappa(P_\theta) := \theta$ has the gradient (see Proposition 5.3.1)

$$\kappa^\cdot(\cdot,P_\theta) = \Lambda(\theta)\ell^{(\cdot)}(\cdot,\theta) .$$

If the family is restricted to $\mathfrak{P}_0$, the tangent space is restricted from $T(P_\theta,\mathfrak{P}) = [\ell^{(j)}(\cdot,\theta)\colon j = 1,\dots,k]$ to $T(P_\theta,\mathfrak{P}_0) = \{a'\ell^{(\cdot)}(\cdot,\theta)\colon a \in \mathbb{R}^k,$ $D(\theta)a = 0\}$. The canonical gradient $\kappa^*(\cdot,P_\theta)$, obtained by projection of $\kappa^\cdot(\cdot,P_\theta)$ into $T(P_\theta,\mathfrak{P}_0)$, is $C(\theta)\ell^{(\cdot)}(\cdot,\theta)$ with $C := \Lambda - \Lambda D'BD\Lambda$. (Hint: Determine C such that $DC' = 0$ and $(\Lambda-C)La = 0$ for all a with $Da = 0$.) From this we obtain the minimal as. covariance matrix

$$P_\theta(\kappa^*(\cdot,P_\theta)\kappa^*(\cdot,P_\theta)') = C(\theta)L(\theta)C(\theta)'$$
$$= \Lambda(\theta)-\Lambda(\theta)D(\theta)'B(\theta)D(\theta)\Lambda(\theta).$$

By (13.3.9), this minimal as. covariance matrix is attained by the estimator-sequence $\underline{\theta}^n$ obtained by projection of an as. efficient estimator-sequence into the subfamily $\mathfrak{P}_0$.

An estimator-sequence with the same as. behavior was obtained by Aitchison and Silvey (1958) by an ad hoc method, using Lagrangian multipliers: They suggest (see p. 814) to determine the estimator for the sample size n as a solution in θ of the following system of equations

$$\sum_{\nu=1}^{n} \ell^{(\cdot)}(x_\nu,\theta) + D(\theta)'\lambda = 0 ,$$
$$F(\theta) = 0 ,$$

and prove that the estimator-sequence thus obtained is as. normal with covariance matrix (13.3.9) (see Aitchison and Silvey, p. 824, Theorem 2). The as. optimality of this estimator-sequence is not discussed by these authors.

Since the rank of D is at most m, the covariance matrix (13.3.9) is singular. We hope that this will not irritate the reader.

14. RANDOM NUISANCE PARAMETERS

14.1. Introduction

Consider a parametric family $\mathfrak{P} = \{P_{\theta,\eta}: (\theta,\eta) \in \Theta \times H\}$ with $\Theta \subset \mathbb{R}^p$ and H arbitrary. We are interested in estimating the (structural) parameter θ. The value of the nuisance parameter η changes from observation to observation, being a random variable, distributed according to some p-measure Γ on $(H,\mathscr{B})$, i.e., the observation x_ν is a realization governed by P_{θ,η_ν}, and η_ν is a realization governed by Γ.

In the following sections we consider first the case that the realizations η_ν are known to the experimenter, second the case that they remain unknown.

The p-measure $Q_{\theta,\Gamma}$ governing the relations (x_ν,η_ν) is uniquely defined on $\mathscr{A}\times\mathscr{B}$ by

$$Q_{\theta,\Gamma}(A\times B) := \int_B P_{\theta,\eta}(A)\,\Gamma(d\eta), \qquad A \in \mathscr{A},\ B \in \mathscr{B}.$$

Let

$$\mathfrak{P}_\Gamma := \{Q_{\theta,\Gamma}: \theta \in \Theta\}, \qquad \mathfrak{Q}_\theta := \{Q_{\theta,\Gamma}: \Gamma \in \mathscr{G}\}.$$

According to Proposition 2.5.2,

$$[(x,\eta) \to \ell^{(i)}(x,\theta,\eta): i = 1,\dots,p] \subset T(Q_{\theta,\Gamma}, \mathfrak{P}_\Gamma),$$

and

$$\{(x,\eta) \to k(\eta): k \in \mathscr{L}_*(\Gamma)\} \subset T(Q_{\theta,\Gamma}, \mathfrak{Q}_\Gamma)$$

if we assume that the family $\mathscr{G}$ is 'full'. Furthermore,

(14.1.1) $$T_o(Q_{\theta,\Gamma},\Omega) := \{(x,\eta) \to a_i \ell^{(i)}(x,\theta,\eta)+k(\eta):\ a \in \mathbb{R}^p,\ k \in {}_*(\Gamma)\} \subset T(Q_{\theta,\Gamma},\Omega)\ .$$

14.2. Estimating a structural parameter in the presence of a known random nuisance parameter

For $i = 1,\dots,p$ let

(14.2.1) $$\kappa_i(Q_{\theta,\Gamma}) := \theta_i\ .$$

This definition presumes that θ is identifiable, i.e. that

$$\theta' \neq \theta'' \text{ implies } Q_{\theta',\Gamma'} \neq Q_{\theta'',\Gamma''} \text{ for all } \Gamma',\Gamma'' \in \mathcal{G}.$$

By Definition 4.1.1 a gradient $\kappa_i^{\bullet}(\cdot,\cdot;\theta,\Gamma)$ of κ_i in $\mathscr{L}_*(Q_{\theta,\Gamma})$ fulfills for every $a \in \mathbb{R}^p$ the relations

$$a_i = t^{-1}(\kappa_i(Q_{\theta+ta,\Gamma}) - \kappa_i(Q_{\theta,\Gamma}))$$
$$= \int \kappa_i^{\bullet}(x,\eta;\theta,\Gamma)a_j \ell^{(j)}(x,\theta,\eta)Q_{\theta,\Gamma}(d(x,\eta)) + o(t^{\circ}),$$

and for every path Γ_t, $t \downarrow 0$, with derivative $k \in \mathscr{L}_*(\Gamma)$ the relation

$$0 = t^{-1}(\kappa_i(Q_{\theta,\Gamma_t}) - \kappa_i(Q_{\theta,\Gamma}))$$
$$= \int \kappa_i^{\bullet}(x,\eta;\theta,\Gamma)k(\eta)Q_{\theta,\Gamma}(d(x,\eta)) + o(t^{\circ})\ .$$

Hence

(14.2.2) $$\int\int \kappa_i^{\bullet}(x,\eta;\theta,\Gamma)\ell^{(j)}(x,\theta,\eta)P_{\theta,\eta}(dx)\Gamma(d\eta) = \delta_{ij} \quad \text{for } j=1,\dots,p,$$

(14.2.3) $$\int \kappa_i^{\bullet}(x,\eta;\theta,\Gamma)P_{\theta,\eta}(dx) = 0 \qquad \text{for } \Gamma\text{-a.a. } \eta \in H.$$

The projection of $\kappa_i^{\bullet}(\cdot,\cdot;\theta,\Gamma)$ into $T(Q_{\theta,\Gamma},\Omega)$ may be written as

$$\kappa_i^*(x,\eta;\theta,\Gamma) = a_{ij}^*\ell^{(j)}(x,\theta,\eta) + k_i^*(\eta)\ .$$

(Since (θ,Γ) remains fixed, we refrain from indicating the dependence of a_{ij}^* and k_i^* on (θ,Γ).)

Relation (14.2.3) implies $k_i^*(\eta) = 0$ for Γ-a.a. $\eta \in H$, and from (14.2.2) we obtain

(14.2.4) $$a_{ij}^* = \Lambda_{ij}(\theta,\Gamma), \qquad i,j = 1,\dots,p,$$

where the matrix $\Lambda(\theta,\Gamma)$ is the inverse of the matrix $L(\theta,\Gamma)$ with elements

(14.2.5) $$L_{i,j}(\theta,\Gamma) = \int \ell^{(i)}(x,\theta,\eta)\ell^{(j)}(x,\theta,\eta)P_{\theta,\eta}(dx)\Gamma(d\eta)\,, \quad i,j = 1,\ldots,p\,.$$

Hence the projection of $\kappa_i^{\cdot}(\cdot,\cdot;\theta,\Gamma)$ into $T_o(Q_{\theta,\Gamma},\mathfrak{Q})$ is

(14.2.6) $$\kappa_i^*(x,\eta;\theta,\Gamma) = \Lambda_{ij}(\theta,\Gamma)\ell^{(j)}(x,\theta,\eta)\,.$$

The conclusion: An as. lower bound for the covariance matrix is $\Lambda(\theta,\Gamma)$.

Let $p(\cdot,\theta,\eta)$ denote a μ-density of $P_{\theta,\eta}$, and assume that $\Gamma \in \mathscr{G}$ has a ν-density γ. Then $Q_{\theta,\Gamma}$ has a $\mu\times\nu$-density $(x,\eta) \to p(x,\theta,\eta)\gamma(\eta)$. From this particular type of density it is obvious how an as. efficient estimator-sequence can be obtained: Since the unknown 'shape' γ occurs only as a factor not depending on θ, an as. efficient estimator for θ can be obtained by the maximum likelihood method, applied to $\theta \to \prod_{\nu=1}^{n} p(x_\nu,\theta,\eta_\nu)$. (For the computation of this estimate, it is irrelevant whether Γ is known or not.) To determine the distribution of this estimator, one has to bear in mind that x_ν is a realization governed by P_{θ,η_ν}, and η_ν a realization governed by Γ. So far, a theorem asserting that the maximum likelihood estimator for θ has, in fact, the minimal as. covariance matrix, is not available. There are, however, results available on the as. distribution of the maximum likelihood estimator obtained from observations x_ν, governed by $P_{\nu,\theta}$, $\nu = 1,\ldots,n$. According to Hoadley (1971, p. 1983, Theorem 2), this distribution is as. normal, its covariance matrix being the inverse of

$$\Big(\lim_{n\to\infty} \frac{1}{n}\sum_{\nu=1}^{n} P_{\nu,\theta}(\ell_\nu^{(i)}(\cdot,\theta)\ell_\nu^{(j)}(\cdot,\theta))\Big)_{i,j=1,\ldots,p}\,,$$

(where $\ell_\nu(\cdot,\theta)$ pertains to $P_{\nu,\theta}$). If we apply this result for $P_{\nu,\theta} = P_{\theta,\eta_\nu}$, the sequence

$$\frac{1}{n}\sum_{\nu=1}^{n} P_{\theta,\eta_\nu}(\ell^{(i)}(\cdot,\eta_\nu,\theta)\ell^{(j)}(\cdot,\eta_\nu,\theta)),\ n \in \mathbb{N},$$

converges with $\Gamma^{\mathbb{N}}$-probability 1 to

$$L_{i,j}(\theta,\Gamma) = \int P_{\theta,\eta}(\ell^{(i)}(\cdot,\theta,\eta)\ell^{(j)}(\cdot,\theta,\eta)\Gamma(d\eta) ,$$

if η_ν, $\nu = 1,\dots,n$, are independent realizations governed by Γ. Hence the sequence of maximum likelihood estimators attains the as. variance bound $\Lambda(\theta,\Gamma)$.

We mention that Hoadley's result was obtained under slightly different regularity conditions by Rogge (1970, p. 67, Satz (4.2)).

<u>14.2.7. Remark</u>. If Γ is known, the space $T_o(Q_{\theta,\Gamma},\mathfrak{Q})$ reduces to $[(x,\eta) \to \ell^{(i)}(x,\theta,\eta): i = 1,\dots,p]$. Since κ_i^* belongs to this restricted space, the knowledge of Γ does not help to reduce the as. lower bound for the covariance matrix (see Section 9.5). Hence estimators which are as. optimal for unknown Γ remain as. optimal if Γ is known. The knowledge of Γ does not help to improve the estimators for θ. (This is intuitively plausible. The realizations η_ν, $\nu = 1,\dots,n$, contain enough information about Γ, so that the knowledge of the true Γ is dispensable.)

14.3. Estimating a structural parameter in the presence of an unknown random nuisance parameter

If the realizations η_ν, $\nu = 1,\dots,n$, of the nuisance parameter remain unknown, then the observations, consisting of x_ν, $\nu = 1,\dots,n$, are governed by $Q'_{\theta,\Gamma}$, the first marginal of $Q_{\theta,\Gamma}$, defined as $Q'_{\theta,\Gamma} := Q_{\theta,\Gamma} * \pi_1$, with $\pi_1(x,\eta) := x$. Let $\mathfrak{Q}' := \{Q'_{\theta,\Gamma}: \theta \in \Theta, \Gamma \in \mathscr{G}\}$,

(14.3.1) $$\bar{\ell}^{(i)}(x) := \int \ell^{(i)}(x,\theta,\eta)p(x,\theta,\eta)\Gamma(d\eta)/\int p(x,\theta,\eta)\Gamma(d\eta) ,$$

(14.3.2) $$\mathscr{K} := \{x \to \int k(\eta)p(x,\theta,\eta)\Gamma(d\eta)/\int p(x,\theta,\eta)\Gamma(d\eta): k \in \mathscr{L}_*(\Gamma)\}.$$

According to Proposition 2.5.2,

$$T_o(Q'_{\theta,\Gamma},\mathfrak{Q}') := \{a_i\bar{\ell}^{(i)} + \bar{k}: a \in \mathbb{R}^p, \bar{k} \in \mathscr{K}\} \subset T(Q'_{\theta,\Gamma},\mathfrak{Q}').$$

Assuming the stronger identifiability condition

$$\theta' \neq \theta'' \text{ implies } Q'_{\theta',\Gamma'} \neq Q'_{\theta'',\Gamma''} \text{ for all } \Gamma',\Gamma'' \in \mathscr{G} ,$$

we may consider $\kappa_i(Q_{\theta,\Gamma}) := \theta_i$ also as a functional on $\mathfrak{Q}' = \{Q'_{\theta,\Gamma} : \theta \in \Theta, \Gamma \in \mathscr{G}\}$. A gradient of κ_i in $\mathscr{L}_*(Q'_{\theta,\Gamma})$, say $\bar{\kappa}_i(x,\theta,\Gamma)$, can be obtained from

$$a_i = t^{-1}(\kappa_i(Q'_{\theta+at,\Gamma}) - \kappa_i(Q'_{\theta,\Gamma}))$$
$$= \int \bar{\kappa}_i(x,\theta,\Gamma) a_j \bar{\ell}^{(j)}(x) Q'_{\theta,\Gamma}(dx) + o(t^o) ,$$
$$o = t^{-1}(\kappa_i(Q'_{\theta,\Gamma_t}) - \kappa_i(Q'_{\theta,\Gamma}))$$
$$= \int \bar{\kappa}_i(x,\theta,\Gamma) \bar{k}(x) Q'_{\theta,\Gamma}(dx) + o(t^o) ,$$

where $a \in \mathbb{R}^p$ and Γ_t, $t \downarrow 0$, is a path with derivative k, and

$$\bar{k}(x) := \int k(\eta) p(x,\theta,\eta) \Gamma(d\eta) / \int p(x,\theta,\eta) \Gamma(d\eta) .$$

This implies

(14.3.3) $\quad \int \bar{\kappa}_i(x,\theta,\Gamma) \bar{\ell}^{(j)}(x) Q'_{\theta,\Gamma}(dx) \qquad$ for $j = 1,\ldots,p$,

(14.3.4) $\quad \int \bar{\kappa}_i(x,\theta,\Gamma) \bar{k}(x) Q'_{\theta,\Gamma}(dx) = 0 \qquad$ for $\bar{k} \in \mathscr{K}$.

If we apply (14.3.3) and (14.3.4) for the projection of $\bar{\kappa}_i$ into $T_o(Q'_{\theta,\Gamma},\mathfrak{Q}')$, say

(14.3.5) $\quad \hat{\kappa}_i(\cdot,\theta,\Gamma) = a^o_{ij} \bar{\ell}^{(j)} + \bar{k}_o$

with $\bar{k}_o \in \mathscr{K}$, we obtain

(14.3.6) $\quad \int (a^o_{ir} \bar{\ell}^{(r)}(x) + \bar{k}_o(x)) \bar{\ell}^{(j)}(x) Q'_{\theta,\Gamma}(dx) = 0 \qquad$ for $j = 1,\ldots,p$,

(14.3.7) $\quad \int (a^o_{ir} \bar{\ell}^{(r)}(x) + \bar{k}_o(x)) \bar{k}(x) Q'_{\theta,\Gamma}(dx) = 0 \qquad$ for $\bar{k} \in \mathscr{K}$.

(14.3.7) implies that $\bar{k}_o$ is the projection of $-a^o_{ir} \bar{\ell}^{(r)}$ into $\mathscr{K}$, i.e.,

$$\bar{k}_o = -a^o_{ir} \hat{\ell}^{(r)} \qquad Q'_{\theta,\eta}\text{-a.e.},$$

where $\hat{\ell}^{(r)}$ denotes the projection of $\bar{\ell}^{(r)}$ into $\mathscr{K}$. Hence (14.3.6) implies

(14.3.8) $\quad a^o_{ir} Q'_{\theta,\Gamma}((\bar{\ell}^{(r)} - \hat{\ell}^{(r)}) \bar{\ell}^{(j)}) = \delta_{ij} \qquad$ for $j = 1,\ldots,p$.

Since $\hat{\ell}^{(r)}$ is the projection of $\bar{\ell}^{(r)}$ into $\mathscr{K}$, we have

$Q'_{\theta,\Gamma}((\bar{\ell}^{(r)} - \hat{\ell}^{(r)})\hat{\ell}^{(j)}) = 0$ for $j = 1,\dots,p$, so that (14.3.8) may be rewritten as

$$a^o_{ir} D_{rj} = \delta_{ij}, \qquad j = 1,\dots,p, \tag{14.3.9}$$

with

$$D_{rj} = Q'_{\theta,\Gamma}((\bar{\ell}^{(r)}-\hat{\ell}^{(r)})(\bar{\ell}^{(j)}-\hat{\ell}^{(j)})), \quad r,j = 1,\dots,p. \tag{14.3.10}$$

Since the relations (14.3.9) hold for $i = 1,\dots,p$, $A^o = (a^o_{ir})_{i,r=1,\dots,p}$ is the inverse of the matrix $D = (D_{rj})_{r,j=1,\dots,p}$.

By (14.3.5),

$$\hat{\kappa}(\cdot,\theta,\Gamma) = D^{-1}(\bar{\ell}^{(\cdot)} - \hat{\ell}^{(\cdot)}),$$

and therefore

$$Q'_{\theta,T}(\hat{\kappa}(\cdot,\theta,\Gamma)\hat{\kappa}(\cdot,\theta,\Gamma)') = D^{-1}.$$

To summarize: D^{-1} is an as. lower bound for the covariance matrix.

This presumes that D is nonsingular. In particular, it excludes cases where the functions $\bar{\ell}^{(i)}$, $i = 1,\dots,p$, are linearly dependent, or where one of these functions belongs to $\mathscr{K}$.

If Γ is known (i.e. $k \equiv 0$ throughout our considerations), we obtain the inverse of $(Q'_{\theta,\Gamma}(\bar{\ell}^{(\cdot)}\bar{\ell}^{(\cdot)\prime}))$ as an as. bound for the covariance matrix. In general, this bound will be smaller, excepting the case that $\bar{\ell}^{(i)}$ is orthogonal to $\mathscr{K}$ for $i = 1,\dots,p$.

With $\bar{\ell}^{(i)}$ being conditional expectations, and $\hat{\ell}^{(i)}$ projections of these conditional expectations, the as. variance bound D^{-1}, with D given by (14.3.10), is not very transparent. We therefore consider in the following a more special model where D can be computed explicitly, and where estimator-sequences attaining D are known.

<u>Special model</u>. Let $\Theta \subset \mathbb{R}^p$ and $H \subset \mathbb{R}^q$. Moreover, there exists a statistic $T|X$, with values in an arbitrary set, which is complete and sufficient for η in the following sense: For each $\theta \in \Theta$,

(i) T is sufficient for the family $\{P_{\theta,\eta}: \eta \in H\}$,

(ii) the family of induced measures $\{P_{\theta,\eta}*T: \eta \in U\}$ is complete for any set $U \subset H$ of positive Lebesgue measure.

Under these assumptions there exists a factorization

(14.3.11) $$p(x,\theta,\eta) = q(x,\theta)p_o(T(x),\theta,\eta) .$$

Throughout the following we shall assume that $p_o(\cdot,\theta,\eta)$ is a density of $P_{\theta,\eta}*T$ with respect to an appropriate measure ν. If a factorization (14.3.11) exists, this can always be assumed w.l.g.

Let $q_o(\cdot,\theta)$ be a conditional expectation of $q(\cdot,\theta)$, given T, with respect to μ. Then another factorization is $p(x,\theta,\eta)=\bar{q}(x,\theta)\bar{p}_o(T(x),\theta,\eta)$, with $\bar{q}(x,\theta) = q(x,\theta)/q_o(T(x),\theta)$ and $\bar{p}_o(t,\theta,\eta) = q_o(t,\theta)p_o(t,\theta,\eta)$. Since $\bar{p}_o(\cdot,\theta,\eta)$ is a conditional expectation of $p(\cdot,\theta,\eta)$, given T, with respect to μ, it is a density of $P_{\theta,\eta}*T$ with respect to $\mu*T$. In particular examples it will, however, be more convenient to work with Lebesgue densities rather than $\mu*T$-densities. Hence we consider the more general case that $p_o(\cdot,\theta,\eta)$ is the density of $P_{\theta,\eta}*T$ with respect to some arbitrary dominating measure.

14.3.12. Lemma. *For any function* $f \in \mathscr{L}_*(Q'_{\theta,\Gamma})$, *the projection into* $\mathscr{K}$ *is a conditional expectation of* f, *given* T, *with respect to* $P_{\theta,\eta}$.

Proof. Every function $\bar{k} \in \mathscr{K}$ is of the form

$$\bar{k}(x) = \frac{\int k(\eta)q(x,\theta)p_o(T(x),\theta,\eta)\Gamma(d\eta)}{\int q(x,\theta)p_o(T(x),\theta,\eta)\Gamma(d\eta)} \qquad \text{with } k \in \mathscr{L}_*(\Gamma).$$

Since

$$\bar{k}(x) = \frac{\int k(\eta)p_o(T(x),\theta,\eta)\Gamma(d\eta)}{\int p_o(T(x),\theta,\eta)\Gamma(d\eta)} ,$$

all functions in $\mathscr{K}$ are contractions of T. Hence the projection of f into $\mathscr{K}$ can be expressed as a function of T, say $f_o \circ T$. (Since (θ,Γ)

remains fixed, we refrain from indicating the dependence of f_o on (θ,Γ).) By definition of the projection, we have for all $k \in \mathscr{L}_*(\Gamma)$,

$$\begin{aligned} 0 &= \int (f(x) - f_o(T(x)))\bar{k}(x) Q'_{\theta,\Gamma}(dx) \\ &= \int (f(x) - f_o(T(x)))(\int k(\eta)p(x,\theta,\eta)\Gamma(d\eta))\mu(dx) \\ &= \int (\int (f(x) - f_o(T(x)))P_{\theta,\eta}(dx))k(\eta)\Gamma(d\eta) , \end{aligned}$$

so that

$$(14.3.13) \quad \int (f(x) - f_o(T(x)))P_{\theta,\eta}(dx) = 0 \qquad \text{for } \Gamma\text{-a.a. } \eta \in H.$$

Since T is sufficient for η, there exists a version of the conditional expectation of f, given T, with respect to $P_{\theta,\eta}$, which does not depend on η, say $f_* \circ T$. Hence (14.3.13) implies

$$P_{\theta,\eta} * T(f_* - f_o) = 0 \qquad \text{for } \Gamma\text{-a.a. } \eta \in H,$$

and therefore on some open set, say U. Together with the assumption that $\{P_{\theta,\eta}*T\colon \eta \in U\}$ is complete for any open set U, this implies that $f_o = f_*$ $P_{\theta,\eta}*T$-a.e. Hence the projection $f_o \circ T$ is a conditional expectation, given T, with respect to $P_{\theta,\eta}$ for any $\eta \in H$.

Moreover, we have (see (14.3.1))

$$\begin{aligned} (14.3.14) \quad \bar{\ell}^{(i)}(x) &= \int \ell^{(i)}(x,\theta,\eta)p(x,\theta,\eta)\Gamma(d\eta)/\int p(x,\theta,\eta)\Gamma(d\eta) \\ &= q^{(i)}(x,\theta)/q(x,\theta) + d_i(T(x),\theta,\Gamma) \end{aligned}$$

with

$$(14.3.15) \quad d_i(t) := \int p_o^{(i)}(t,\theta,\eta)\Gamma(d\eta)/\int p_o(t,\theta,\eta)\Gamma(d\eta) .$$

Assuming that the order between integration with respect to ν and differentiation with respect to θ_j is interchangeable, we obtain from $\nu(p_o(\cdot,\theta,\eta)) = 1$ that

$$(14.3.16) \quad P_{\theta,\eta} * T\left(\frac{p_o^{(j)}(\cdot,\theta,\eta)}{p_o(\cdot,\theta,\eta)}\right) = 0 .$$

Assuming that the order between integration with respect to μ and differentiation with respect to θ_j is interchangeable, we obtain from

$\mu(q(\cdot,\theta)p_o(T(\cdot),\theta,\eta)) = 1$ that

$$(14.3.17) \qquad P_{\theta,\eta}\Big(\frac{q^{(j)}(\cdot,\theta)}{q(\cdot,\theta)} - \frac{p_o^{(j)}(T(\cdot),\theta,\eta)}{p_o(T(\cdot),\theta,\eta)}\Big) = 0 \quad \text{for all } \theta\in\Theta,\ \eta\in H.$$

Together with (14.3.16) this implies

$$(14.3.18) \qquad P_{\theta,\eta}\Big(\frac{q^{(j)}(\cdot,\theta)}{q(\cdot,\theta)}\Big) = 0 \qquad \text{for all } \theta\in\Theta,\ \eta\in H.$$

Using the completeness of $P_{\theta,\eta}*T$, $\eta\in H$, we obtain from (14.3.18) that the conditional expectation of $q^{(j)}(\cdot,\theta)/q(\cdot,\theta)$, given T, with respect to $P_{\theta,\eta}$ equals zero $P_{\theta,\eta}$-a.e. (since, by sufficiency of T, this conditional expectation may be chosen independent of η). Hence we obtain from (14.3.14) for $\hat{\ell}^{(i)}$ (the projection of $\bar{\ell}^{(i)}$ into $\mathscr{K}$),

$$(14.3.19) \qquad \hat{\ell}^{(i)}(x) = d_i(T(x),\theta,\Gamma)$$

and therefore

$$\bar{\ell}^{(i)}(x) - \hat{\ell}^{(i)}(x) = q^{(i)}(x,\theta)/q(x,\theta)\ .$$

Hence by (14.3.10) we have

$$(14.3.20) \qquad D = \Big(Q_{\theta,\Gamma}\Big(\frac{q^{(i)}(\cdot,\theta)}{q(\cdot,\theta)}\ \frac{q^{(j)}(\cdot,\theta)}{q(\cdot,\theta)}\Big)\Big)_{i,j=1,\dots,p}\ .$$

According to the general results obtained above, D^{-1} is an as. bound for the covariance matrices. It remains to show that estimator-sequences attaining this bound exist. For this purpose, we can resort to a paper by Andersen (1970). In this paper it is shown that *conditional maximum likelihood* estimators for θ, defined as solutions of

$$\prod_{\nu=1}^{n} q(x_\nu,\theta) = \max\ ,$$

are asymptotically normal with a covariance matrix equal to D^{-1} (see p. 292, Theorem 4, where Andersen's $\varphi(\cdot|\theta,T(\cdot))$ corresponds to our $q(\cdot,\theta)$).

To be more precise, Andersen's theorems refer to a model in which the nuisance parameter η_ν underlying the observation x_ν is some u n k n o w n constant, changing from observation to observation. But his assumptions placed upon the sequence η_ν, $\nu = 1,\ldots,n$, are fulfilled in a stochastic sense if the values η_ν are independent realizations governed by a fixed p-measure Γ. (See also Andersen's Assumption 1.5", p. 290.)

In Sections 7 and 8, Andersen discusses the problem whether conditional maximum likelihood estimators are asymptotically efficient. In this discussion, Andersen restricts himself to comparing the as. variance of these estimators (for $\Theta \subset \mathbb{R}$ and $H \subset \mathbb{R}$) to some Cramêr-Rao bound. Besides the fact that the special kind of this bound presumed in (16) and (17) needs to be justified, the method as such is of limited applicability, since Cramêr-Rao bounds are not sharp in general. Hence Andersen obtains a conclusive result only for certain special cases.

For the case of a random nuisance parameter, the general result obtained above implies the as. optimality of the conditional maximum likelihood estimators under the general conditions of Andersen's model.

Of course, conditional maximum likelihood estimators may not be optimal any more if more informations about the possible 'prior' distributions Γ are available. If Γ is known to belong to a certain parametric family, for instance, this reduces the tangent space of this family at Γ and therefore the tangent space of the family $\mathfrak{Q}$ at $Q_{\theta,\Gamma}$ drastically, thus leading to a lower as. variance bound. But then the model is simply parametric, and to obtain as. optimal estimators becomes routine.

<u>14.3.21. Example</u>. Assume that (x_ν, y_ν) is a realization governed by $N(\mu_\nu, \sigma^2)^2$, where μ_ν is a realization governed by some unknown p-measure Γ. The problem is to estimate σ^2. The λ^2-density of $N(\mu,\sigma^2)^2$ admits a representation (14.3.11) by

$$q(x,y,\sigma^2)p_o(T(x,y),\sigma^2,\mu)$$

with

$$q(x,y,\sigma^2) = \frac{1}{\sqrt{\pi}\sigma} \exp[-(x-y)^2/4\sigma^2] ,$$

$$p_o(t,\sigma^2,\mu) = \frac{1}{\sqrt{\pi}\ 2\sigma} \exp[-(t-2\mu)^2/4\sigma^2] ,$$

$$T(x,y) = x+y .$$

Notice that $p_o(\cdot,\sigma^2,\mu)$ is the λ-density of $N(\mu,\sigma^2)^2 * T = N(2\mu,2\sigma^2)$. Moreover, the family $\{N(2\mu,2\sigma^2): \mu \in U\}$ is complete for any open set U. Hence the results obtained above apply, and the minimal as. variance for σ^2 can be obtained as the inverse of (14.3.20). We have

$$\frac{q'(x,y,\sigma^2)}{q(x,y,\sigma^2)} = \frac{(x-y)^2}{4\sigma^4} - \frac{1}{2\sigma^2}$$

and therefore

$$\int \left(\frac{q'(x,y,\sigma^2)}{q(x,y,\sigma^2)}\right)^2 N(\mu,\sigma^2)(dx)N(\mu,\sigma^2)(dy) = \frac{1}{2\sigma^4} .$$

Since this value does not depend on μ, integration over μ with respect to Γ ceases, so that $2\sigma^4$ is the as. variance bound for estimators of σ^2, irrespective of the value of Γ.

The conditional maximum likelihood estimator for the sample size n is

$$\frac{1}{2n} \sum_{\nu=1}^{n} (x_\nu - y_\nu)^2 .$$

It is easy to check that the as. variance of this estimator-sequence is $2\sigma^4$, i.e., it attains the as. variance bound.

15. INFERENCE FOR SYMMETRIC PROBABILITY MEASURES

15.1. Asymptotic variance bounds for functionals of symmetric distributions

Let $\mathfrak{P}_0$ be the family of all distributions on $\mathbb{B}$ which admit a positive and symmetric Lebesgue density, and $\mathfrak{P} \supset \mathfrak{P}_0$ a full family of distributions with positive Lebesgue density. Let p denote the Lebesgue density of P, $\ell(x,P) := \log p(x)$, and $\ell'(x,P) := (d/dx)\ell(x,P)$.

According to Propositions 2.3.1 and 2.3.3, we have

$$T(P,\mathfrak{P}_0) = [\ell'(\cdot,P)] \oplus \Psi(P) ,$$

where $\Psi(P)$ denotes the class of all functions in $\mathscr{L}_*(P)$ which are symmetric about the same center as P, say $m(P)$. Since $\ell(\cdot,P)$ is symmetric about $m(P)$, and therefore $\ell'(\cdot,P)$ skew-symmetric about $m(P)$, the two components, $[\ell'(\cdot,P)]$ and $\Psi(P)$, are orthogonal.

Let $\kappa: \mathfrak{P} \to \mathbb{R}$ be a differentiable functional and $\kappa^{\cdot}(\cdot,P)$ its gradient in $\mathscr{L}_*(P)$. Since $[\ell'(\cdot,P)]$ and $\Psi(P)$ are orthogonal, the projection of $\kappa^{\cdot}(\cdot,P)$ into $T(P,\mathfrak{P}_0)$ is the sum of the projections into $[\ell'(\cdot,P)]$ and $\Psi(P)$, which are

$$x \to \ell'(x,P)P(\kappa^{\cdot}(\cdot,P)\ell'(\cdot,P))/P(\ell'(\cdot,P)^2)$$

and

$$x \to \tfrac{1}{2}\kappa^{\cdot}(x,P) + \tfrac{1}{2}\kappa^{\cdot}(2m(P) - x,P)$$

(see Propositions 4.2.3 and 4.3.2).

Hence the canonical gradient of κ at $P \in \mathfrak{P}_0$ in $T(P,\mathfrak{P}_0)$ is

(15.1.1) $$\kappa^*(x,P) = \ell'(x,P)P(\kappa^\cdot(\cdot,P)\ell'(\cdot,P))/P(\ell'(\cdot,P)^2) + \frac{1}{2}\kappa^\cdot(x,P) + \frac{1}{2}\kappa^\cdot(2m(P)-x,P).$$

If κ is any location functional (i.e., translation equivariant), we have $\int\kappa^\cdot(\xi,P)\ell'(\xi,P)P(d\xi) \equiv -1$. Hence

(15.1.2) $$\kappa^*(x,P) = -\ell'(x,P)/P(\ell'(\cdot,P)^2) + \frac{1}{2}\kappa^\cdot(x,P) + \frac{1}{2}\kappa^\cdot(2m(P)-x,P).$$

Consider in particular a minimum contrast functional, based on a contrast function $(x,t) \to f(x-t)$ with f symmetric and unimodal (see Section 5.2). Restricted to a family of symmetric unimodal distributions, these functionals become identical with the median. Correspondingly, their canonical gradients in $T(P,\mathfrak{P}_0)$ become identical. (By Remark 5.2.4, the gradient in $\mathscr{L}_*(P)$ is

$$\kappa^\cdot(x,P) = -f'(x-\kappa(P))/\int f'(\xi-\kappa(P))\ell'(\xi,P)P(d\xi),$$

which is skew-symmetric about $\kappa(P)$. Hence (15.1.2) reduces to $\kappa^*(x,P) = -\ell'(x,P)/P(\ell'(\cdot,P)^2)$.)

As another instance consider now the estimation of the β-quantile, say κ_β. By Proposition 5.4.2, the gradient of κ_β in $\mathscr{L}_*(P)$ is

$$\kappa^\cdot_\beta(x,P) = (\beta - 1_{(-\infty,\kappa_\beta(P))}(x))/p(\kappa_\beta(P)).$$

The application of (15.1.2) yields the canonical gradient of κ_β in $T(P,\mathfrak{P}_0)$,

(15.1.3) $$\kappa^*_\beta(x,P) = -\ell'(x,P)/P(\ell'(\cdot,P)^2) + [2\beta-1+1_{(-\infty,\kappa_{1-\beta}(P))}(x) - 1_{(-\infty,\kappa_\beta(P))}(x)]/2p(\kappa_\beta(P))$$

If $T(P,\mathfrak{P}) = \mathscr{L}_*(P)$, then the as. variance bound for κ_β on $\mathfrak{P}$ is

(15.1.4) $$\sigma^2_\beta(P) = P(\kappa^\cdot_\beta(\cdot,P)^2) = \beta(1-\beta)/p(\kappa_\beta(P))^2.$$

This as. variance bound is attained by the β-quantile of the sample. In other words:

If nothing is known about the true p-measure (except that it has a positive Lebesgue density), one cannot do better asymptotically than use the sample quantile for estimating the quantile of the true p-measure.

This is not as obvious as it sounds. One cannot exclude in advance the possibility that better estimators can be obtained if the density is sufficiently smooth (e.g., as quantiles of sufficiently smooth density estimators). Such additional informations about the density of the true p-measure have, in fact, a positive effect on the accuracy of the best possible estimators, but this effect is of smaller order (see Reiss, 1980).

If it is known that the true p-measure belongs to $\mathfrak{P}_o$, then the as. variance bound reduces to

(15.1.5) $$\bar{\sigma}_\beta^2(P) := P(\kappa_\beta^*(\cdot,P)^2) = 1/P(\ell'(\cdot,P)^2) + a(\beta)/p(\kappa_\beta(P))^2$$

with $a(\beta) = |\frac{1}{2} - \beta| \cdot (\frac{1}{2} - |\frac{1}{2} - \beta|)$.

Of particular interest is the case of the median, corresponding to $\beta = \frac{1}{2}$. In this case, (15.1.5) specializes to

(15.1.6) $$\bar{\sigma}_{1/2}^2(P) = 1/P(\ell'(\cdot,P)^2),$$

which equals the as. variance bound for the location parameter family $\mathfrak{P}_1 = \{P*(x \to x+c): c \in \mathbb{R}\}$ generated by P. In other words: If we have to estimate the median of a symmetric distribution, then the as. variance bound is not increased by knowing the shape of the distribution.

The reason behind this phenomenon is that $\kappa_{1/2}^*(x,P) = -\ell'(x,P)/P(\ell'(\cdot,P)^2)$, i.e. the canonical gradient of $\kappa_{1/2}$ in $T(P,\mathfrak{P}_o)$, even belongs to the smaller tangent space $T(P,\mathfrak{P}_1)$.

For $\beta \neq \frac{1}{2}$, this is not the case any more. For a β-quantile other than the median the situation presents itself as follows: If we know that the distribution is symmetric, we can find a better estimator than the sample quantile, but this estimator will not be as good as the estimator obtainable if the shape of the symmetric distribution is known.

The situation is totally different for functionals κ for which the gradient $\kappa^{\cdot}(\cdot,P) \in \mathscr{L}_*(P)$ is symmetric about $m(P)$. In this case, (15.1.1) yields $\kappa^*(\cdot,P) = \kappa^{\cdot}(\cdot,P)$, i.e., $\kappa^{\cdot}(\cdot,P)$ is also the canonical gradient in $T(P,\mathfrak{P}_o)$. Hence knowing that the true p-measure is symmetric does not enable us to obtain as. better estimators. Natural examples of functionals with symmetric gradients are measures of dispersion, such as $\kappa(P) := \kappa_{1-\beta}(P) - \kappa_\beta(P)$, or $\kappa(P) := \frac{1}{2}\int k_o(x_1-x_2)P(dx_1)P(dx_2)$, with k_o symmetric about 0 (see Section 5.1).

15.1.7. Remark. The results obtained above suggest a natural example for a function of two inefficient estimators which is as. efficient. For symmetric P,

$$\kappa_{1/2}(P) = \frac{1}{2}\kappa_\beta(P) + \frac{1}{2}\kappa_{1-\beta}(P) .$$

If κ_α^n is an estimator-sequence with as. variance $\bar{\sigma}_\alpha^2(P)$, then $\kappa^n = \frac{1}{2}\kappa_\beta^n + \frac{1}{2}\kappa_{1-\beta}^n$ is an estimator-sequence for $\kappa_{1/2}(P)$ with as. variance $\bar{\sigma}_{1/2}^2(P)$. (This follows immediately from the fact that the as. covariance between $n^{1/2}(\kappa_\beta^n - \kappa_\beta(P))$ and $n^{1/2}(\kappa_{1-\beta}^n - \kappa_{1-\beta}(P))$ is $1/P(\ell'(\cdot,P)^2) - a(\beta)/p(\kappa_\beta(P))^2$.) If p is known up to location, then κ_α^n is inefficient for the α-quantile. In spite of this, the function $\frac{1}{2}\kappa_\beta^n + \frac{1}{2}\kappa_{1-\beta}^n$ of the two inefficient estimators κ_β^n, $\kappa_{1-\beta}^n$, is as. efficient for $\kappa_{1/2}$.

15.2. Asymptotically efficient estimators for functionals of symmetric distributions

In order to obtain an estimator-sequence which is as. efficient for the family of all symmetric distributions on $\mathbb{B}$, we start from an estimator-sequence which is as. efficient for the family of all p-measures which are equivalent to the Lebesgue measure. Let $p_n(\underline{x},\cdot)$ denote the Lebesgue density of such an estimator-sequence. (According to Example 10.2.6, such estimators can be obtained by the kernel method.)

The desired estimator-sequence which is as. efficient for the family of all symmetric distributions can be obtained by projecting $p_n(\underline{x},\cdot)$ into this family. According to Section 7.8, this projection is

(15.2.1) $$\bar{p}_n(\underline{x},\xi) = \tfrac{1}{2}p_n(\underline{x},\xi) + \tfrac{1}{2}p_n(\underline{x},2M_n(\underline{x}) - \xi) ,$$

where M_n, the estimator for the median, is obtained as the solution in M of

(15.2.2) $$\int \frac{p_n(\underline{x},\xi)-p_n(\underline{x},2M-\xi)}{p_n(\underline{x},\xi)+p_n(\underline{x},2M-\xi)}\, p_n'(\underline{x},\xi)d\xi = 0 .$$

Since $\int(p_n(\underline{x},\xi)^2/p(\xi))d\xi$ will not be finite in general, Theorem 10.4.8, establishing the as. efficiency of the projection $\bar{p}_n$, is not applicable here. Hence we indicate in the following how the as. efficiency of $\bar{p}_n$ can be proved directly. To transform this sketch into a precise proof, it may be necessary for technical reasons to take for p_n a modified version of the kernel estimator.

Let p denote the Lebesgue density of the true p-measure, and M its median. At first we shall show that

(15.2.3) $$n^{1/2}(M_n(\underline{x})-M) = -\tilde{\ell}'(\underline{x},P)/P(\ell'(\cdot,P)^2) + o_P(n^o) .$$

(This stochastic expansion implies in particular that the estimator-sequence M_n is as. efficient for the median in the family of all symmetric distributions.)

To simplify our notations, we assume w.l.g. M = O, i.e. that p is symmetric about O. Using the relation

$$p(2M-\xi) = p(\xi) - 2Mp'(\xi) + O(M^2) ,$$

and the notations

$$\Delta_n(\underline{x},\xi) := p_n(\underline{x},\xi)-p(\xi), \quad \Delta_n'(\underline{x},\xi) := p_n'(\underline{x},\xi)-p'(\xi) ,$$

we obtain the following equalities $\doteq$ up to terms of order $o_P(n^{-1/2})$:

$$p_n(\underline{x},\xi)-p_n(\underline{x},2M_n(\underline{x})-\xi) \doteq 2M_n(\underline{x})p'(\xi)+\Delta_n(\underline{x},\xi)-\Delta_n(\underline{x},2M_n(\underline{x})-\xi),$$

$$p_n(\underline{x},\xi)+p_n(\underline{x},2M_n(\underline{x})-\xi) \doteq 2p(\xi)\big(1-M_n(\underline{x})\ell'(\xi,P)+\Delta_n(\underline{x},\xi)/2p(\xi) + \Delta_n(\underline{x},2M_n(\underline{x})-\xi)/2p(\xi)\big),$$

(15.2.4) $$\frac{p_n(\underline{x},\xi)-p_n(\underline{x},2M_n(\underline{x})-\xi)}{p_n(\underline{x},\xi)+p_n(\underline{x},2M_n(\underline{x})-\xi)}\, p_n'(\underline{x},\xi)$$

$$\doteq \frac{M_n(\underline{x})p'(\xi)+\Delta_n(\underline{x},\xi)/2-\Delta_n(\underline{x},2M_n(\underline{x})-\xi)/2}{p(\xi)(1-M_n(\underline{x})\ell'(\xi,P)+[\Delta_n(\underline{x},\xi)+\Delta_n(\underline{x},2M_n(\underline{x})-\xi)]/2p(\xi))}(p'(\xi)+\Delta_n'(\underline{x},\xi))$$

$$\doteq M_n(\underline{x})\ell'(\xi,P)^2p(\xi) + \tfrac{1}{2}\ell'(\xi,P)(\Delta_n(\underline{x},\xi)-\Delta_n(\underline{x},2M_n(\underline{x})-\xi)) + M_n(\underline{x})\ell'(\xi,P)\Delta_n'(\underline{x},\xi).$$

Since p_n is as. efficient (see 10.1.1),

$$\int\ell'(\xi,P)\Delta_n(\underline{x},2M_n(\underline{x})-\xi)d\xi = \int\ell'(2M_n(\underline{x})-\eta,P)\Delta_n(\underline{x},\eta)d\eta$$

$$\doteq -\int\ell'(\eta,P)\Delta_n(\underline{x},\eta)d\eta + 2M_n(\underline{x})\int\ell''(\eta,P)\Delta_n(\underline{x},\eta)d\eta$$

$$\doteq -n^{-1/2}\tilde{\ell}'(\underline{x},P) + 2M_n(\underline{x})n^{-1/2}\tilde{\ell}''(\underline{x},P)$$

and by partial integration

$$\int\ell'(\xi,P)\Delta_n'(\underline{x},\xi)d\xi = -\int\ell''(\xi,P)\Delta_n(\underline{x},\xi)d\xi = o_P(n^{-1/2}) .$$

Together with (15.2.2) and (15.2.4) these relations imply

$$M_n(\underline{x})P(\ell'(\cdot,P)^2) \doteq -n^{-1/2}\tilde{\ell}'(\underline{x},P) ,$$

which proves (15.2.3).

It remains to be shown that any symmetrized density (15.2.1) is as. efficient for the family of all p-measures with symmetric density, provided M_n fulfills (15.2.3). Assuming, as above, that p is symmetric about zero, we have to show that

$$(15.2.5) \qquad n^{1/2}\int f(\xi)p_n(\underline{x},\xi)d\xi = \tilde{f}(\underline{x}) + o_P(n^o)$$

for any $f \in T(P,\mathfrak{P}_o)$, i.e. for any $f = c\ell'(\cdot,P) + \psi$, where ψ is symmetric about 0. This is possible only under additional regularity conditions on ψ.

The consequence: If we use the p-measure with Lebesgue density $\bar{p}_n(\underline{x},\cdot)$, say $\bar{P}_n(\underline{x},\cdot)$, to compute the estimate $\kappa(\bar{P}_n(\underline{x},\cdot))$, this estimate will be as. efficient for the family of all symmetric distributions only if these additional regularity conditions are fulfilled for the symmetric component of the gradient $\kappa^{\cdot}$.

We have

$$(15.2.6) \qquad n^{1/2}\int f(\xi)\bar{p}_n(\underline{x},\xi)d\xi$$

$$= n^{1/2}\int f(\xi)p_n(\underline{x},\xi)d\xi + \tfrac{1}{2}n^{1/2}\int f(\xi)(p_n(\underline{x},2M_n(\underline{x})-\xi)-p_n(\underline{x},\xi))d\xi$$

$$= \tilde{f}(\underline{x}) + \tfrac{1}{2}n^{1/2}\int(f(2M_n(\underline{x})-\xi)-f(\xi))p_n(\underline{x},\xi)d\xi + o_P(n^o) .$$

The case $f = c\ell'(\cdot,P)$ is treated as follows. Appropriate smoothness assumptions on $\ell''(\cdot,P)$ guarantee

$$\int\ell''(\xi,P)p_n(\underline{x},\xi)d\xi = P(\ell''(\cdot,P))+o_P(n^o) = -P(\ell'(\cdot,P)^2)+o_P(n^o).$$

Since

$$\ell'(2M_n(\underline{x})-\xi,P)-\ell'(\xi,P) = -2\ell'(\xi,P) + 2M_n(\underline{x})\ell''(\xi,P) + o_P(n^{-1/2}),$$

we obtain

$$\tfrac{1}{2}n^{1/2}\int(\ell'(2M_n(\underline{x})-\xi,P)-\ell'(\xi,P))p_n(\underline{x},\xi)d\xi$$

$$= -n^{1/2}\int\ell'(\xi,P)p_n(\underline{x},\xi)d\xi + n^{1/2}M_n(\underline{x})\int\ell''(\xi,P)p_n(\underline{x},\xi)d\xi + o_P(n^o)$$

$$= -\tilde{\ell}'(\underline{x},P)-n^{1/2}M_n(\underline{x})P(\ell'(\cdot,P)^2) + o_P(n^o) = o_P(n^o) .$$

The last equality follows from (15.2.3) (with $M = 0$).

The case $f = \psi$, symmetric about $m(P)$, is more delicate, because here smoothness assumptions on ψ rule out certain interesting applications. (Recall, for instance, that the canonical gradient of the β-quantile has, according to (15.1.3), the symmetric component $\psi = 1_{(\kappa_\beta(P),\kappa_{1-\beta}(P))} + 2\beta - 1$ for $\beta \leq 1/2$.)

Because of (15.2.6), applied for $f = \psi$, it remains to be shown that

$$n^{1/2}\int(\psi(2M_n(\underline{x})-\xi)-\psi(\xi))p_n(\underline{x},\xi)d\xi = o_p(n^o) .$$

Let

$$r(x,M) := \psi(2M-x)-\psi(x).$$

To illustrate the basic idea, we first consider M as a real-valued parameter. If P_n, $n \in \mathbb{N}$, is as. efficient, there is a fair chance that

$$n^{1/2}\int r(\xi,M_n)p_n(\underline{x},\xi)d\xi = n^{-1/2}\sum_{\nu=1}^{n} r(x_\nu,M_n) + o_p(n^o) .$$

(Notice that this relation requires slightly more than as. efficiency, since $\xi \to r(\xi,M_n)$ depends on n.) Hence

$$n^{1/2}\int r(\xi,M_n)p_n(\underline{x},\xi)d\xi = o_p(n^o)$$

is in this case equivalent to

(15.2.7) $$n^{-1/2}\sum_{\nu=1}^{n} r(x_\nu,M_n) = o_p(n^o) .$$

Using the symmetry of p and ψ about 0 we obtain

$$\int r(\xi,M)p(\xi)d\xi = \int\psi(\xi)(p(\xi-2M)-p(\xi))d\xi$$

$$= -2M\int\psi(\xi)p'(\xi)d\xi + O(M^2) = O(M^2) .$$

Moreover, Lemma 19.1.4 implies

$$\int r(\xi,M)^2p(\xi)d\xi = o(M^o) .$$

Hence (15.2.7) follows from the degenerate convergence lemma if $M_n = o(n^{-1/4})$.

We have $M_n(\underline{x}) = O_p(n^{-1/2})$. This is more than we need to establish

$$n^{-1/2} \sum_{\nu=1}^{n} r(x_\nu, M_n(\underline{x})) = o_p(n^o) ,$$

except for the stochastic dependence between x_ν and $M_n(\underline{x})$. This problem is the same as that mentioned in Section 11.5, and can, perhaps, be solved along the lines indicated there.

<u>15.2.8. Remark</u>. In the considerations indicated above, we have only used that the estimator M_n fulfills (15.2.3). Hence we could have used for M_n as well the estimator obtained by minimizing the Hellinger distance between $p_n(\underline{x},\xi)$ and $\frac{1}{2}p_n(\underline{x},\xi) + \frac{1}{2}p_n(\underline{x},2M-\xi)$ as a function of M. This is, for instance, carried through in Beran (1978, p. 299, Theorem 3), taking for p_n a modified kernel estimator.

In Remark 10.5.7 we mentioned that projections of minimizations using essentially different distance functions, for instance distance functions based on distribution functions, lead in general to inefficient estimators. B.V.Rao, Schuster and Littell (1975) determine an estimator for the median of a symmetric distribution using the Kolmogorov distance. Starting from the empirical distribution function $t \to G_n(\underline{x},t)$, $\frac{1}{2}G_n(\underline{x},t) + \frac{1}{2}(1- G_n(\underline{x},2M-t))$ is a symmetrized version, having M as its center of symmetry. The difference between the original and the symmetrized distribution function is then $\frac{1}{2}(G_n(\underline{x},t) + G_n(\underline{x},2M-t)-1)$. Minimizing

$$\sup\{|G_n(\underline{x},t) + G_n(\underline{x},2M-t) - 1| : t \in \mathbb{R}\}$$

as a function of M yields an estimator M_n. If the underlying p-measure is symmetric, M_n is a reasonable estimator for the median. It is, however, as. inefficient (see Rao, Schuster and Littell, 1975, p. 866, Theorem 4). A fortiori: The symmetric p-measure pertaining to the distribution function $\frac{1}{2}G_n(\underline{x},t) + \frac{1}{2}(1-G_n(\underline{x},2M(\underline{x})-t))$ is as. inefficient as an estimator for the true p-measure.

15.2.9. Remark. Among the functionals on the family of symmetric p-measures, the median is particularly simple. Its canonical gradient becomes (see (15.1.3) for $\beta = \frac{1}{2}$)

$$\kappa^*_{1/2}(x,P) = -\ell'(x,P)/P(\ell'(\cdot,P)^2) .$$

In this case, the symmetric component of the gradient - which may cause some technical difficulties in the general case - disappears, and the problem of obtaining an as. efficient estimator-sequence becomes particularly simple. Starting from the basic paper of Stein (1956), this problem has attracted the attention of many authors, among them van Eeden (1970), Takeuchi (1971), Beran (1974), Sacks (1975), and Stone (1975). The most satisfactory solution now seems to be Theorem 3 in Beran (1978).

15.3. Symmetry in two-dimensional distributions

Let $\mathfrak{P}$ be the family of all p-measures $P|\mathbb{B}^2$ with symmetric first marginal. Our problem is to estimate the center of symmetry, say $\kappa(P)$.

It seems intuitively clear that the best one can do in this situation is to use the first marginal of the empirical p-measure to obtain an estimator which is as. efficient in the family of all symmetric p-measures (see Sections 13.1 and 13.2). Yet intuition seems to be misleading in this case.

For the following we assume that P admits a positive and continuous λ^2-density, say $(x,y) \to p(x,y)$. Then the first marginal has λ-density $p_1(x) = \int p(x,\eta)d\eta$. Under suitable regularity conditions,

$$T(P,\mathfrak{P}) = [\ell^{(1)}(\cdot,P)] \oplus G(P) ,$$

where G(P) is the class of all functions $g \in \mathscr{L}_*(P)$ such that $x \to \int g(x,\eta)p(x,\eta)d\eta$ is symmetric about $\kappa(P)$, and where $\ell(\cdot,P) := \log p$

and $\ell^{(1)}(x,y,P) := (\partial/\partial x)\ell(x,y,P)$.

Since $\kappa(P)$ is the median of the first marginal of P, κ has by Proposition 5.4.2 the gradient

$$\kappa^{\cdot}(x,y,P) = (\tfrac{1}{2} - 1_{(-\infty,\kappa(P))}(x))/p_1(\kappa(P)) .$$

Since the function $x \to \kappa^{\cdot}(x,y,P)$ is skew-symmetric about $\kappa(P)$, it is orthogonal to $G(P)$. Its projection into $T(P,\mathfrak{P})$ is therefore

$$\kappa^*(x,y,P) = \ell^{(1)}(x,y,P)/P(\ell^{(1)}(\cdot,P)^2) . \tag{15.3.1}$$

Hence the as. variance bound for estimators of κ is

$$\sigma_o(P)^2 := P(\kappa^*(\cdot,P)^2) = 1/P(\ell^{(1)}(\cdot,P)^2) . \tag{15.3.2}$$

The minimal as. variance for estimators based on the first marginal only is (see (15.1.6) and (15.2.9))

$$\sigma(P)^2 = 1/P(\ell_1'(\cdot,P)^2) \tag{15.3.3}$$

with $\ell_1(\cdot,P) = \log p_1$ and $\ell_1'(x,P) = (d/dx)\ell_1(x,P)$.

Since $\sigma_o(P)^2$ is the as. variance bound for estimators of κ, we necessarily have $\sigma_o(P)^2 \le \sigma(P)^2$. In the following this will be checked from the explicit expressions (15.3.2) and (15.3.3), and we show that the inequality is strict unless x and y are independent. By the Schwarz inequality, for every $x \in X$,

$$(\int p^{(1)}(x,\eta)d\eta)^2 \le \int(\frac{p^{(1)}(x,\eta)}{p(x,\eta)})^2 p(x,\eta)d\eta \int p(x,\eta)d\eta .$$

Hence

$$(\frac{p_1'(x)}{p_1(x)})^2 p_1(x) \le \int(\frac{p^{(1)}(x,\eta)}{p(x,\eta)})^2 p(x,\eta)d\eta , \tag{15.3.4}$$

i.e.,

$$\ell_1'(x,P)^2 p_1(x) \le \int \ell^{(1)}(x,\eta,P)^2 p(x,\eta)d\eta .$$

Integration over x leads to $\sigma_o(P)^2 \le \sigma(P)^2$, where the inequality is strict unless equality holds in (15.3.4) for λ-a.a. $x \in \mathbb{R}$. This is possible only if $p^{(1)}(x,\eta)/p(x,\eta) = c(x)$ for λ-a.a. $\eta \in \mathbb{R}$. Since $p^{(1)}$ is continuous, this implies $p(x,y) = A(x)B(y)$, from which independence follows easily.

We remark that knowing p up to the median $\kappa(P)$ does not help to improve the as. variance bound σ_o^2, since the canonical gradient κ^*, given by (15.3.1), even belongs to the tangent space of this one-parameter family. A fortiori, it does not help to know something more about the symmetry of p (for instance that $x \rightarrow p(x,y)$ is symmetric about $\kappa(P)$ for every $y \in \mathbb{R}$, or that $(x,y) \rightarrow p(x,y)$ is centric symmetric or even circular symmetric about a certain point with first component $\kappa(P)$).

Such additional informations can, however, be of great help in obtaining as. efficient estimators. In other words: Estimators using such additional information may be distinctively superior for small sample sizes.

In fact, it is hard to guess how an as. efficient estimator-sequence can be obtained, knowing only that the first marginal is symmetric. Whether we wish to obtain an as. efficient estimator for P, or to apply the improvement procedure to a preliminary estimator for $\kappa(P)$, in any case we need an estimator for the density which belongs to our basic family. It suggests itself to obtain such an estimator by applying some symmetrization procedure to a preliminary density estimator. But how can this be done without knowing more about P than just the symmetry of its first marginal?

16. INFERENCE FOR MEASURES ON PRODUCT SPACES

16.1. Introduction

For $i \in \{1,\dots,m\}$ let $(X_i,\mathscr{A}_i)$ be measurable spaces. In the following, sums Σ and products $\times,\Pi$ over i always run from 1 to m. Let $\mathfrak{P}$ be a family of p-measures on $\times\mathscr{A}_i$, and $\kappa: \mathfrak{P} \to \mathbb{R}$ a functional. Our problem is to estimate $\kappa(P)$ under various conditions on $\mathfrak{P}$.

As an illustration, consider the von Mises functional

$$\kappa(P) := \int k(x_1,\dots,x_m)P(d(x_1,\dots,x_m)) .$$

It suggests itself to estimate κ by replacing P by the empirical p-measure, thus obtaining the estimate

$$\kappa^n(\underline{x}) := \frac{1}{n}\sum_{\nu=1}^{n} k(x_{1\nu},\dots,x_{m\nu}) .$$

If we know that P is a product measure, a reasonable estimate is

$$\bar{\kappa}^n(\underline{x}) := \frac{1}{n^m}\sum_{\nu_1=1}^{n}\dots\sum_{\nu_m=1}^{n} k(x_{1\nu_1},\dots,x_{m\nu_m}) .$$

We know from Remark 11.2.4 that κ^n is as. efficient for κ if $\mathfrak{P}$ is sufficiently large. But is $\bar{\kappa}^n$ as. efficient if $\mathfrak{P}$ is a (large) family of product measures? Do we perhaps have to construct our estimator more carefully? If the components P_i of $P = \times P_i$ have Lebesgue densities, we could, for instance, use the i-th components $x_{i\nu}$, $\nu = 1,\dots,n$, to obtain a density estimator for P_i, say $p_{in}(\underline{x}_i,\cdot)$, and could then construct the estimator

$$\underline{x} \to \int k(\xi_1,\dots,\xi_m)\Pi p_{in}(\underline{x}_i,\xi_i)d\xi_1\dots d\xi_m .$$

Is such a more circumspect procedure perhaps necessary to obtain an as. efficient estimator-sequence? In the following we shall see that the answer is no.

16.2. Variance bounds

Assume that the functional κ is differentiable with gradient $\kappa^{\cdot}(\cdot,P) \in \mathscr{L}_*(P)$. If $T(P,\mathfrak{P}) = \mathscr{L}_*(P)$, then the as. variance bound is

(16.2.1) $\quad \sigma^2(P) := P(\kappa^{\cdot}(\cdot,P)^2)$.

If we know that the true p-measure is a *product* measure, then the as. variance bound will be smaller. To be more formal, let $P = \times Q_i$. Given $T(Q_i,\mathfrak{Q}_i)$, we have (see Proposition 2.4.1)

$$T(\times Q_i,\times\mathfrak{Q}_i) = \{(x_1,\dots,x_m) \to \Sigma g_j(x_j): g_j \in T(Q_j,\mathfrak{Q}_j),\ j = 1,\dots,m\}.$$

The projection of $\kappa^{\cdot}(\cdot,\times Q_i)$ into this tangent space, say

$$\kappa^*(x_1,\dots,x_m,\times Q_i) = \Sigma\bar{g}_j(x_j,\times Q_i) ,$$

is uniquely determined by the relation (see Proposition 4.3.2)

(16.2.2) $\quad \int(\kappa^{\cdot}(x_1,\dots,x_m,\times Q_i) - \Sigma\bar{g}_j(x_j,\times Q_i))g_k(x_k)\times Q_\ell(dx_\ell) = 0$

for all $g_k \in T(Q_k,\mathfrak{Q}_k)$, $k = 1,\dots,m$.

Let $\bar{\kappa}_k$ be defined by

(16.2.3) $\quad \bar{\kappa}_k(x_k,\times Q_i) = \int\kappa^{\cdot}(x_1,\dots,x_m,\times Q_i) \underset{j\neq k}{\times} Q_j(dx_j)$

and let $\kappa_k^{\cdot}(\cdot,\times Q_i)$ be the projection of $\bar{\kappa}_k(\cdot,\times Q_i)$ into $T(Q_k,\mathfrak{Q}_k)$. With these notations, (16.2.2) implies

(16.2.4) $\quad \int(\kappa_k^{\cdot}(\xi,\times Q_i) - \bar{g}_k(\xi,\times Q_i))g_k(\xi)Q_k(d\xi) = 0$

for all $g_k \in T(Q_k,\mathfrak{Q}_k)$, $k = 1,\dots,m$.

This implies

$$\bar{g}_k = \kappa_k^{\cdot} , \qquad k = 1,\dots,m.$$

Hence the canonical gradient is

(16.2.5) $\quad \kappa^*(x_1,\dots,x_m,\times Q_i) = \Sigma\kappa_j^{\cdot}(x_j,\times Q_i)$.

The pertaining lower bound for the as. variance is

(16.2.6) $$\bar{\sigma}^2(\times Q_i) = \Sigma\int \kappa_j^{\bullet}(\xi,\times Q_i)^2 Q_j(d\xi) ,$$

a quantity smaller than $\sigma^2(\times Q_i)$, given by (16.2.1).

The variance bound can be further reduced if $(X_i,\mathscr{A}_i) = (X_1,\mathscr{A}_1)$ for $i = 2,\ldots,m$, and if it is known that the true p-measure is a product of *identical* components, i.e. if $P = Q^m$. Let $\mathfrak{P}_o = \{Q^m: Q \in \mathfrak{Q}\}$. We have (see Proposition 2.4.1)

$$T(Q^m,\mathfrak{P}_o) = \{(x_1,\ldots,x_m) \rightarrow \Sigma g(x_i): g \in T(Q,\mathfrak{Q})\}.$$

The projection of $\kappa^{\bullet}$ (or κ^*) into $T(Q^m,\mathfrak{P}_o)$, say

$$\kappa^{**}(x_1,\ldots,x_m,Q^m) = \Sigma\bar{\bar{g}}(x_i,Q^m),$$

is uniquely determined by (see Proposition 4.3.2)

$$\int(\kappa^*(x_1,\ldots,x_m,Q^m)-\Sigma\bar{\bar{g}}(x_i,Q^m))\Sigma g(x_k)\times Q(dx_\ell) = 0 \quad \text{for all } g \in T(Q,\mathfrak{Q}).$$

From this we obtain

$$\bar{\bar{g}}(x,Q^m) = \frac{1}{m}\Sigma\kappa_j^{\bullet}(x,Q^m) .$$

Hence the canonical gradient is

(16.2.7) $$\kappa^{**}(x_1,\ldots,x_m,Q^m) = \frac{1}{m}\sum_{i=1}^{m}\sum_{j=1}^{m}\kappa_j^{\bullet}(x_i,Q^m) .$$

The resulting lower bound for the as. variance is

(16.2.8) $$\bar{\bar{\sigma}}^2(Q^m) = \frac{1}{m}\int(\Sigma\kappa_i^{\bullet}(\xi,Q^m))^2 Q(d\xi) .$$

Since

$$\Sigma\int\kappa_i^{\bullet}(\xi,Q^m)^2 Q(d\xi) = \Sigma\int(\kappa_i^{\bullet}(\xi,Q^m) - \frac{1}{m}\Sigma\kappa_j^{\bullet}(\xi,Q^m))^2 Q(d\xi)$$
$$+ \frac{1}{m}\int(\Sigma\kappa_j^{\bullet}(\xi,Q^m))^2 Q(d\xi),$$

we have $\bar{\bar{\sigma}}^2(Q^m) < \bar{\sigma}^2(Q^m)$ unless the functions $\kappa_j^{\bullet}$, $j = 1,\ldots,m$, are identical.

16.3. Asymptotically efficient estimators for product measures

Section 16.2 contains as. variance bounds under various assumptions on the true p-measure ((i) arbitrary, (ii) product measure, (iii) product of identical components). In Section 11.2 it was suggested that an as. efficient estimator-sequence for a functional can be obtained by applying the functional to an as. efficient estimator-sequence of the p-measure.

In this section we shall show that an as. efficient estimator-sequence for a product measure can be obtained as a product of as. efficient estimator-sequences of the components.

16.3.1. Proposition. *Assume we are given families of p-measures* $\mathfrak{P}_i|\mathscr{A}_i$, $i = 1,\dots,m$. *Let* $P_{in}(\underline{x}_i,\cdot) \in \mathfrak{P}_i$ *be as. efficient on* $\mathfrak{P}_i$. *Then* $\times P_{in}(\underline{x}_i,\cdot)$ *is as. efficient on* $\mathfrak{P} = \{\times P_i\colon P_i \in \mathfrak{P}_i,\ i = 1,\dots,m\}$

Proof. Let $g \in T(\times P_i,\mathfrak{P})$ be arbitrary. By Proposition 2.4.1, $g(\xi_1,\dots,\xi_m) = \Sigma g_i(\xi_i)$ with $g_i \in T(P_i,\mathfrak{P}_i)$. We have

$$n^{1/2}\int g(\xi_1,\dots,\xi_m)\times P_{in}(\underline{x}_i,d\xi_i) = \Sigma n^{1/2}\int g_i(\xi_i)P_{in}(\underline{x}_i,d\xi_i)$$

$$= \Sigma\tilde{g}_i(\underline{x}_i) + o_P(n^0) = \tilde{g}((x_{1\nu},\dots,x_{m\nu})_{\nu=1,\dots,n}) + o_P(n^0)\ .$$

Hence the assertion follows from Definition 10.1.1.

Now we consider products of *identical* components.

16.3.2. Proposition. *Assume that* $\mathfrak{P}_i|\mathscr{A}_i = \mathfrak{P}_1|\mathscr{A}_1$, $i = 2,\dots,m$, *and* $\mathfrak{P} = \{P^m\colon P \in \mathfrak{P}_1|\mathscr{A}_1\}$. *Let* $P_n(\underline{x}_i,\cdot) \in \mathfrak{P}_1$ *be as. efficient on* $\mathfrak{P}_1$, *and define*

$$P_{on}((x_{1\nu},\dots,x_{m\nu})_{\nu=1,\dots,n},\cdot) = \frac{1}{m}\Sigma P_n(\underline{x}_i,\cdot).$$

Then $P_{on}((x_{1\nu},\dots,x_{m\nu})_{\nu=1,\dots,n},\cdot)$ *is as. efficient for* P *(and therefore* $(P_{on})^m$ *as. efficient for* P^m*).*

__Proof.__ P_{on} is, in fact, an estimator for P based on a sample of size mn. Let $g \in T(P,\mathfrak{P}_1)$ be arbitrary. We have

$$(mn)^{1/2}\int g(\xi)P_{on}(\underline{x},d\xi) = (mn)^{1/2}\frac{1}{m}\sum_{i=1}^{m}\int g(\xi)P_n(\underline{x}_i,d\xi)$$

$$= m^{-1/2}\sum_{i=1}^{m} n^{-1/2}\sum_{\nu=1}^{n} g(x_{i\nu}) + o_P(n^o)$$

$$= (mn)^{-1/2}\sum_{i=1}^{m}\sum_{\nu=1}^{n} g(x_{i\nu}) + o_P(n^o) .$$

By Definition 10.1.1 this proves the assertion.

16.4. Estimators for von Mises functionals

In Section 16.3 we have shown that an as. efficient estimator-sequence for a product measure can be obtained from as. efficient estimator-sequences of the components. In this section we consider, in particular, von Mises functionals. For these functionals, the empirical p-measures can be used instead of as. efficient estimator-sequences (despite the fact that they do not belong to the basic family).

Let

(16.4.1) $\quad \kappa(P) = \int k(x_1,\dots,x_m)P(d(x_1,\dots,x_m)).$

If $T(P,\mathfrak{P}) = \mathscr{L}_*(P)$, we have

$$\kappa^\bullet(x_1,\dots,x_m,P) = k(x_1,\dots,x_m) - \kappa(P).$$

An estimator for $\kappa(P)$ can be obtained by replacing P in (16.4.1) by the empirical p-measure. This leads to the estimator-sequence

(16.4.2) $\quad \kappa^n((x_{1\nu},\dots,x_{m\nu})_{\nu=1,\dots,n}) = \frac{1}{n}\sum_{\nu=1}^{n} k(x_{1\nu},\dots,x_{m\nu}).$

Obviously, this estimator-sequence attains the as. variance bound given by (16.2.1), i.e.,

$$\int (k(\xi_1,\ldots,\xi_m) - \kappa(P))^2 P(d(\xi_1,\ldots,\xi_m)).$$

If the true p-measure is known to be a *product measure*, a better estimator for P is available, namely the product of the empirical p-measures $Q_n((x_{i\nu})_{\nu=1,\ldots,n},\cdot)$ for $i = 1,\ldots,m$. This yields for $\kappa(P)$ the estimator

$$(16.4.3)\qquad \bar{\kappa}^n((x_{1\nu},\ldots,x_{m\nu})_{\nu=1,\ldots,n}) = \frac{1}{n^m}\sum_{\nu_1=1}^{n}\ldots\sum_{\nu_m=1}^{n} k(x_{1\nu_1},\ldots,x_{m\nu_m}).$$

Let

$$k_j(x_j,\times Q_i) := \int k(x_1,\ldots,x_m) \underset{i\neq j}{\times} Q_i(dx_i).$$

Since $\bar{\kappa}^n$ is a generalized von Mises statistic (see, e.g., Sen, 1981, pp. 71ff.), $(\times Q_i)^n * n^{1/2}(\bar{\kappa}^n-\kappa(P))$ is as. normal with variance

$$(16.4.4)\qquad \Sigma\int (k_j(\xi,\times Q_i)-\kappa(\times Q_i))^2 Q_j(d\xi).$$

If $\mathfrak{P} = \{\times Q_i : Q_i \in \mathfrak{Q}_i\}$ with $T(Q_i,\mathfrak{Q}_i) = \mathscr{L}_*(Q_i)$, then $\kappa_k^{\bullet}(\cdot,\times Q_i)$, defined in (16.2.3) as the projection of $\bar{\kappa}_k(\cdot,\times Q_i)$ into $T(Q_k,\mathfrak{Q}_k)$, becomes identical with $\bar{\kappa}_k(\cdot,\times Q_i)$. In our particular case, this yields

$$\kappa_j^{\bullet}(\xi,\times Q_i) = k_j(\xi,\times Q_i) - \kappa(\times Q_i).$$

Hence the as. variance given by (16.4.4) coincides with the as. variance bound (16.2.6). In other words: If the basic family is the product of 'full' families, then the estimator-sequence (16.4.3) is as. efficient.

If the true p-measure is known to be a product of *identical* components, the estimator for P can be further improved to

$$\frac{1}{m}\Sigma Q_n((x_{i\nu})_{\nu=1,\ldots,n},\cdot).$$

This yields for $\kappa(P)$ the estimator

$$(16.4.5)\qquad \bar{\bar{\kappa}}^n((x_{1\nu},\ldots,x_{m\nu})_{\nu=1,\ldots,n})$$

$$= \frac{1}{(mn)^m}\sum_{i_1=1}^{m}\ldots\sum_{i_m=1}^{m}\sum_{\nu_1=1}^{n}\ldots\sum_{\nu_m=1}^{n} k(x_{i_1\nu_1},\ldots,x_{i_m\nu_m}).$$

Since $\overline{\overline{\kappa}}^n$ can be viewed as a von Mises statistic based on a sample of size mn, it is as. normal (see, e.g., Sen, 1981, p. 59) with variance

(16.4.6) $\quad \frac{1}{m}\int(\Sigma(k_i(\xi,Q^m) - \kappa(Q^m))Q(d\xi)$.

If $\overline{\overline{\mathfrak{P}}} = \{Q^m: Q \in \mathfrak{Q}\}$ with $T(Q,\mathfrak{Q}) = \mathscr{L}_*(Q)$, then this as. variance coincides with the as. variance bound given by (16.2.8), so that the estimator-sequence (16.4.5) is in this case as. efficient.

In the particular case that $k(x_1,\ldots,x_m)$ is invariant under permutations of $(x_1,\ldots,x_m)$, the functions $\kappa_i^{\bullet}(\xi,\times Q_j) = k_i(\xi,\times Q_j) - \kappa(\times Q_j)$ are identical for $i = 1,\ldots,m$, in which case the as. variance bounds (16.2.6) and (16.2.8) (and - correspondingly - the estimators (16.4.3) and (16.4.5)) become identical. Hence the knowledge that the true p-measure is the product of *identical* components does not help to obtain better estimators if k is permutation invariant.

16.5. A special example

For $P|\mathbb{B}^2$ consider the problem of estimating $\kappa(P) := P\{(x,y) \in \mathbb{R}^2: x > y\}$. This is a special von Mises functional, $\kappa(P) = \int k(x,y)P(d(x,y))$, with $k(x,y) = 1_{(0,\infty)}(x-y)$. Hence all the results of Section 16.4 apply.

If nothing is known about P, the best possible estimator for κ is (see (16.4.2))

(16.5.1) $\quad \frac{1}{n}\sum_{\nu=1}^{n} 1_{(0,\infty)}(x_\nu - y_\nu)$,

its as. variance being

(16.5.2) $\quad \kappa(P)(1-\kappa(P))$.

If it is known that (x,y) are stochastically independent, i.e. $P = Q_1 \times Q_2$, then a better estimator is available, namely (see (16.4.3)) the Wilcoxon two-sample statistic

(16.5.3) $$\frac{1}{n^2}\sum_{\mu=1}^{n}\sum_{\nu=1}^{n} 1_{(0,\infty)}(x_\mu - y_\nu) ,$$

its as. variance being

(16.5.4) $$\int Q_2(-\infty,x)^2 Q_1(dx) + \int Q_1(y,\infty)^2 Q_2(dy) - 2\kappa(Q_1 \times Q_2)^2 .$$

This variance is always smaller than $\kappa(Q_1 \times Q_2)(1-\kappa(Q_1 \times Q_2))$ since the gradient $(x,y) \to 1_{(0,\infty)}(x-y) - \kappa(Q_1 \times Q_2)$ differs from its projection into $T(Q_1 \times Q_2, \mathfrak{Q}_1 \times \mathfrak{Q}_2)$, $(x,y) \to Q_2(-\infty,x) - \kappa(P) + Q_1(y,\infty) - \kappa(P)$.

Assume now that Q_1 is known to be a member of a parametric family, say $\mathfrak{P} = \{P_\theta : \theta \in \Theta\}$ with $\Theta \subset \mathbb{R}^k$, whereas nothing is known about Q_2 except that it has a Lebesgue density over $\mathbb{R}$. It is intuitively clear how a better estimator can be obtained in this case.

We have

$$\kappa(P_\theta \times Q) = \int P_\theta(y,\infty) Q(dy) .$$

If $\theta^n(\underline{x})$ is an as. efficient estimator-sequence for θ, $P_{\theta^n(\underline{x})}$ is an as. efficient estimator-sequence for P_θ (see Proposition 10.3.2). By Proposition 16.3.1, as. efficient estimator-sequences for product measures can be obtained as products of as. efficient estimator-sequences of the components. Hence it suggests itself to try $\kappa(P_{\theta^n(\underline{x})} \times Q_n(\underline{y},\cdot))$ as an estimator-sequence for $\kappa(P_\theta \times Q)$, with Q_n being the empirical p-measure. Written explicitly, this estimator-sequence is

(16.5.5) $$\kappa^n((x_\nu,y_\nu)_{\nu=1,\dots,n}) = n^{-1}\sum_{\nu=1}^{n} P_{\theta^n(\underline{x})}(y_\nu,\infty) .$$

Since the empirical p-measure is not a strict estimator and does, therefore, not fulfill the conditions of Definition 10.1.1, Theorem 11.2.1 is not applicable. Hence it appears advisable to check the as. efficiency of (16.5.5) directly.

Since

$$n^{-1/2}\sum_{\nu=1}^{n}(P_{\theta^n(\underline{x})}(y_\nu,\infty) - \kappa(P_\theta \times Q)) = n^{-1/2}\sum_{\nu=1}^{n}(P_\theta(y_\nu,\infty) - \kappa(P_\theta \times Q))$$

$$+ n^{1/2}(\theta_i^n(\underline{x}) - \theta_i)\frac{1}{n}\sum_{\nu=1}^{n}\frac{\partial}{\partial\theta_i} P_\theta(y_\nu,\infty) + o_{P_\theta \times Q}(n^0)$$

and

$$n^{1/2}(\theta_i^n(\underline{x})-\theta_i)\frac{1}{n}\sum_{\nu=1}^{n}\frac{\partial}{\partial\theta_i}P_\theta(y_\nu,\infty)$$

$$= \Lambda_{ij}(\theta)\tilde{\ell}^{(j)}(\underline{x},\theta)\frac{\partial}{\partial\theta_i}\int P_\theta(y,\infty)Q(dy) + o_{P_\theta\times Q}(n^0) ,$$

the estimator-sequence (16.5.5) is as. normal with variance

(16.5.6) $\Lambda_{ij}(\theta)\frac{\partial}{\partial\theta_i}\kappa(P_\theta\times Q)\frac{\partial}{\partial\theta_j}\kappa(P_\theta\times Q) + \int P_\theta(y,\infty)^2 Q(dy) - \kappa(P_\theta\times Q)^2.$

To see that this as. variance is, in fact, minimal, we proceed as follows. The gradient of κ in $\mathscr{L}_*(P_\theta\times Q)$ is

$$\kappa^\cdot(x,y,P_\theta\times Q) = 1_{(0,\infty)}(x-y) - \kappa(P_\theta\times Q) .$$

Moreover (see (16.2.3)),

$$\bar{\kappa}_1(x,P_\theta\times Q) = \int\kappa^\cdot(x,\eta,P_\theta\times Q)Q(d\eta) = Q(-\infty,x) - \kappa(P_\theta\times Q) ,$$

$$\bar{\kappa}_2(y,P_\theta\times Q) = \int\kappa^\cdot(\xi,y,P_\theta\times Q)P_\theta(d\xi) = P_\theta(y,\infty) - \kappa(P_\theta\times Q).$$

According to (16.2.5) we have to determine the projections of these functions into $T(P_\theta,\mathfrak{P})$ (with $\mathfrak{P} = \{P_\theta: \theta\in\Theta\}$) and $T(Q,\mathfrak{Q})$, respectively.

Since $T(P_\theta,\mathfrak{P}) = [\ell^{(i)}(\cdot,\theta): i = 1,\dots,k]$, we obtain

(16.5.7) $\kappa_1^\cdot(x,P_\theta\times Q) = a_i(\theta,Q)\ell^{(i)}(x,\theta)$

with $a_i(\theta,Q) := \Lambda_{ij}(\theta)\int Q(-\infty,\xi)\ell^{(j)}(\xi,\theta)P_\theta(d\xi) = \Lambda_{ij}(\theta)\frac{\partial}{\partial\theta_j}\kappa(P_\theta\times Q).$

If $T(Q,\mathfrak{Q}) = \mathscr{L}_*(Q)$, we have

(16.5.8) $\kappa_2^\cdot(y,P_\theta\times Q) = P_\theta(y,\infty) - \kappa(P_\theta\times Q).$

The as. variance bound (16.2.6), computed with (16.5.7) and (16.5.8), becomes equal to (16.5.6). Hence the estimator-sequence (16.5.5) is, in fact, as. efficient.

17. DEPENDENCE - INDEPENDENCE

17.1. Measures of dependence

In this section we introduce certain measures of dependence. We concentrate on measures which take the direction of dependence (positive or negative) into account. Our outline will be restricted to the simplest case, that of a family $\mathfrak{P}$ of p-measures $P|\mathbb{B}^2$.

A minimal requirement for any *measure of dependence* κ is

(17.1.1) $\quad \kappa(P_1 \times P_2) = 0 \qquad$ for any $P_i|\mathbb{B}$, $i = 1,2$.

For $P|\mathbb{B}^2$ let F_P denote the distribution function, defined by $F_P(x,y) = P(-\infty,x] \times (-\infty,y]$, and $P_i|\mathbb{B}$ the i-th marginal measure, with F_{P_i} the pertaining distribution function. We have $F_{P_1}(x) = F_P(x,\infty)$, and $F_{P_2}(y) = F_P(\infty,y)$. To simplify our notation, we assume that there exists an open rectangle in $\mathbb{R}^2$, say $I_1 \times I_2$, on which each $P \in \mathfrak{P}$ is concentrated and has positive Lebesgue density. This implies in particular that F_{P_i} is continuous and increasing on I_i.

At first we consider a class of measures of dependence which remain unchanged under monotone transformations, i.e.,

(17.1.2) $\quad \kappa(P*(m_1,m_2)) = \kappa(P) \quad$ for any $P \in \mathfrak{P}$ and $m_i: I_i \to \mathbb{R}$ increasing.

Under our assumptions, this implies in particular that

(17.1.3) $\quad \kappa(P) = \kappa(P*(F_{P_1},F_{P_2}))$,

i.e., the functional κ depends on P only through the induced measure $P*(F_{P_1},F_{P_2})$ on the Borel algebra of $(0,1)^2$, the so-called *copula*.

The term 'copula' was introduced by Sklar (1959), but the idea of such a representation is so natural that it arises independently in several places (see, e.g., Lehmann, 1953, and Kruskal, 1958). Its properties are extensively studied in Schweizer and Sklar (1974). See also Deheuvels, who uses the concept of a copula under the name of a 'dependence function' in a series of papers, starting with Deheuvels (1979).

Two widely used measures of dependence which are functionals of the copula are *Spearman's* ρ

(17.1.4) $$\rho(P) = 12(\iint F_P(x,y)P_1(dx)P_2(dy) - \tfrac{1}{4})$$

and the *quadrant correlation coefficient*

(17.1.5) $$\varrho(P) = \iint \operatorname{sign}(x_1-x_2)\operatorname{sign}(y_1-y_2)P(d(x_1,y_1))P(d(x_2,y_2)).$$

The quadrant correlation coefficient was introduced by Sheppard (1899; see also Blomqvist, 1950). Written in a different form, it became known as Kendall's τ, i.e.,

(17.1.6) $$\tau(P) = 4(\iint F_P(x,y)P(d(x,y)) - 1).$$

See Schweizer and Wolff (1981) for other measures of dependence which can be defined in terms of the copula. It is easy to check that both ϱ and ρ fulfill (17.1.1) and (17.1.2).

We remark that the quadrant correlation coefficient is a special case of the general measure of dependence introduced in 5.1.3 as an example of a von Mises functional. Therefore, Proposition 5.1.5 yields the gradient of the quadrant correlation coefficient in $\mathscr{L}_*(P)$,

(17.1.7) $$\varrho^\cdot(x,y,P) = 8(F_P(x,y) - \int F_P(\xi,\eta)P(d(\xi,\eta)) - \tfrac{1}{2}F_{P_1}(x) - \tfrac{1}{2}F_{P_2}(y) + \tfrac{1}{2}).$$

From this we obtain the as. variance bound for the quadrant correlation coefficient,

(17.1.8) $$P(\varrho^\cdot(\cdot,P^2) = 64[\int F_P(x,y)^2 P(d(x,y)) - (\int F_P(x,y)P(d(x,y)))^2$$
$$+ \int F_P(x,y)P(d(x,y)) - \int F_{P_1}(x)F_P(x,y)P(d(x,y))$$
$$- \int F_{P_2}(y)F_P(x,y)P(d(x,y)) + \tfrac{1}{2}\int F_{P_1}(x)F_{P_2}(y)P(d(x,y)) - \tfrac{1}{12}].$$

A direct computation (using partial integration) yields for Spearman's ρ the following gradient in $\mathscr{L}_*(P)$

(17.1.9) $$\rho^{\bullet}(x,y,P) = 12[F_{P_1}(x)F_{P_2}(y)-\int F_{P_1}(\xi)F_{P_2}(\eta)P(d(\xi,\eta))$$
$$+ \int F_P(x,\eta)P_2(d\eta) + \int F_P(\xi,y)P_1(d\xi) - 2\int\int F_P(\xi,\eta)P_1(d\xi)P_2(d\eta)$$
$$- F_{P_1}(x) - F_{P_2}(y) + 1].$$

17.2. Estimating measures of dependence

Since the functionals ϱ and ρ are also defined for the empirical p-measure, our general considerations in Section 11.2 suggest that as. efficient estimators for these functionals can be obtained by applying them to the empirical p-measure.

In case of the quadrant correlation coefficient this leads to

(17.2.1) $$\varrho(Q_n((x_\nu,y_\nu)_{\nu=1,\ldots,n},\cdot)) = \frac{1}{n^2}\sum_{\nu=1}^{n}\sum_{\mu=1}^{n}\text{sign}(x_\nu-x_\mu)\text{sign}(y_\nu-y_\mu).$$

(With n^2 replaced by $n(n-1)$, this is Kendall's τ for the sample $(x_\nu,y_\nu)_{\nu=1,\ldots,n}$.) Kendall (1955, pp. 81ff.) gives the as. variance for τ (which coincides with the as. variance of the estimator (17.2.1)) for discrete p-measures $P|\mathbb{B}^2$, having their probability concentrated in N points. By letting $N \to \infty$ one obtains an expression for the as. variance of the estimator-sequence (17.2.1) which coincides with (17.1.8). This supports our conjecture that (17.2.1) is an as. efficient estimator for $\varrho(P)$.

Since the empirical distribution function $G_n(\underline{x},\cdot)$ fulfills

$$G_n(\underline{x},x_\nu) = n^{-1}\sum_{\mu=1}^{n} 1_{(-\infty,x_\nu]}(x_\mu) = n^{-1}R_\nu(\underline{x}),$$

where $R_\nu(\underline{x})$ denotes the rank of x_ν in the sample $\underline{x} = (x_1,\ldots,x_n)$, $\rho(Q_n((x_\nu,y_\nu)_{\nu=1,\ldots,n},\cdot))$ may be written as

$$12(n^{-3}\sum_{\nu=1}^{n} R_\nu(\underline{x})R_\nu(\underline{y}) - \tfrac{1}{4}) .$$

This is as. equivalent to Spearman's ρ, which may be written

$$12[\frac{1}{n(n^2-1)}\sum_{\nu=1}^{n} R_\nu(\underline{x})R_\nu(\underline{y}) - \frac{n+1}{4(n-1)}] .$$

17.3. Tests for independence

Nonparametric tests for independence are often based on measures of dependence, such as Spearman's ρ or Kendall's τ. Measures of dependence have the property of being zero for product measures, but not only for these. Tests based on an estimator κ^n for a measure of dependence, say $\kappa(P)$, are, in fact, tests for the hypothesis $\mathfrak{P}_1 = \{P \in \mathfrak{P}: \kappa(P) = 0\}$. The hypothesis of independence, however, $\mathfrak{P}_o = \{P \in \mathfrak{P}: P = P_1 \times P_2\}$, is usually much smaller: Whereas $T^\perp(P;\mathfrak{P}_1,\mathfrak{P})$ is one-dimensional, $T^\perp(P;\mathfrak{P}_o,\mathfrak{P})$ is infinite-dimensional if $\mathfrak{P}$ is a full family of p-measures.

Tests for the hypothesis $\mathfrak{P}_1$ are, of course, a legitimate tool of inference provided the measure of dependence as such is of interest because it seizes on an essential feature of the substance matter in question. Tests for measures of dependence as a substitute for a test of independence are unjustified. If they are as. optimal for testing the hypothesis $\mathfrak{P}_1 = \{P \in \mathfrak{P}: \kappa(P) = 0\}$, they are particularly unreliable: They are as. optimal for testing $\mathfrak{P}_o$ against alternatives which deviate from the true p-measure $P_1 \times P_2 \in \mathfrak{P}_o$ in direction $\kappa^\cdot(\cdot, P_1 \times P_2)$ and have, therefore (see Theorem 8.5.3), for alternatives in direction g an as. power function with slope

$$P_1 \times P_2(g\kappa^\cdot(\cdot,P_1 \times P_2))/(P_1 \times P_2(\kappa^\cdot(\cdot,P_1 \times P_2)^2))^{1/2} . \tag{17.3.1}$$

This implies in particular that they have as. power zero for directions

$g \in T(P_1 \times P_2, \mathfrak{P}_1)$, part of which belong to $T^{\perp}(P_1 \times P_2; \mathfrak{P}_o, \mathfrak{P})$ and correspond, therefore, to alternatives of $\mathfrak{P}_o$.

As an example consider tests based on Spearman's ρ or the quadrant correlation coefficient. Since $\varrho(Q_n((x_\nu, y_\nu)_{\nu=1,\dots,n}, \cdot))$ and $\rho(Q_n((x_\nu, y_\nu)_{\nu=1,\dots,n}, \cdot))$ are as. efficient estimators of $\varrho(P)$ resp. $\rho(P)$, tests for the hypothesis $\{P \in \mathfrak{P}: \varrho(P) = 0\}$ resp. $\{P \in \mathfrak{P}: \rho(P) = 0\}$ are as. most powerful against alternatives in direction $\varrho^{\bullet}(\cdot, P)$ resp. $\rho^{\bullet}(\cdot, P)$. For $P \in \mathfrak{P}_o$, we obtain from (17.1.7) resp. (17.1.9)

$$(17.3.2) \qquad \varrho^{\bullet}(x, y, P_1 \times P_2) = 8(F_{P_1}(x) - \tfrac{1}{2})(F_{P_2}(y) - \tfrac{1}{2})$$

resp.

$$(17.3.3) \qquad \rho^{\bullet}(x, y, P_1 \times P_2) = 12(F_{P_1}(x) - \tfrac{1}{2})(F_{P_2}(y) - \tfrac{1}{2}) .$$

(Notice that these directions are orthogonal to $T(P_1 \times P_2, \mathfrak{P}_o)$, because $\int (F_{P_1}(x) - \frac{1}{2})(F_{P_2}(\eta) - \frac{1}{2}) P_2(d\eta) \equiv 0$ and $\int (F_{P_1}(\xi) - \frac{1}{2})(F_{P_2}(y) - \frac{1}{2}) P_1(d\xi) \equiv 0$.) Since these gradients coincide up to a constant factor, we obtain the following result: Critical regions based on $\varrho(Q_n((x_\nu, y_\nu)_{\nu=1,\dots,n}, \cdot))$ or $\rho(Q_n((x_\nu, y_\nu)_{\nu=1,\dots,n}, \cdot))$, considered as tests for $\mathfrak{P}_o$, have the *same as. power*. If we specialize (17.3.1) for this case, we obtain for directions $g \in T^{\perp}(P_1 \times P_2; \mathfrak{P}_o, \mathfrak{P})$ the slope

$$(17.3.4) \qquad 12 \iint g(\xi, \eta) F_{P_1}(\xi) F_{P_2}(\eta) P_1(d\xi) P_2(d\eta) .$$

(Hint: Since $g \perp T(P_1 \times P_2, \mathfrak{P}_o)$, we have (see Example 8.1.5) $\int g(x, \eta) P_2(d\eta) \equiv 0$ and $\int g(\xi, y) P_1(d\xi) \equiv 0$.)

The as. equivalence of these two tests was first proved by Daniels (1944). See also Hoeffding (1948, p. 321).

The result that tests based on Spearman's ρ are as. most powerful against alternatives in direction

$$(17.3.5) \qquad (x, y) \to (F_{P_1}(x) - \tfrac{1}{2})(F_{P_2}(y) - \tfrac{1}{2})$$

is equivalent to the result of Lehmann (1953, p. 39) that these tests are locally as. optimal against alternatives with distribution function

$$(x, y) \to (1-t) F_{P_1}(x) F_{P_2}(y) + t F_{P_1}(x)^2 F_{P_2}(y)^2 .$$

These alternatives have $P_1 \times P_2$-density

$$(x,y) \to 1 + t(4F_{P_1}(x)F_{P_2}(y) - 1) ,$$

i.e. they approach $P_1 \times P_2$ from the direction

$$(17.3.6) \qquad (x,y) \to 4F_{P_1}(x)F_{P_2}(y) - 1 .$$

This direction is different from (17.3.5). The reason is that these alternatives are not represented in a canonical way, using densities with respect to the closest element in $\mathfrak{P}_o$ and - consequently - a direction orthogonal to the tangent space. In order to show the equivalence we have to prove (see Theorem 8.5.3) that (17.3.6) is in the space spanned by (17.3.5) and $T(P_1 \times P_2, \mathfrak{P}_o)$. This follows from

$$4F_{P_1}(x)F_{P_2}(y) - 1 = 4(F_{P_1}(x)-\tfrac{1}{2})(F_{P_2}(y)-\tfrac{1}{2}) - 2(F_{P_1}(x)+F_{P_2}(y)),$$

since $(x,y) \to F_{P_1}(x) + F_{P_2}(y) \in T(P_1 \times P_2, \mathfrak{P}_o)$ by Proposition 2.4.1.

Of course, there are tests for the hypothesis of independence, $\mathfrak{P}_o$, which have positive slope for all directions in $T^{\perp}(P_1 \times P_2; \mathfrak{P}_o, \mathfrak{P})$, but the choice of such a test poses principal problems which lead beyond our treatise (see Section 8.2).

The exception: If the co-space of the hypothesis is one-dimensional. This is, for instance, the case in the model discussed in Section 2.4. Under this model,

$$T(P_1 \times P_2, \mathfrak{P}_o) \;=\; \{(x,y) \to g_1(x)+g_2(y):\; g_i \in \mathscr{L}_*(P_i),\; i = 1,2\}$$

and

$$T(P_1 \times P_2, \mathfrak{P}) \;=\; \{(x,y) \to g_1(x)+g_2(y) + a\ell'(x,P_1)\ell'(y,P_2):\; g_i \in \mathscr{L}_*(P_i),\; i = 1,2, \text{ and } a \in \mathbb{R}\},$$

hence

$$T^{\perp}(P_1 \times P_2; \mathfrak{P}_o, \mathfrak{P}) \;=\; \{(x,y) \to a\ell'(x,P_1)\ell'(y,P_2):\; a \in \mathbb{R}\}.$$

According to Section 12.2, as. efficient tests for the hypothesis $\mathfrak{P}_o$ can be based on the test-statistic

$$n^{-1/2} \sum_{\nu=1}^{n} \ell_n'(x_\nu, \underline{x})\,\ell_n'(y_\nu, \underline{y}) ,$$

where $\ell_n'(\cdot,\underline{z})$ is an estimate for $\ell'(\cdot,Q)$, based on a sample $\underline{z} = (z_1,\ldots,z_n)$ from Q^n. Estimates appropriate for this purpose are suggested by Rieder (1980, p. 816, Proposition). (The test suggested by Hâjek and Šidák (1967, p. 75, Theorem 4.11) is based on $n^{-1/2}\sum_{\nu=1}^{n} \ell'(x_\nu,P_1)\ell'(y_\nu,P_2)$ and presumes P_1,P_2 to be known.)

17.3.7. Remark. If we restrict the basic family $\mathfrak{P}$ to the family of all two-dimensional normal distributions, say $\mathfrak{Q}$, then $\mathfrak{P}_o$ reduces to $\mathfrak{Q}_o$, the family of all two-dimensional normal distributions with $\rho = 0$. According to Example 8.6.8 (specialized to $\rho_o = 0$) we have

$$T^{\perp}(N(\mu_1,\sigma_1^2)\times N(\mu_2,\sigma_2^2);\mathfrak{Q}_o,\mathfrak{Q}) = \{(x,y)\to a(x-\mu_1)(y-\mu_2):\ a\in\mathbb{R}\}.$$

Since $\ell'(z,N(\mu,\sigma^2)) = -(z-\mu)/\sigma^2$, we have

$$T^{\perp}(N(\mu_1,\sigma_1^2)\times N(\mu_2,\sigma_2^2);\mathfrak{P}_o,\mathfrak{P}) = T^{\perp}(N(\mu_1,\sigma_1^2)\times N(\mu_2,\sigma_2^2);\mathfrak{Q}_o,\mathfrak{Q}) .$$

According to Proposition 8.6.5 this implies that for the model introduced in Section 2.4 a restriction to the family of all two-dimensional normal distributions does not lead to a better as. envelope power function.

18. TWO-SAMPLE PROBLEMS

18.1. Introduction

Let $(x_1,...,x_m)$ and $(y_1,...,y_n)$ be two independent samples (of i.i.d. variables). A natural problem occuring in this framework is to test that the x_i and the y_j come from the same distribution. Another problem is to estimate a functional measuring the difference (if any) between these distributions.

Since our general theory is restricted to the one-sample problem, only the particular case $m = n$ can be handled within our framework. The generalization of our approach necessary for the general two-sample problem offers no principal difficulties but we avoid such a breach to preserve homogeneity.

If $m = n$, we may consider the two samples $(x_1,...,x_n)$ and $(y_1,...,y_n)$ as n independent realizations of the two-dimensional random variable (x_ν,y_ν), $\nu = 1,...,n$, distributed according to a p-measure in

$$\mathfrak{P} = \{P_1 \times P_2 \colon P_i \in \mathfrak{P}_i,\ i = 1,2\}.$$

The hypothesis that x and y have the same distribution is equivalent to the hypothesis

$$\mathfrak{P}_o = \{P \times P \colon P \in \mathfrak{P}_1 \cap \mathfrak{P}_2\}.$$

If $\mathfrak{P}_2 = \mathfrak{P}_1$ and if $T(P,\mathfrak{P}_1)$ is a linear space, we have (see Proposition 2.4.1)

(18.1.1) $T(P \times P, \mathfrak{P}) = \{(x,y) \to g_1(x)+g_2(y): g_i \in T(P,\mathfrak{P}_1), i=1,2\}$

and

(18.1.2) $T(P \times P, \mathfrak{P}_o) = \{(x,y) \to g(x) + g(y): g \in T(P,\mathfrak{P}_1)\}$,

hence

(18.1.3) $T^{\perp}(P \times P; \mathfrak{P}_o, \mathfrak{P}) = \{(x,y) \to g(x)-g(y): g \in T(P,\mathfrak{P}_1)\}$.

(Hint: $\int (g_1(x)+g_2(y))(g(x)+g(y))P(dx)P(dy) = 0$ for all $g \in T(P,\mathfrak{P}_1)$ implies

$$\int (g_1(x)+g_2(x))g(x)P(dx) = 0 \qquad \text{for all } g \in T(P,\mathfrak{P}_1).$$

Since $g_1+g_2 \in T(P,\mathfrak{P}_1)$, this implies $g_1+g_2 = 0$ P-a.e.)

Since the co-space $T^{\perp}(P \times P; \mathfrak{P}_o, \mathfrak{P})$ is infinite-dimensional in this general model, the problem of finding an optimal test for the hypothesis is indeterminate, unless we have a weighting system for the different alternatives, or there is an inherent relationship between x and y so that deviations between the two distributions are restricted to certain directions.

18.2. Inherent relationships between x and y

In this section we present some models for an inherent relationship between the two components of the variable (x,y). Generally speaking, we assume that each p-measure in $\mathfrak{P}_*$ is the 'nucleus' of a parametric family. More formally: There is a parameter set Θ and a map assigning to each pair $(\theta,P) \in \Theta \times \mathfrak{P}_*$ a p-measure θP. Assume that for a certain $\varepsilon \in \Theta$ we have $\varepsilon P = P$ for all $P \in \mathfrak{P}_*$

Our basic assumption is: If x is distributed according to a certain p-measure $P \in \mathfrak{P}_*$, then the distribution of y is necessarily an element of $\{\theta P: \theta \in \Theta\}$. More formally: Our basic family is $\mathfrak{P} = \{P \times \theta P: P \in \mathfrak{P}_*, \theta \in \Theta\}$. Our problem is to estimate a functional of θ, or to test a hypothesis about θ. Of particular interest is the hypothesis $\theta = \varepsilon$, i.e. the hypothesis that x and y have the same distribution.

To illustrate the flexibility of this model, we mention a few special cases.

18.2.1. Example. Each $\theta \in \Theta$ defines a *transformation* on X, say $x \to \theta x$, and $\theta P := P*(x \to \theta x)$. As a special example we mention the *linear transformation* on $\mathbb{R}$. To each element $(a,b) \in \Theta = \mathbb{R} \times (0,\infty)$ corresponds the transformation $x \to a+bx$. The unit element is $\varepsilon = (0,1)$.

18.2.2. Example. For $P|\mathcal{B}$ let F_P denote its distribution function, defined by $F_P(x) := P(-\infty,x]$. For $\theta \in \Theta \subset \mathbb{R}$ let $B(\cdot,\theta)$: $[0,1] \to [0,1]$ be an increasing function with $B(0,\theta) = 0$ and $B(1,\theta) = 1$ for all $\theta \in \Theta$. Assume that $B(u,\varepsilon) \equiv u$ for a certain $\varepsilon \in \Theta$. If necessary, we assume further regularity conditions (like differentiability in u and θ). Let θP be the p-measure with distribution function $x \to B(F_P(x),\theta)$.

A model of this type has first been considered by Lehmann (1953, Sections 4 and 6). A popular special case is

$$B(u,\theta) = (1-\theta)u + \theta B_0(u), \qquad \theta \in [0,1], \tag{18.2.3}$$

where B_0: $[0,1] \to [0,1]$ is increasing with $B_0(0) = 0$ and $B_0(1) = 1$.

Another interesting case is the *proportional failure rate* model, given by

$$B(u,\theta) = 1 - (1-u)^\theta, \qquad \theta \in (0,\infty). \tag{18.2.4}$$

Since $1 - F_{\theta P}(x) = (1 - F_P(x))^\theta$, the Lebesgue density of θP is $p(x,\theta) = \theta(1 - F_P(x))^\theta p(x)$, and therefore

$$\frac{p(x,\theta)}{1-F_{\theta P}(x)} = \theta \frac{p(x)}{1-F_P(x)} .$$

With this motivation, this model was used by Cox (1972, p. 189) and many other authors since.

18.3. The tangent spaces

Starting from the basic family $\mathfrak{P} = \{P \times \theta P: P \in \mathfrak{P}_*, \theta \in \Theta\}$, we are interested in the as. envelope power for tests on a hypothesis about θ, and in the as. variance bound for estimators of (functionals of) θ. In either case we need the tangent space $T(P\times\theta P, \mathfrak{P})$. Our primary interest is in nonparametric problems, where $\mathfrak{P}_*$ is a full family of p-measures, so that $T(P,\mathfrak{P}_*) = \mathscr{L}_*(P)$.

Under appropriate regularity conditions, the tangent space can be written as (see Section 2.6)

$$T(P\times\theta P,\mathfrak{P}) = T(P\times\theta P,\mathfrak{Q}_\theta) + T(P\times\theta P,\mathfrak{P}_P)$$

where

$$\mathfrak{Q}_\theta = \{Q\times\theta Q: Q \in \mathfrak{P}_*\},$$

$$\mathfrak{P}_P = \{P\times\tau P: \tau \in \Theta\} .$$

Observe that

$$T(P\times\theta P,\mathfrak{P}_P) = \{(x,y) \to b(y): b \in T(\theta P,\{\tau P: \tau \in \Theta\}) .$$

In the following, we use this decomposition to determine the tangent space in each of the examples.

Let us first consider the *transformation model* 18.2.1. Let $\Theta \subset \mathbb{R}^k$, $X \subset \mathbb{R}^m$. We assume that the transformations form a group, and that the dominating measure μ is covariant, i.e., $\theta\mu(A) := \mu(\theta^{-1}A) = \Delta(\theta^{-1})\mu(A)$, $A \in \mathscr{A}$, so that $d\theta\mu/d\mu = \Delta(\theta^{-1})$. Since $p \in dP/d\mu$ implies $x \to p(\theta^{-1}x)$ $\in d\theta P/d\theta\mu$, and since μ is covariant, we obtain

$$(18.3.1) \quad x \to \Delta(\theta^{-1})p(\theta^{-1}x) \in d\theta P/d\mu .$$

Let P_t, $t \downarrow 0$, be a differentiable path in $\mathfrak{P}_*$ with μ-density $x \to p(x)(1+t(g(x)+r_t(x)))$. The transformed path θP, $t \downarrow 0$, has μ-density

$$y \to \Delta(\theta^{-1})p(\theta^{-1}y)(1 + t(g(\theta^{-1}y) + r_t(\theta^{-1}y))) .$$

Hence the path $P_t \times \theta P_t$, $t \downarrow 0$, in $\mathfrak{Q}_\theta$ has derivative

(18.3.2) $\quad (x,y) \to g(x) + g(\theta^{-1}y)$,

so that

(18.3.3) $\quad T(P\times\theta P, \mathfrak{Q}_\theta) = \{(x,y) \to g(x) + g(\theta^{-1}y): g \in T(P, \mathfrak{P}_*)\}$.

If the transformations are differentiable, then the path $P\times(\theta+ta)P$, $t \downarrow 0$, in $\mathfrak{P}_P$, with $a \in \mathbb{R}^k$ fixed, has derivative

(18.3.4) $\quad (x,y) \to a_i c_{ij}(\theta) f_j(\theta^{-1}y,P)$,

where

(18.3.5) $\quad f_j(z,P) = h_{jq}(z)\ell^{(q)}(z,P) - P(h_{jq}\ell^{(q)}(\cdot,P))$

with $\ell^{(q)}(z,P) = (\partial/\partial z_q)\log p(z)$.

(If the group is differentiable, there exists for every $\theta \in \Theta$ a $k\times k$-matrix $C(\theta)$ such that for all $a \in \mathbb{R}^k$ and $t \downarrow 0$,

$$(\theta+ta)^{-1} = (\varepsilon + C(\theta)'a)\theta^{-1} + o(t) .$$

Moreover, for every $x \in X$ there exists a $k\times m$-matrix $H(x)$ such that for all $b \in \mathbb{R}^k$ and $t \downarrow 0$,

$$(\varepsilon+tb)x = x + tH(x)'b + o(t) .$$

Therefore,

$$p((\theta+ta)^{-1}z) = p((\varepsilon+tC(\theta)'a)\theta^{-1}z) + o(t)$$

$$= p(\theta^{-1}z)(1+a'C(\theta)H(\theta^{-1}z)\ell^{(\cdot)}(\theta^{-1}z,P)) + o(t) .$$

Since the μ-density of θP is $z \to \Delta(\theta^{-1})p(\theta^{-1}z)$, the derivative of the path $(\theta+ta)P$, $t \downarrow 0$, is $z \to a'C(\theta)H(\theta^{-1}z)\ell^{(\cdot)}(\theta^{-1}z,P)$, up to an additive term depending on θ only.)

From (18.3.4) we obtain

(18.3.6) $\quad T(P\times\theta P, \mathfrak{P}_P) = \{(x,y) \to a_i c_{ij}(\theta) f_j(\theta^{-1}y,P): a \in \mathbb{R}^k\}$.

Next we consider the model 18.2.2. We have $F_{\theta P}(x) = B(F_P(x),\theta)$ and therefore, for $p(\cdot,\theta) \in d\theta P/d\lambda$,

$$p(x,\theta) = \beta(F_P(x),\theta)p(x) ,$$

where $\beta(u,\theta) := (\partial/\partial u)B(u,\theta)$. Moreover, we need $\beta_1(u,\theta) := (\partial/\partial u)\beta(u,\theta)$ and $\beta_2(u,\theta) := (\partial/\partial\theta)\beta(u,\theta)$.

Let P_t, $t \downarrow 0$, be a differentiable path in $\mathfrak{P}_*$ with λ-density $x \to p(x)(1 + t(g(x) + r_t(x)))$. The transformed path θP, $t \downarrow 0$, has λ-density

$$y \to \beta(F_P(y),\theta)p(y)\Big(1+t[g(y) + \frac{\beta_1(F_P(y),\theta)}{\beta(F_P(y),\theta)} \int_{-\infty}^{y} g(\eta)P(d\eta) + \bar{r}_t(\eta)]\Big) .$$

Hence the path $P_t \times \theta P_t$, $t \downarrow 0$, in $\mathfrak{Q}_\theta$ has derivative

$$(18.3.7) \qquad (x,y) \to g(x) + g(y) + \frac{\beta_1(F_P(y),\theta)}{\beta(F_P(y),\theta)} \int_{-\infty}^{y} g(\eta)P(d\eta) ,$$

so that

$$(18.3.8) \qquad T(P\times\theta P,\mathfrak{Q}_\theta) = \{(x,y) \to g(x) + g(y) + \frac{\beta_1(F_P(y),\theta)}{\beta(F_P(y),\theta)} \int_{-\infty}^{y} g(\eta)P(d\eta) : g \in T(P,\mathfrak{P}_*)\} .$$

Furthermore, the path $(\theta+ta)P$, $t \downarrow 0$, with $a \in \mathbb{R}$ fixed, has λ-density

$$y \to \beta(F_P(y),\theta)p(y)\Big(1 + t[a \frac{\beta_2(F_P(y),\theta)}{\beta(F_P(y),\theta)} + \bar{\bar{r}}_t(y)]\Big) .$$

Hence the path $P\times(\theta+ta)P$, $t \downarrow 0$, in $\mathfrak{P}_P$ has derivative

$$(18.3.9) \qquad (x,y) \to a \frac{\beta_2(F_P(y),\theta)}{\beta(F_P(y),\theta)}$$

so that

$$(18.3.10) \qquad T(P\times\theta P,\mathfrak{P}_P) = \{(x,y) \to a \frac{\beta_2(F_P(y),\theta)}{\beta(F_P(y),\theta)} : a \in \mathbb{R}\} .$$

The different structure of the tangent spaces in Examples 18.2.1 and 18.2.2 demonstrates that general results cannot be expected in the two-sample problem. The situation is somewhat more favorable if we consider the hypothesis $\theta = \varepsilon$, i.e. $\mathfrak{P}_o = \{P\times P: P \in \mathfrak{P}_*\}$. This case will be treated in Section 18.4. In Sections 18.5 and 18.6 we consider the problem of estimating θ (or testing a general hypothesis about θ) for the transformation model and the proportional failure rate model, respectively.

18.4. Testing for equality

In the case of a general hypothesis $\theta = \theta_o$, say, the tangent space $T(P\times\theta_o P, \mathfrak{P})$ is complicated by the fact that a path $P_t \times \theta_o P_t$, $t \downarrow 0$, leads to a direction $(x,y) \to g(x) + g(y,\theta_o)$, with $g(\cdot,\theta_o) \neq g$ (see (18.3.2) and (18.3.7)). The situation is different in case of the hypothesis $\theta = \varepsilon$. Since $\varepsilon P = P$, we have $\mathfrak{P}_o = \{P\times P: P \in \mathfrak{P}_*\}$ and therefore

$$T(P\times P, \mathfrak{P}_o) = \{(x,y) \to g(x) + g(y): g \in T(P, \mathfrak{P}_*)\}.$$

If the decomposition discussed in Section 2.6 is valid, we have

$$T(P\times P, \mathfrak{P}) = T(P\times P, \mathfrak{P}_o) + T(P\times P, \mathfrak{P}_P)$$

with $\mathfrak{P}_P = \{P\times\tau P: \tau \in \Theta\}$. Furthermore,

$$T(P\times P, \mathfrak{P}_P) = \{(x,y) \to b(y): b \in T(P, \{\tau P: \tau \in \Theta\})\}.$$

If $\Theta \subset \mathbb{R}$, then $T(P, \{\tau P: \tau \in \Theta\})$ is one-dimensional, say $[b(\cdot,P)]$, and

(18.4.1) $\quad T(P\times P, \mathfrak{P}) = \{(x,y) \to g(x) + g(y) + ab(y,P): g \in T(P, \mathfrak{P}_*),\ a \in \mathbb{R}\}.$

18.4.2. Remark. (i) For the transformation model (i.e., $X \subset \mathbb{R}^m$, $\theta P = P*(x \to \theta x)$) we obtain from (18.3.5) (applied for $k = 1$) that (18.4.1) holds with $b(y,P) = h_i(y)\ell^{(i)}(y,P) - P(h_i \ell^{(i)}(\cdot,P))$.

(ii) For Lehmann's alternatives (see Example 18.2.2), relation (18.4.1) holds with $b(y,P) = \beta_2(F_P(y),\varepsilon)$, since $B(u,\varepsilon) \equiv u$ implies $\beta(u,\varepsilon) \equiv 1$.

To determine the as. envelope power function for alternatives $P\times(\varepsilon+n^{-1/2}t)P$, we have to determine the distance of $P\times(\varepsilon+n^{-1/2}t)P$ from $\mathfrak{P}_o$ (see Remark 8.4.5). According to Proposition 6.2.18 we have to determine the component of the function $y\to b(y,P)$ which is orthogonal to $T(P\times P,\mathfrak{P}_o)$.

Let $(x,y)\to g_o(x)+g_o(y)$ denote the projection of $y\to b(y,P)$ into $T(P\times P,\mathfrak{P}_o)$. By definition,

$$\int(g_o(x)+g_o(y)-b(y,P))(g(x)+g(y))P(dx)P(dy)=0 \text{ for all } g\in T(P,\mathfrak{P}_*),$$

which implies

$$(18.4.3)\qquad \int(2g_o(y)-b(y,P))g(y)P(dy) = 0 \qquad \text{for all } g\in T(P,\mathfrak{P}_*).$$

If $T(P,\mathfrak{P}_*) = \mathscr{L}_*(P)$, relation (18.4.3) implies

$$g_o(y) = \tfrac{1}{2}b(y,P) .$$

Hence the projection of $y\to b(y,P)$ into $T(P\times P,\mathfrak{P}_o)$ is $(x,y)\to\frac{1}{2}b(x,P) + \frac{1}{2}b(y,P)$, and the orthogonal component is

$$(18.4.4)\qquad (x,y)\to\tfrac{1}{2}b(y,P)-\tfrac{1}{2}b(x,P) .$$

Since $\int(\frac{1}{2}b(y,P)-\frac{1}{2}b(x,P))^2P(dx)P(dy) = \frac{1}{2}P(b(\cdot,P)^2)$, the slope of the as. envelope power function under alternatives $P\times(\varepsilon+n^{-1/2}t)P$ is, according to Remark 8.4.5,

$$(18.4.5)\qquad 2^{-1/2}P(b(\cdot,P)^2)^{1/2} .$$

If we change from $\mathfrak{P}_*$ to a smaller family, say $\mathfrak{Q}_*$, then the hypothesis $\mathfrak{P}_o = \{P\times P\colon P\in\mathfrak{P}_*\}$ is replaced by the smaller hypothesis $\mathfrak{Q}_o = \{P\times P\colon P\in\mathfrak{Q}_*\}$, and the tangent space $T(P\times P,\mathfrak{P}_o)$ by the smaller tangent space $T(P\times P,\mathfrak{Q}_o)$. If we restrict our attention to alternatives in direction $b(\cdot,P)$, we obtain:

Conclusion: The restriction from the full family $\mathfrak{P}_*$ to a smaller family $\mathfrak{Q}_*$ will not increase the as. envelope power function against alternatives in direction $b(\cdot,P)$ iff $b(\cdot,P)\in T(P,\mathfrak{Q}_*)$.

(Since the projection of $b(\cdot,P)$ into $T(P\times P,\mathfrak{P}_o)$ is $(x,y) \to \frac{1}{2}b(x,P) + \frac{1}{2}b(y,P)$, this projection belongs to

$$T(P\times P,\mathfrak{Q}_o) = \{(x,y) \to g(x) + g(y): g \in T(P,\mathfrak{Q}_*)\}$$

iff $b(\cdot,P) \in T(P,\mathfrak{Q}_*)$. Hence the conclusion follows from Remark 8.6.12.)

Let us now consider the proportional failure rate model (18.2.4). In this case we have $\beta_2(u,\varepsilon) = 1 + \log(1-u)$, hence

(18.4.6) $\qquad b(x,P) = 1 + \log(1 - F_P(x))$.

We consider the restriction to the family $\mathfrak{Q}_* = \{P_\lambda: \lambda > 0\}$, where P_λ is the exponential distribution (with Lebesgue density $x \to \lambda\exp[-\lambda x]$, $x > 0$). In this case, $T(P_\lambda,\mathfrak{Q}_*) = \{x \to c(1-\lambda x): c \in \mathbb{R}\}$. Since $F_{P_\lambda}(x) = 1 - \exp[-\lambda x]$, we have from (18.4.6) that $b(x,P_\lambda) = 1 - \lambda x$, so that $b(\cdot,P_\lambda) \in T(P_\lambda,\mathfrak{Q}_*)$. Hence knowing that P is an exponential distribution does not help to obtain better tests for the hypothesis that the two samples come from the same distribution. The rank test based on (18.4.10), which becomes in this case

$$n^{-1/2} \sum_{\nu=1}^{n} \log \frac{1-R''_{n\nu}/2n}{1-R'_{n\nu}/2n} ,$$

is as. efficient even if the distributions are known to be exponential.

If we consider a family $\mathfrak{Q}_* = \{P_\lambda: \lambda > 0\}$ with $F_{P_\lambda}(x) = 1-\exp[-T(\lambda x)]$, where T is an increasing twice differentiable function with $\lim_{x\downarrow 0} T(x)=0$ and $\lim_{x\uparrow\infty} T(x) = \infty$, then $b(x,P_\lambda) = 1 - T(\lambda x)$ will, in general, not belong to $T(P_\lambda,\mathfrak{Q}_*)$, so that the restriction from a full family $\mathfrak{P}_*$ to the family $\mathfrak{Q}_*$ leads to an as. envelope power function with steeper slope.

Let us now consider the problem of obtaining as. efficient test-sequences for the hypothesis $\mathfrak{P}_o = \{P\times P: P \in \mathfrak{P}_*\}$ if $\mathfrak{P}_*$ is a full family. By (18.4.4) we have

(18.4.7) $\qquad T^\perp(P\times P;\mathfrak{P}_o,\mathfrak{P}) = \{(x,y) \to c(b(y,P)-b(x,P)): c \in \mathbb{R}\}$.

The heuristic principle (12.2.1) suggests to use the test-statistic

$$(18.4.8) \qquad \underline{z} \to n^{-1/2} \sum_{\nu=1}^{n} (b_n(y_\nu,\underline{z}) - b_n(x_\nu,\underline{z}))$$

with $\underline{z} = (x_\nu,y_\nu)_{\nu=1,\dots,n}$, where $b_n(\cdot,\underline{z})$ is an appropriate estimator for $b(\cdot,P)$. Whether such estimators are easy to obtain or not depends very much on the particular structure of $b(\cdot,P)$.

Since x_ν,y_ν, $\nu = 1,\dots,n$, are independently and identically distributed, an estimate $P_n(\underline{z},\cdot)$ of P can be obtained (see (10.6.1)) as

$$(18.4.9) \qquad P_n(\underline{z},\cdot) := \tfrac{1}{2}M_n(\underline{x},\cdot) + \tfrac{1}{2}M_n(\underline{y},\cdot) ,$$

where $M_n(\underline{x},\cdot)$ is an appropriate estimate of P, based on $\underline{x}$. But to find such an estimate will not always be easy.

If we consider the model (18.2.2), we have $b(x,P) = \beta_2(F_P(x),\varepsilon)$, so that $b(x,P)$ depends on P only through $F_P(x)$. In this case, it suffices to estimate the distribution function. Let now $G_n(\underline{x},\cdot)$ denote the empirical distribution function, defined by

$$G_n(\underline{x},\xi) := n^{-1} \sum_{\nu=1}^{n} 1_{(-\infty,\xi]}(x_\nu) .$$

Then our estimate for $F_P(\xi)$, derived from (18.4.9), is

$$F_n(\underline{z},\xi) := \frac{1}{2}\, G_n(\underline{x},\xi) + \frac{1}{2}\, G_n(\underline{y},\xi) ;$$

hence

$$F_n(\underline{z},x_\nu) = \frac{1}{2n} R'_{n\nu}(\underline{z})$$

and

$$F_n(\underline{z},y_\nu) = \frac{1}{2n} R''_{n\nu}(\underline{z}) ,$$

where $R'_{n\nu}(\underline{z})$ resp. $R''_{n\nu}(\underline{z})$ denotes the rank of x_ν resp. y_ν in the combined sample $\underline{z}$.

If we use $\beta_2(F_n(\underline{z},\xi),\varepsilon)$ as an estimate for $b(\xi,P) = \beta_2(F_P(\xi),\varepsilon)$, we obtain from (18.4.8) the test statistic

$$(18.4.10) \qquad \underline{z} \to n^{-1/2} \sum_{\nu=1}^{n} (\beta_2(R''_{n\nu}(\underline{z})/2n,\varepsilon) - \beta_2(R'_{n\nu}(\underline{z})/2n,\varepsilon)) .$$

Hence our heuristic principle, applied to Example 18.2.2, leads to rank tests. That tests based on (18.4.10) are, in fact, as. efficient, follows easily from Theorem 2.1(b) in Behnen (1972, p. 1842).

There are other cases in which the estimation of $b(\cdot,P)$ is rather difficult, so for instance in the shift model 18.2.1, where $b(\cdot,P) = \ell'(\cdot,P)$. Stein (1956, Section 5) and Takeuchi (1970) indicate how as. efficient tests can be constructed in this case.

18.5. Estimation of a transformation parameter

For $\theta = (\theta_1,\dots,\theta_k) \in \Theta \subset \mathbb{R}^k$ let $x \to \theta x$ be a transformation of the sample space X. Assume that

$$\mathfrak{P} = \{P\times\theta P\colon P \in \mathfrak{P}_*,\ \theta \in \Theta\}$$

where $\mathfrak{P}_*$ is a full family of p-measures.

We are interested in as. variance bounds for estimators of one of the components of θ, say θ_1, or in the as. envelope power function for tests about θ_1. For this purpose, we define a functional $\kappa|\mathfrak{P}$ by

(18.5.1) $\quad \kappa(P\times\theta P) := \theta_1 \,.$

By Definition 4.1.1, the canonical gradient of this functional, say $\kappa^\bullet(\cdot,P\times\theta P) \in T(P\times\theta P,\mathfrak{P})$, fulfills the relations (see (18.3.2) and (18.3.4))

(18.5.2) $\quad \int\kappa^\bullet(x,y,P\times\theta P)(g(x)+g(\theta^{-1}y))P(dx)\theta P(dy) = 0$ for all $g \in T(P,\mathfrak{P}_*)$,

(18.5.3) $\quad \int\kappa^\bullet(x,y,P\times\theta P)a_i c_{ij}(\theta)f_j(\theta^{-1}y,P)P(dx)\theta P(dy) = a_1$ for all $a \in \mathbb{R}^k$.

By (18.3.3) and (18.3.6) the canonical gradient admits a representation

(18.5.4) $\quad \kappa^\bullet(x,y,P\times\theta P) = \bar{g}(x) + \bar{g}(\theta^{-1}y) + \bar{a}_i c_{ij}(\theta)f_j(\theta^{-1}y,P)$

with $\bar{g} \in T(P,\mathfrak{P}_*)$ and $\bar{a} \in \mathbb{R}^k$. Hence (18.5.2) and (18.5.3) imply

(18.5.5) $\quad \int(2\bar{g}(x) + \bar{a}_i c_{ij}(\theta)f_j(x,P))g(x)P(dx) = 0 \qquad$ for all $g \in T(P,\mathfrak{P}_*)$,

(18.5.6) $\quad \int(\bar{g}(x) + \bar{a}_i c_{ij}(\theta)f_j(x,P))c_{qr}(\theta)f_r(x,P)P(dx) = \delta_{1q}$ for $q=1,\dots,k$.

If $T(P,\mathfrak{P}_*) = \mathscr{L}_*(P)$, relation (18.5.5) implies

(18.5.7) $\quad \bar{g}(x) = -\tfrac{1}{2}\bar{a}_i c_{ij}(\theta)f_j(x,P) \qquad$ for P-a.a. $x \in X$.

Together with (18.5.6) this implies

(18.5.8) $\quad \frac{1}{2}\bar{a}_i c_{ij}(\theta) c_{qr}(\theta) P(f_j(\cdot,P) f_r(\cdot,P)) = \delta_{1q} \qquad$ for $q = 1,\dots,k$.

From (18.5.7) we obtain the solution $\bar{a}(\theta,P)$ as two times the first row of the inverse of

$$(c_{ij}(\theta) c_{qr}(\theta) P(f_j(\cdot,P) f_r(\cdot,P)))_{i,q=1,\dots,k}\,.$$

From (18.5.4) and (18.5.7) we obtain

(18.5.9) $\quad \kappa^{\cdot}(x,y,P\times\theta P) = \frac{1}{2}\bar{a}_i(\theta,P) c_{ij}(\theta)(f_j(\theta^{-1}y,P) - f_j(x,P))$.

Hence the as. variance bound is (use (18.5.9) and (18.5.8))

(18.5.10) $\quad P\times\theta P(\kappa^{\cdot}(\cdot,P\times\theta P)^2) = \frac{1}{2}P((\bar{a}_i(\theta,P) c_{ij}(\theta) f_j(\cdot,P))^2) = \bar{a}_1(\theta,P)$.

To obtain an interpretation for $\bar{a}_1(\theta,P)$, assume now that P is known and that we have to estimate θ from a sample $y_1,\dots,y_n$ governed by θP. From (18.5.3) with $g \equiv 0$ we arrive at (18.5.6) with $g \equiv 0$, which is (18.5.8) without the factor $\frac{1}{2}$. This leads to the as. variance bound $\frac{1}{2}\bar{a}_1(\theta,P)$.

To summarize: If P is unknown, we need another sample $x_1,\dots,x_n$ governed by P (in addition to the sample $y_1,\dots,y_n$ governed by θP); the resulting estimator for θ based on these two samples has an as. variance bound twice the as. variance bound of the estimator for θ based on $y_1,\dots,y_n$ in case P is known.

Finally, we consider the restriction from the full family $\mathfrak{P}_*$ to the parametric subfamily $\mathfrak{Q}_* = \{\tau\bar{Q}: \tau \in \Theta\}$, where $\bar{Q}$ is a fixed p-measure. We have

(18.5.11) $\quad T(\tau\bar{Q}\times\theta(\tau\bar{Q}), \{\tau\bar{Q}\times\eta\bar{Q}: \tau,\eta \in \Theta\})$

$$= \{(x,y) \to a_i f_i(x,\tau\bar{Q}) + b_i f_i(\theta^{-1}y,\tau\bar{Q}): a,b \in \mathbb{R}^k\}.$$

Since this reduced tangent space contains the canonical gradient (18.5.9), we obtain the following

Conclusion: The reduction from a full family $\mathfrak{P}_*$ to any parametric subfamily $\mathfrak{Q}_* = \{\tau\bar{Q}: \tau \in \Theta\}$ does not reduce the as. variance bound for estimators of θ_1. This implies a corresponding result for tests of hypotheses on θ_1 (as a consequence of our general Remark 8.4.9).

For Θ being the group of shifts on $\mathbb{R}$ or the group of dilations on $\mathbb{R}$, this conclusion was arrived at by Stein (1956, Section 5), where he also raised this question for arbitrary linear transformations on $\mathbb{R}$.

For the construction of as. efficient estimators of a shift parameter see van Eeden (1970) and Beran (1974). Weiss and Wolfowitz (1970) consider simultaneous estimation of location and scale parameters, and Wolfowitz (1974) constructs an as. efficient estimator of a scale parameter.

18.6. Estimation in the proportional failure rate model

Let $\mathfrak{P} = \{P\times\theta P: P \in \mathfrak{P}_*, \theta \in \Theta\}$, where $\mathfrak{P}_*$ is a family of p-measures over $(0,\infty)$, and $\Theta \subset \mathbb{R}$, and where θP is the p-measure with distribution function $x \to 1-(1-F_P(x))^\theta$. In this case the failure rate of θP is θ times the failure rate of P (see the lines following (18.2.4)). Our problem is to estimate the proportionality factor θ.

The results on paths and tangent spaces needed for an application of our general theory can be obtained from Section 18.3 by specializing for $B(u,\theta) = 1-(1-u)^\theta$. We have $\beta(u,\theta) = \theta(1-u)^{\theta-1}$, $\beta_1(u,\theta) = -\theta(\theta-1)(1-u)^{\theta-2}$, $\beta_2(u,\theta) = (1-u)^{\theta-1}(1+\theta\log(1-u))$.

Let P_t, $t \downarrow 0$, be a path in $\mathfrak{P}_*$ with derivative g. The transformed path $P_t\times\theta P_t$, $t\downarrow 0$, in $\mathfrak{Q}_\theta$ has derivative

$$(18.6.1)\qquad (x,y) \to g(x)+g(y) - \frac{\theta-1}{1-F_P(y)} \int_0^y g(\eta)P(d\eta)\,.$$

A path $P\times(\theta+ta)P$, $t\downarrow 0$, in $\mathfrak{P}_P$, with $a\in\mathbb{R}$ fixed, has derivative

$$(18.6.2)\qquad (x,y) \to a(\theta^{-1} + \log(1-F_P(y)))\,.$$

Hence

(18.6.3) $\quad T(P\times\theta P,\mathfrak{P}) = \{(x,y) \rightarrow g(x)+g(y) - \frac{\theta-1}{1-F_P(y)} \int\limits_0^y g(\eta)P(d\eta)$

$+ a(\theta^{-1} + \log(1-F_P(y))): g \in T(P,\mathfrak{P}_*),\ a \in \mathbb{R}\}.$

To avoid technicalities, we assume that for any $P \in \mathfrak{P}_*$ the distribution function F_P is strictly increasing. In this case, we may use the representation $g(x) = h(F_P(x))$, where h is defined on (0,1). The condition $\int g(x)P(dx) = 0$ is equivalent to the condition $\int\limits_0^1 h(u)du = 0$, so that $g \in \mathscr{L}_*(P)$ iff $h \in \mathscr{L}_*(E)$, where E is the uniform distribution over (0,1). Let $S(P,\mathfrak{P}_*)$ denote the class of functions h corresponding to the functions g in $T(P,\mathfrak{P}_*)$.

For $h \in \mathscr{L}_*(E)$, $a \in \mathbb{R}$ and $\theta \in \Theta$ we define

$$H(u,v;h,\theta) := h(u) + h(v) - \frac{\theta-1}{1-v}\int\limits_0^v h(\xi)d\xi ,$$

$$K(v;a,\theta) \quad := a(\theta^{-1} + \log(1-v)) .$$

Then the tangent space (18.6.3) may be written as

(18.6.4) $\quad T(P\times\theta P,\mathfrak{P}) = \{(x,y) \rightarrow H(F_P(x),F_P(y);h,\theta) + K(F_P(y);a,\theta):$

$h \in S(P,\mathfrak{P}_*),\ a \in \mathbb{R}\}.$

Now we define a functional κ on $\mathfrak{P}$ by

(18.6.5) $\quad \kappa(P\times\theta P) = \theta, \qquad P \in \mathfrak{P}_*,\ \theta \in \Theta.$

Let $h_o \in S(P,\mathfrak{P}_*)$ and $a_o \in \mathbb{R}$ determine the canonical gradient of κ in $T(P\times\theta P,\mathfrak{P})$, i.e.,

(18.6.6) $\quad \kappa^{\bullet}(x,y;P\times\theta P) = H(F_P(x),F_P(y);h_o,\theta) + K(F_P(y);a_o,\theta)$.

By definition of the gradient (see 4.1.1) we obtain from (18.6.1) and (18.6.2) the conditions

(18.6.7) $\quad \int \kappa^{\bullet}(x,y;P\times\theta P)H(F_P(x),F_P(y);h,\theta)P(dx)\theta P(dy) = 0$

for all $h \in S(P,\mathfrak{P}_*)$,

(18.6.8) $\quad \int \kappa^{\bullet}(x,y;P\times\theta P)K(F_P(y);a,\theta)P(dx)\theta P(dy) = a \qquad$ for all $a \in \mathbb{R}$.

Intuitively speaking, (18.6.7) expresses that $\kappa^{\bullet}$ is orthogonal to $T(P\times\theta P,\mathfrak{Q}_\theta)$ (because κ remains constant on $\mathfrak{Q}_\theta$).

From (18.6.6), (18.6.7) and (18.6.8), applied for $h=h_o$ and $a=a_o$, we obtain the following as. variance bound for estimators of θ:

$$\sigma^2(P\times\theta P) := \int \kappa^{\bullet}(x,y;P\times\theta P)^2 P(dx)\theta P(dy) \tag{18.6.9}$$
$$= \int \kappa^{\bullet}(x,y;P\times\theta P)[H(F_P(x),F_P(y);h_o,\theta)+K(F_P(y);a_o,\theta)]P(dx)\theta P(dy)$$
$$= a_o \; .$$

To determine a_o and h_o, we proceed as follows. Because of (18.6.6), conditions (18.6.7) and (18.6.8) can be written as

$$\int_o^1 [h_o(u) + h_o(v) - \frac{\theta-1}{1-v}\int_o^v h_o(\xi)d\xi + a_o(\theta^{-1} + \log(1-v))] \tag{18.6.10}$$
$$[h(u) + h(v) - \frac{\theta-1}{1-v}\int_o^v h(\xi)d\xi](1-v)^{\theta-1}dudv = 0$$
$$\text{for all } h\in S(P,\mathfrak{P}_*),$$

$$\int_o^1 [h_o(u) + h_o(v) - \frac{\theta-1}{1-v}\int_o^v h_o(\xi)d\xi + a_o(\theta^{-1} + \log(1-v))] \tag{18.6.11}$$
$$[\theta^{-1} + \log(1-v)]\theta(1-v)^{\theta-1}dudv = 1 \; .$$

Carrying through the integration over u and using partial integration with respect to v we obtain from (18.6.10) and (18.6.11)

$$\int_o^1 [h_o(v)(1+\theta(1-v)^{\theta-1}) + \theta(\theta-1)\int_o^v h_o(\xi)(1-\xi)^{\theta-2}d\xi \tag{18.6.12}$$
$$- \frac{\theta(\theta-1)}{1-v}\int_o^v h_o(\xi)(1-\xi)^{\theta-1}d\xi$$
$$+ a_o(\theta(1-v)^{\theta-1}-1)/(\theta-1)]h(v)dv = 0 \qquad \text{for all } h\in S(P,\mathfrak{P}_*),$$

$$\int_o^1 [h_o(v) - \frac{\theta-1}{1-v}\int_o^v h_o(\xi)d\xi + \frac{a_o}{\theta}(1+\theta\log(1-v))] \tag{18.6.13}$$
$$[1+\theta\log(1-v)](1-v)^{\theta-1}dv = 1 \; .$$

It is easy to see that any solution h_o of (18.6.12) is also a solution of (18.6.10). This implies that, for any a_o, (18.6.12) has at most one solution. For let h_o' and h_o'' be two solutions. Then (18.6.10) holds with $a_o = 0$ and $h_o = h_o'-h_o''$. Applied for $h = h_o'-h_o''$ this shows that $h_o' = h_o''$.

If $\mathfrak{P}_*$ is a full family, i.e. $T(P,\mathfrak{P}_*) = \mathscr{L}_*(P)$ and therefore $S(P,\mathfrak{P}_*) = \mathscr{L}_*(E)$, relation (18.6.12) implies

$$(18.6.14)\quad \begin{aligned} &h_o(v)(1+\theta v^{\theta-1}) + \theta(\theta-1)\int_o^v h_o(\xi)(1-\xi)^{\theta-2}d\xi \\ &- \frac{\theta(\theta-1)}{1-v}\int_o^v h_o(\xi)(1-\xi)^{\theta-1}d\xi + a_o[\theta(1-v)^{\theta-1}-1]/(\theta-1) = 0 \end{aligned}$$

for E-a.a. $v \in (0,1)$.

By a straightforward calculation one can verify that

$$(18.6.15)\quad h_o(v,\theta) := a_o(\theta)\left(-\frac{(1-v)^{\theta-1}}{1+\theta(1-v)^{\theta-1}} + \frac{\log[(1+\theta)/(1+\theta(1-v)^{\theta-1})]}{\theta(\theta-1)}\right)$$

is a function in $\mathscr{L}_*(E)$ which fulfills (18.6.14).

So far, we tacitly assumed that $\theta \neq 1$. The degenerate case $\theta = 1$ needs a separate treatment and leads to

$$(18.6.15')\quad h_o(v,1) := -\tfrac{1}{2}a_o(\theta)(1 + \log(1-v)) .$$

Using (18.6.13) and (18.6.15) (resp. (18.6.15')) we obtain

$$a_o(\theta) = \theta^2 / \int_o^1 (1+\theta z^{(\theta-1)/\theta})^{-1}\, dz .$$

Because of (18.6.9) this implies

$$(18.6.16)\quad \sigma^2(P\times\theta P) = \theta^2 / \int_o^1 (1+\theta z^{(\theta-1)/\theta})^{-1}\, dz .$$

The as. variance bound (18.6.16) has recently been obtained by Begun and Wellner (1981, Theorem 1) for the more general case of unequal sample sizes. (The reader should not be confused by the factor $\frac{1}{2}$ occuring in their theorem. The standardization of their estimator with $(mn/(m+n))^{1/2}$ corresponds for $m = n$ to a standardization with $n^{1/2}/2^{1/2}$.) Working with 'least favorable' contiguous p-measures they obtain for this model an analogue to Hájek's convolution theorem (corresponding to our general Theorem 9.3.1). The as. variance bound (18.6.16) is sharp, because it is attained, for instance, by Cox's 'partial likelihood estimator' (see Efron, 1977, Section 4).

We remark that the as. variance bound given by (18.6.16) does not depend on the p-measure P. This peculiarity should not be confounded with the question whether a more precise knowledge about P can be used to obtain better estimators for θ. This is, indeed, the case. Assume, for example, that $\mathfrak{P}_*$ is the family of exponential distributions P_λ, $\lambda > 0$, with Lebesgue densities $x \to \lambda \exp[-\lambda x]$, $x > 0$. Then the tangent space of P_λ in this family, i.e. $T(P_\lambda, \mathfrak{P}_*)$, is the linear space generated by $1-\lambda x$. Hence $S(P_\lambda, \mathfrak{P}_*)$ is the linear space generated by $v \to 1 + \log(1-v)$. This space does not contain the function h_o given by (18.6.15), unless $\theta = 1$. Hence a projection of the canonical gradient into the smaller tangent space of this model leads, for $\theta \neq 1$, to a lower as. variance bound. It is, however, not necessary to carry this through, since now we are back to a parametric model, and the canonical gradient can be obtained from Proposition 5.3.1. This leads to the as. variance bound $2\theta^2$. The same holds true if we wish to test a hypothesis about θ, excepting the case $\theta = 1$.

18.7. Dependent samples

In this section we consider observations $(x,y) \in X^2$ which are - perhaps - dependent. If we think of y as an observation under treatment, and x as the control, then the hypothesis that the treatment has no effect is equivalent to the hypothesis that the distribution $P|\mathscr{A}^2$ of (x,y) is (bivariate) symmetric in the sense that $P*((x,y) \to (y,x)) = P$. (Bivariate symmetry of P implies that the two marginals are identical, but is a stronger property.)

A function $f|X^2$ will be called symmetric if $f(y,x) = f(x,y)$ for all $(x,y) \in X^2$. If P has a $\mu\times\mu$-density, then symmetry of P is equivalent to symmetry μ^2-a.e. of its density.

Let $\mathfrak{P}_o$ be a full family of symmetric distributions on $\mathscr{A}^2$. Then we have

(18.7.1) $T(P,\mathfrak{P}_o) = \mathscr{S}(P)$,

where $\mathscr{S}(P)$ is the class of all symmetric functions in $\mathscr{L}_*(P)$.

It is straightforward to show that any skew-symmetric function in $\mathscr{L}_*(P)$ (i.e. $f(y,x) = -f(x,y)$) is orthogonal to $\mathscr{S}(P)$. Any function $f \in \mathscr{L}_*(P)$ may be written as

(18.7.2) $f(x,y) = \frac{1}{2}(f(x,y) + f(y,x)) + \frac{1}{2}(f(x,y) - f(y,x))$,

with the first term being the projection of f into $\mathscr{S}(P)$, and the second term being orthogonal to $\mathscr{S}(P)$.

This implies that the co-space of $\mathscr{S}(P)$ in $\mathscr{L}_*(P)$ is the family of all skew-symmetric functions. Since this space is infinite-dimensional, the problem of finding a test for $\mathfrak{P}_o$ which is as. optimal in a full family $\mathfrak{P}$ is indeterminate. Even if we restrict ourselves to certain types of alternatives for which y is stochastically larger than x (see Yanagimoto and Sibuya, 1972a, Schaafsma, 1976, Snijders, 1981), the co-space remains too large. To get a meaningful problem we have to assume that there is a close inherent relationship between the variables x and y. A natural restriction of this kind is the following.

Assume that the family $\mathfrak{P}$ of alternatives is generated from $\mathfrak{P}_o$ by a map $(\theta,P) \to \theta P$, $\theta \in \Theta \subset \mathbb{R}$, $P \in \mathfrak{P}_o$, i.e.,

(18.7.3) $\mathfrak{P} = \{\theta P: P \in \mathfrak{P}_o , \theta \in \Theta\}$.

Assume the existence of $\varepsilon \in \Theta$ such that $\varepsilon P = P$ for all $P \in \mathfrak{P}_o$. Since Θ is one-dimensional, it is natural to assume that (under appropriate smoothness conditions on $\theta \to \theta P$) the path $(\varepsilon+t)P$, $t \downarrow 0$, has derivative $b(\cdot,P) \in \mathscr{L}_*(P)$, so that

(18.7.4) $T(P,\{\theta P: \theta \in \Theta\}) = [b(\cdot,P)]$.

The considerations of Section 2.6 suggest that

(18.7.5) $T(P,\mathfrak{P}) = \mathscr{S}(P) + [b(\cdot,P)]$.

According to (18.7.2) and Remark 8.4.5, the slope of the as. envelope power function for testing the hypothesis $\mathfrak{P}_o$ (of bivariate symmetry) against alternatives $(\varepsilon+n^{-1/2}t)P$ is

(18.7.6) $$\tfrac{1}{2}(\int (b(x,y,P) - b(y,x,P))^2 P(d(x,y)))^{1/2}$$
$$= 2^{-1/2}[\int b(x,y,P)^2 P(d(x,y)) - \int b(x,y,P)b(y,x,P)\, P(d(x,y))]^{1/2}.$$

As a particular instance, we mention the *transformation model*, where Θ is a transformation group acting on $X \subset \mathbb{R}^m$, and

(18.7.7) $$\theta P = P*((x,y) \to (x,\theta y)) .$$

In this case we obtain by the same arguments as for (18.3.4) that (18.7.4) holds with

(18.7.8) $$b(x,y,P) = h_\alpha(y)\ell^{(\alpha)}(x,y,P) - \int h_\alpha(\eta)\ell^{(\alpha)}(\xi,\eta,P)P(d(\xi,\eta)),$$

where $\ell^{(\alpha)}(x,y,P) = (\partial/\partial y_\alpha)\log p(x,y)$.

To motivate the transformation model, think of an experiment in which the two eyes of a test animal are infected. One of the eyes gets treatment A, the other one treatment B. For test animal ν let x_ν denote the curing time under treatment A, y_ν the curing time under treatment B. The hypothesis 'treatments A and B have the same effect' is equivalent to the hypothesis that the joint distribution of (x_ν,y_ν) is symmetric in its arguments. To find a probabilistic model for the possible alternatives, assume there is some variation of the curing time without treatment between the individuals, represented by the p-measure M, and there are treatment effects which act independently and multiplicatively, so that the joint distribution of the observed curing time has a density

$$(x,y) \to \int_0^\infty q_A(x\xi)q_B(y\xi)\xi^2 M(d\xi) ,$$

where the densities q_A and q_B correspond to treatments A and B, respectively.

If both treatments have the same effect, we have $q_A = q_B$, so that

$$p(x,y) := \int_0^\infty q_A(x\xi)q_A(y\xi)\xi^2 M(d\xi)$$

is symmetric in its arguments.

If treatment A is superior to treatment B, the random variable with density q_B is stochastically larger than the random variable with density q_A. A particularly simple assumption describing such an effect is $q_B(y) = \theta^{-1}q_A(\theta^{-1}y)$ for some $\theta > 1$. Then the joint distribution of the curing time has density

$$p(x,y,\theta) := \theta^{-1} \int_0^\infty q_A(x\xi)q_A(\theta^{-1}y\xi)\xi^2 M(d\xi) ,$$

so that $p(x,y,\theta) = \theta^{-1}p(x,\theta^{-1}y)$. Hence the variables (x,y) obey the transformation model with $\theta y = \theta \cdot y$.

Following the general ideas outlined in Section 8.6 we discuss now two particular instances of subfamilies which are much smaller than the family $\mathfrak{P}$ (defined by (18.7.3)), without admitting as. better tests for the hypothesis of symmetry.

At first we consider the case that the variables (x,y) are independent, i.e. that θP is a product measure for every $P \in \mathfrak{P}_o$, $\theta \in \Theta$. In particular: the b i v a r i a t e s y m m e t r i c $P \in \mathfrak{P}_o$ is a product of i d e n t i c a l components, say $P = Q \times Q$. More precisely, we have $\mathfrak{Q}_o = \{Q \times Q : Q \in \mathfrak{P}_*\}$, where $\mathfrak{P}_*$ is a full family of p-measures on $\mathscr{A}$. According to Proposition 2.4.1,

(18.7.9) $\quad T(Q \times Q, \mathfrak{Q}_o) = \{(x,y) \to g(x) + g(y) : g \in \mathscr{L}_*(Q)\}$.

If $\theta(Q \times Q)$ is a product measure for every $Q \in \mathfrak{P}_*$, $\theta \in \Theta$, we necessarily have

(18.7.10) $\quad b(x,y,Q \times Q) = b_1(x,Q) + b_2(y,Q)$.

The assumptions that $\theta(Q \times Q)$ is a product measure for every $\theta \in \Theta$ is in particular true in case of the transformation model $\theta P = P * ((x,y) \to (x,\theta y))$. Since $(\partial/\partial y_\alpha)\log q(x)q(y) = \ell^{(\alpha)}(y,Q)$, relation (18.7.8) yields

$$b(x,y,Q \times Q) = h_i(y)\ell^{(i)}(y,Q) - Q(h_i\ell^{(i)}(\cdot,Q)) .$$

The projection of b into $T(Q \times Q, \mathfrak{P}_o) = \mathscr{S}(Q \times Q)$ is (see (18.7.2))

$$\begin{aligned}(x,y) \to\ & \tfrac{1}{2}(b_1(x,Q) + b_2(y,Q)) + \tfrac{1}{2}(b_1(y,Q) + b_2(x,Q)) \\ &= b_o(x,Q) + b_o(y,Q)\end{aligned}$$

with $b_o(z,Q) = \frac{1}{2}b_1(z,Q) + \frac{1}{2}b_2(z,Q)$. Since this projection belongs to $T(Q\times Q,\mathfrak{Q}_o)$ (see (18.7.9)), we obtain from Remark 8.6.12 the following

Conclusion: If we test the hypothesis of bivariate symmetry of P against alternatives θP, the as. envelope power function does not increase if we assume, in addition, independence of the variables (under the hypothesis as well as under all alternatives).

Of course, estimating the function $(x,y) \to b(x,y,P)$, assuming symmetry of P only, may be much more difficult than estimating the function $y \to b_o(y,Q)$. Hence our conclusion, being a s y m p t o t i c in nature, may not be representative for small samples.

Let now Θ be a transformation group acting on $X \subset \mathbb{R}^m$. We replace the family $\mathfrak{P} = \{\theta P\colon P \in \mathfrak{P}_o, \theta \in \Theta\}$, with $\theta P = P*((x,y) \to (x,\theta y))$, by the parametric family $\mathfrak{Q} = \{\bar{Q}_{\delta,\tau}\colon \delta,\tau \in \Theta\}$, where $\bar{Q} \in \mathfrak{P}_o$ is a fixed bivariate symmetric p-measure on $\mathscr{A}^2$, and

$$\bar{Q}_{\delta,\tau} := \bar{Q}*((x,y) \to (\delta x,\tau y)) .$$

We have $\mathfrak{Q} \subset \mathfrak{P}$, since $\bar{Q}_{\delta,\tau} = \tau\delta^{-1}\bar{Q}_{\delta,\delta}$, where $\bar{Q}_{\delta,\delta} \in \mathfrak{P}_o$ and $\tau\delta^{-1} \in \Theta$.

Let $\bar{q}$ be a density of $\bar{Q}$ with respect to $\mu\times\mu$, where μ may be assumed covariant w.l.g. Then $\bar{Q}_{\delta,\tau}$ has $\mu\times\mu$-density

$$(x,y) \to \Delta(\delta^{-1})\Delta(\tau^{-1})\bar{q}(\delta^{-1}x,\tau^{-1}y) .$$

We obtain similarly as in (18.3.4) that the path $\bar{Q}_{\delta+ta_1,\tau}$, $t \downarrow 0$, has derivative

$$a_1c(\delta)f_1(\delta^{-1}x,\tau^{-1}y;\bar{Q})$$

with

$$f_1(x,y;\bar{Q}) = h_i(x)\ell^{(i)}(x,y,\bar{Q}) - \int h_i(\xi)\ell^{(i)}(\xi,\eta,\bar{Q})\bar{Q}(d(\xi,\eta)), \tag{18.7.11}$$

where $\ell^{(i)}(x,y,\bar{Q}) = (\partial/\partial x_i)\log \bar{q}(x,y)$. Analogously, the path $\bar{Q}_{\delta,\tau+a_2t}$, $t \downarrow 0$, has derivative

$$a_2c(\tau)f_2(\delta^{-1}x,\tau^{-1}y;\bar{Q}) .$$

Due to the symmetry of $\bar{Q}$ we have

$$f_2(x,y;\bar{Q}) = f_1(y,x;\bar{Q}) .$$

We obtain

(18.7.12) $$T(\bar{Q}_{\delta,\delta},\mathfrak{Q}) = [(x,y) \to f_1(\delta^{-1}x,\delta^{-1}y;\bar{Q}),\ (x,y) \to f_1(\delta^{-1}y,\delta^{-1}x;\bar{Q})],$$

(18.7.13) $$T(\bar{Q}_{\delta,\delta},\mathfrak{Q}_o) = [(x,y) \to f_1(\delta^{-1}x,\delta^{-1}y;\bar{Q}) + f_1(\delta^{-1}y,\delta^{-1}x;\bar{Q})].$$

Hence

$$T(\bar{Q}_{\delta,\delta},\mathfrak{Q}) = T(\bar{Q}_{\delta,\delta},\mathfrak{Q}_o) + [(x,y) \to f_1(\delta^{-1}y,\delta^{-1}x;\bar{Q})].$$

For the general transformation model (18.7.7), we obtain from (18.7.1), (18.7.5) and (18.7.8) that

$$T(P,\mathfrak{P}) = \mathscr{S}(P) + [(x,y) \to h_\alpha(y)\ell^{(\alpha)}(x,y,P) - \int h_\alpha(\eta)\ell^{(\alpha)}(\xi,\eta,P)P(d(\xi,\eta))].$$

Applied for $P = Q_{\delta,\delta}$ this yields

$$T(\bar{Q}_{\delta,\delta},\mathfrak{P}) = \mathscr{S}(\bar{Q}_{\delta,\delta}) + [(x,y) \to f_1(\delta^{-1}y,\delta^{-1}x;\bar{Q})].$$

The projection of $(x,y) \to f_1(\delta^{-1}y,\delta^{-1}x;\bar{Q})$ into $T(\bar{Q}_{\delta,\delta},\mathfrak{P}_o) = \mathscr{S}(\bar{Q}_{\delta,\delta})$ is (see (18.7.2))

$$(x,y) \to \tfrac{1}{2}f_1(\delta^{-1}y,\delta^{-1}x;\bar{Q}) + \tfrac{1}{2}f_1(\delta^{-1}x,\delta^{-1}y;\bar{Q}) .$$

Since this projection belongs to $T(\bar{Q}_{\delta,\delta},\mathfrak{Q}_o)$ (see (18.7.13)), we obtain from Remark 8.6.12 the following

Conclusion: If we restrict in the transformation model the basic family from $\mathfrak{P}$ to a parametric family $\mathfrak{Q} = \{\bar{Q}_{\delta,\tau}: \delta,\tau \in \Theta\}$, this does not improve the slope of the as. envelope power function for testing the hypothesis of bivariate symmetry.

Presuming the existence of as. efficient tests, we may illustrate our conclusions by the following example: Assume that the variables x,y are only allowed to differ by a shift. If it is known that the two variables are independent and normally distributed with equal variance, the t-test is the optimal similar test for equality of means. Asymptotically, the slope of this test agrees with the slope of an as. efficient as. similar test for bivariate symmetry which presumes neither independence nor a particular shape of the distribution.

<u>18.7.14. Remark</u>. In case of the shift model $\theta P = P*((x,y) \to (x,y+\theta))$, it is usual to base tests of the hypothesis $\theta = 0$ (i.e. bivariate symmetry) on statistics depending on (x,y) through $y-x$ only. If (x,y) is distributed with density $(x,y) \to p(x,y-\theta)$ then $y-x$ is distributed with density $z \to p_o(z-\theta)$, where

$$(18.7.15) \qquad p_o(z) = \int p(x,z+x)dx .$$

Since p is symmetric in its arguments, p_o is symmetric about zero. It follows from Example 8.1.1 that the as. envelope power function for the hypothesis of symmetry about zero has under alternatives $p_o(z-n^{-1/2}t)$ the slope

$$(\int \ell_o'(z)^2 p_o(z)dz)^{1/2} ,$$

where $\ell_o'(z) = (\partial/\partial z)\log p_o(z)$.

Since the as. envelope power function for tests based on y-x cannot be better than the as. envelope power function for tests based on (x,y), we have from (18.7.6), applied for $b(x,y,P) = \ell_2(x,y,P)$

$$(18.7.16) \qquad \int \ell_o'(z)^2 p_o(z)dz \le \tfrac{1}{4}\int (\ell_2(x,y,P)-\ell_2(y,x,P))^2 p(x,y)dxdy .$$

In this special case, (18.7.16) can also be seen directly. From (18.7.15), $p_o'(z) = \int p_2(x,x+z)dx$, where $p_2(x,y) = (\partial/\partial y)p(x,y)$. Moreover, symmetry of p implies

$$\int p_2(x,x+z)dx = -\int p_2(x,x-z)dx = -\int p_2(x+z,x)dx,$$

hence

$$p_o'(z) = \tfrac{1}{2}\int (p_2(x,x+z) - p_2(x+z,x))dx .$$

By Schwarz's inequality

$$(18.7.17) \qquad (\tfrac{1}{2}\int (p_2(x,x+z) - p_2(x+z,x))dx)^2$$
$$\le \tfrac{1}{4}\int \frac{(p_2(x,x+z) - p_2(x+z,x))^2}{p(x,x+z)} dx \int p(x,x+z)dx .$$

This implies

$$\ell_o'(z)^2 p_o(z) \le \tfrac{1}{4}\int (\ell_2(x,x+z)-\ell_2(x+z,x))^2 p(x,x+z)dx$$

for all $z \in \mathbb{R}$.

Relation (18.7.16) follows by integration over z.

Notice that inequality (18.7.16) is strict unless equality holds in (18.7.17) for Lebesgue-a.a. $z \in \mathbb{R}$, which can be true only if $x \to (p_2(x,x+z) - p_2(x+z,x))/p(x,x+z)^{1/2}$ is proportional to $x \to p(x,x+z)^{1/2}$ for Lebesgue-a.a. $z \in \mathbb{R}$, i.e.,

$$(p_2(x,x+z) - p_2(x+z,x))/p(x,x+z)^{1/2} = a(z)p(x,x+z)^{1/2}$$

for Lebesgue-a.a. $z \in \mathbb{R}$. If p, p_2 are continuous, this implies

(18.7.18) $\quad \ell_2(x,y,P) - \ell_2(y,x,P) = a(y-x)$.

This differential equation is fulfilled for $P = N(\mu,\mu,\sigma^2,\sigma^2,\rho)$, so that in this case the restriction to tests depending on x-y entails no loss of as. efficiency. This normal family is not the only family of bivariate symmetric p-measures for which as. efficient tests may be based on x-y. All p-measures with density $(x,y) \to c \exp[f(x-y)+g(x+y)]$ fulfill an equation of type (18.7.18) (with $a(z) = f'(-z) - f'(z)$).

19. APPENDIX

19.1. Miscellaneous lemmas

By a *null-function* N we understand a function $N: (0,\infty) \to [0,\infty)$ such that $N(s) \downarrow 0$ for $s \downarrow 0$.

19.1.1. Lemma. *Let* $M: (0,\infty)^2 \to [0,\infty)$ *be such that*

(i) $u \to M(u,s)$ *is nonincreasing and right continuous for every* $s > 0$,

(ii) $s \to M(u,s)$ *is a null-function for every* $u > 0$.

Then there exists a null-function N *such that* $M(N(s),s) \to 0$ *for* $s \to 0$.

Proof. Since $u \to M(u,s)$ is nonincreasing, the set $U(s) := \{u > 0: M(u,s) \leq u\}$ is nonempty for every $s > 0$. Let $N(s) := \inf U(s)$. Since $u \to M(u,s)$ is right continuous, we have $U(s) = [N(s),\infty)$; in particular, $M(N(s),s) \leq N(s)$. Since $s \to M(u,s)$ is nondecreasing, $s \to N(s)$ is nonincreasing. It remains to be shown that $N(s) \to 0$ for $s \to 0$. To see this, observe that for every $\varepsilon > 0$ there exists s_ε such that $s < s_\varepsilon$ implies $M(\varepsilon,s) < \varepsilon$. Hence $N(s) < \varepsilon$ for $s < s_\varepsilon$.

The following uniform version of the central limit theorem is a special case of Theorem 18.1 in Bhattacharya and Rao (1976, p. 181).

<u>19.1.2. Theorem</u>. *There exists a constant* $c > 0$ *such that for every* $P|\mathscr{A}$, *every measurable function* $f: X \to \mathbb{R}$ *with* $P(f) = 0$ *and* $P(f^2) = 1$, *and every* $n \in \mathbb{N}$, $t \in \mathbb{R}$, $\varepsilon > 0$,

$$|P^n * \tilde{f}(-\infty,t) - \Phi(t)| \leq c(P(f^2 1_{\{|f| > \varepsilon n^{1/2}\}}) + \varepsilon).$$

We need the following generalization of a result of Bahadur (1964, p. 1549, Lemma 4), due to Droste and Wefelmeyer (1982).

<u>19.1.3. Lemma</u>. *Let* $f_n, f: \mathbb{R}^k \to \mathbb{R}$ *be measurable with* $f_n \to f$ λ^k*-a.e., and let* $y_n \in \mathbb{R}^k$, $n \in \mathbb{N}$, *fulfill* $y_n \to 0$. *Then there exists a subsequence* $\mathbb{N}_o$ of $\mathbb{N}$ *such that* $(f_n(x+y_n))_{n \in \mathbb{N}_o} \to f(x)$ *for* λ^k*-a.a.* $x \in \mathbb{R}^k$.

<u>19.1.4. Lemma</u>. *Let* $1 \leq s < \infty$ *and* $P|\mathbb{B}$ *a p-measure with Lebesgue density* p. *Assume that* p *is locally bounded and bounded away from zero (i.e., for each* $x \in \mathbb{R}$ *there exists a neighborhood* U *of* x *such that* $\sup\{p(\xi): \xi \in U\} < \infty$ *and* $\inf\{p(\xi): \xi \in U\} > 0$*).*

Assume that $f: \mathbb{R} \to \mathbb{R}$ *fulfills the following regularity condition: There exist* $\eta > 0$ *and* $g \in \mathscr{L}_s(P)$ *such that* $|f(\xi)| \leq g(x)$ *for* $\xi, x \in \mathbb{R}$ *with* $|\xi - x| < \eta$.

Then $\lim_{t \to 0} \int |f(\xi+t) - f(\xi)|^s P(d\xi) = 0$.

<u>Proof</u>. Let $\varepsilon > 0$ be arbitrary. Since $g \in \mathscr{L}_s(P)$, there exists a compact K_ε such that

$$\left(\int_{K_\varepsilon^c} g(\xi)^s P(d\xi)\right)^{1/s} < \varepsilon .$$

Let C_ε be a compact neighborhood of K_ε, and define

$$b_\varepsilon := \sup\{p(\xi): \xi \in C_\varepsilon\} .$$

Since p is locally bounded, we have $b_\varepsilon < \infty$. Hence for $|t| < \eta$

$$\left(\int |f(\xi+t) - f(\xi)|^s P(d\xi)\right)^{1/s} < 2\varepsilon + b_\varepsilon^{1/s}\left(\int_{K_\varepsilon} |f(\xi+t) - f(\xi)|^s d\xi\right)^{1/s}.$$

Since p is locally bounded away from zero, we have $f 1_{C_\varepsilon} \in \mathscr{L}_s(\lambda)$. Since the continuous functions are dense in $\mathscr{L}_s(\lambda)$ (see, e.g., Ash, 1972, p. 88, Theorem 2.4.14), there exists a continuous function $f_\varepsilon\colon C_\varepsilon \to \mathbb{R}$ such that

$$\Big(\int_{C_\varepsilon} |f(\xi)-f_\varepsilon(\xi)|^s d\xi\Big)^{1/s} < \varepsilon/b_\varepsilon^{1/s}.$$

Moreover,

$$\begin{aligned}
&\Big(\int_{K_\varepsilon} |f(\xi+t)-f(\xi)|^s d\xi\Big)^{1/s} \\
&\le \Big(\int_{K_\varepsilon} |f(\xi+t)-f_\varepsilon(\xi+t)|^s d\xi\Big)^{1/s} + \Big(\int_{K_\varepsilon} |f_\varepsilon(\xi+t)-f_\varepsilon(\xi)|^s d\xi\Big)^{1/s} \\
&\quad + \Big(\int_{K_\varepsilon} |f(\xi)-f_\varepsilon(\xi)|^s d\xi\Big)^{1/s} \\
&\le 2\varepsilon/b_\varepsilon^{1/s} + \Big(\int_{K_\varepsilon} |f(\xi+t)-f_\varepsilon(\xi)|^s d\xi\Big)^{1/s},
\end{aligned}$$

since

$$\begin{aligned}
\int_{K_\varepsilon} |f(\xi+t)-f_\varepsilon(\xi+t)|^s d\xi &= \int_{K_\varepsilon+t} |f(\xi)-f_\varepsilon(\xi)|^s d\xi \\
&\le \int_{C_\varepsilon} |f(\xi)-f_\varepsilon(\xi)|^s d\xi,
\end{aligned}$$

if t is so small that $K_\varepsilon+t \subset C_\varepsilon$.

Since f_ε is continuous, it is uniformly continuous on C_ε. Hence there exists $\delta_\varepsilon > 0$ such that $|t| < \delta_\varepsilon$ implies

$$|f_\varepsilon(\xi+t)-f_\varepsilon(\xi)| < \varepsilon/b_\varepsilon^{1/s}\lambda(K_\varepsilon)^{1/s} \qquad \text{for all } x \in K_\varepsilon.$$

From this we obtain for $|t| < \delta_\varepsilon$

$$\Big(\int_{K_\varepsilon} |f_\varepsilon(\xi+t)-f_\varepsilon(\xi)|^s d\xi\Big)^{1/s} < \varepsilon/b_\varepsilon^{1/s}.$$

Hence for t sufficiently small,

$$\Big(\int |f(\xi+t)-f(\xi)|^s P(d\xi)\Big)^{1/s} < 5\varepsilon.$$

The following version of the Neyman-Pearson lemma is due to Lehmann and Stein (1948, p. 497).

19.1.5. Lemma. *Let* $f,g: X \to \mathbb{R}$ *be* $\mathscr{A}$*-measurable and finitely* μ*-integrable, and* $\psi: X \to [0,1]$ $\mathscr{A}$*-measurable such that for some* $c \geq 0$,

$$\psi(x) = \begin{cases} 1 & \text{if} \quad g(x) \gtrless cf(x) . \\ 0 & \end{cases}$$

Then for every $\mathscr{A}$*-measurable* $\varphi: X \to [0,1]$,

$$\mu(g(\varphi-\psi)) \leq c\mu(f(\varphi-\psi)).$$

19.2. Asymptotic normality of log-likelihood ratios

In this section we present some results on the as. normality of log-likelihood ratios. Related results can be found in LeCam (1966, 1970) and Hájek (1972).

We write $\| \ \|$ for $\| \ \|_Q$. For our basic Theorem 19.2.7 we need the following assumption.

19.2.1. Assumption. Let N_i, $i = 0,1,2$, be null-functions. A pair of p-measures Q,P fulfills Assumption 19.2.1 if the Q-density of P admits a representation $1+g+r$ such that

(19.2.2) $\quad Q(g^2 1_{\{|g| > c\|g\|\}}) \leq \|g\|^2 N_0(c^{-1}) \quad$ for all $c > 0$,

(19.2.3) $\quad Q(|r| 1_{\{|r| > 1\}}) \leq \|g\|^2 N_1(\|g\|)$,

(19.2.4) $\quad Q(r^2 1_{\{|r| \leq 1\}}) \leq \|g\|^2 N_2(\|g\|)$.

19.2.5. Remark. (i) Condition (19.2.2) implies for every $\varepsilon > 0$ the existence of a null-function $M(\varepsilon,\cdot)$ such that

$$Q(g^2 1_{\{|g| > \varepsilon\}}) \leq \|g\|^2 M(\varepsilon, \|g\|).$$

(ii) Conditions (19.2.3) and (19.2.4) imply for every $\varepsilon > 0$ the existence of a null-function $M(\varepsilon,\cdot)$ such that

$$Q(|r|1_{\{|r|>\varepsilon\}}) \leq \|g\|^2 M(\varepsilon, \|g\|) .$$

Proof. (i) $Q(g^2 1_{\{|g|>\varepsilon\}}) \leq \|g\|^2 N_o(\|g\|/\varepsilon)$.

(ii) If $\varepsilon < 1$ (w.l.g.), then

$$Q(|r|1_{\{|r|>\varepsilon\}}) = Q(|r|1_{\{1\geq|r|>\varepsilon\}}) + Q(|r|1_{\{|r|>1\}})$$

$$\leq \varepsilon^{-1} Q(r^2 1_{\{|r|\leq 1\}}) + Q(|r|1_{\{|r|>1\}})$$

$$\leq \|g\|^2 (N_1(\|g\|) + \varepsilon^{-1} N_2(\|g\|)).$$

19.2.6. Remark. If there exist $c > 0$ and null-functions N_i', $i = 1,2$, such that

$$(19.2.3') \qquad Q(|r|1_{\{|r|>c\}}) \leq \|g\|^2 N_1'(\|g\|) ,$$

$$(19.2.4') \qquad Q(r^2 1_{\{|r|\leq c\}}) \leq \|g\|^2 N_2'(\|g\|) ,$$

then conditions (19.2.3) and (19.2.4) hold.

Proof. As in Remark 19.2.5(ii) it can be shown that (19.2.3') and (19.2.4') imply (19.2.3). Assume w.l.g. that $c < 1$. Then (19.2.3') and (19.2.4') imply

$$Q(r^2 1_{\{|r|\leq 1\}}) = Q(r^2 1_{\{c<|r|\leq 1\}}) + Q(r^2 1_{\{|r|\leq c\}})$$

$$\leq Q(|r|1_{\{|r|>c\}}) + Q(r^2 1_{\{|r|\leq c\}})$$

$$\leq \|g\|^2 (N_1'(\|g\|) + N_2'(\|g\|)) .$$

19.2.7. Theorem. *Let N_i, $i = 0,1,2$, be null-functions. Then there exist sequences $s_n \uparrow \infty$ and $\varepsilon_n \downarrow 0$, $\delta_n \downarrow 0$, with the following properties:*

For all $n \in \mathbb{N}$ and all p-measures Q, P and $\hat{P}$ fulfilling Assumption 19.2.1 and the condition $\|g\| \leq n^{-1/2} s_n$ and $\|\hat{g}\| \leq n^{-1/2} s_n$ (where g pertains to the Q-density of P, and $\hat{g}$ to the Q-density of $\hat{P}$),

$$\hat{P}^n\{\underline{x} \in X^n : |\sum_{\nu=1}^n \log p(x_\nu)/q(x_\nu) - (\sum_{\nu=1}^n g_n(x_\nu) - \tfrac{1}{2} n \|g\|^2)| > \delta_n\} \leq \delta_n$$

with

$$(19.2.8) \qquad g_n := g 1_{\{|g| \leq n^{1/2} \varepsilon_n \|g\|\}} .$$

<u>19.2.9. Corollary</u>. *For all* $n \in \mathbb{N}$ *and all* $t \in \mathbb{R}$,

$$|\hat{P}^n\{\underline{x} \in X^n: \sum_{\nu=1}^{n} g_n(x_\nu) < n^{1/2} t \|g\| + nQ(\hat{g}g)\} - \Phi(t)| \leq \delta_n .$$

<u>Proof of Theorem 19.2.7</u>. (i) We prove the assertion first for $\hat{P} = Q$. By Lemma 19.1.1, applied for $M(u,s) = u^{-2}N_o(s^{-1}u)$, there exist sequences $s_n \uparrow \infty$, $\rho_n \downarrow 0$, such that

$$A_n := \{|g| \leq n^{1/2}\|g\|\rho_n\}$$

fulfills uniformly for $\|g\| \leq n^{-1/2}s_n$

(19.2.10) $\quad Q(A_n^c) = o(n^{-1})$,

(19.2.11) $\quad Q(|g|1_{A_n^c}) = \|g\|o(n^{-1/2}) = o(n^{-1})$,

(19.2.12) $\quad Q(g^2 1_{A_n^c}) = \|g\|^2 o(n^o) = o(n^{-1})$.

By Remark 19.2.5(ii), conditions (19.2.3) and (19.2.4) imply the existence of a function $M: (0,\infty)^2 \rightarrow [0,\infty)$ with the properties required in Lemma 19.1.1, such that

$$Q(|r|1_{\{|r|>\varepsilon\}}) \leq \|g\|^2 M(\varepsilon, \|g\|) .$$

Hence by Lemma 19.1.1, applied for $(u,s) \rightarrow u^{-1}M(u,s)$, there exist sequences $s_n \uparrow \infty$, $\varepsilon_n \downarrow 0$, such that

$$B_n := \{|r| \leq \varepsilon_n/2\}$$

fulfills uniformly for $\|g\| \leq n^{-1/2}s_n$

(19.2.13) $\quad Q(B_n^c) = o(n^{-1})$

(19.2.14) $\quad Q(|r|1_{B_n^c}) = o(n^{-1})$;

and, by condition (19.2.4),

(19.2.15) $\quad Q(r^2 1_{B_n}) = o(n^{-1})$.

Define

$$k_n := g1_{A_n} + r1_{B_n} .$$

The following inequalities hold for $\|g\| \leq n^{-1/2}s_n$ and appropriately chosen $s_n \uparrow \infty$. Assume w.l.g. that $s_n\rho_n \leq \varepsilon_n/2$. Then

(19.2.16) $\quad |k_n| \leq \varepsilon_n$.

By (19.2.12) and (19.2.16),

$$(19.2.17)\qquad Q(k_n^2) = \|g\|^2 - Q(g^2 1_{A_n^c}) + 2Q(gr1_{A_nB_n}) + Q(r^2 1_{B_n}) .$$

(ii) By (19.2.10) and (19.2.13),

$$(19.2.18)\qquad Q^n\{\Sigma \log(p(x_\nu)/q(x_\nu)) \neq \Sigma \log(1 + k_n(x_\nu))\}$$
$$\leq nQ(A_n^c) + nQ(B_n^c) = o(n^o) .$$

A Taylor expansion yields for small $|u|$

$$(19.2.19)\qquad |\log(1+u) - (u - \tfrac{1}{2}u^2)| \leq |u|^3 .$$

We have by (19.2.17)

$$(19.2.20)\qquad Q(|k_n|^3) \leq \varepsilon_n Q(k_n^2) \leq 2\varepsilon_n n^{-1} s_n^2 .$$

From (19.2.19), (19.2.20) and (19.2.17) we obtain

$$(19.2.21)\qquad Q^n\{|\Sigma \log(1+k_n(x_\nu)) - (\Sigma k_n(x_\nu) - \tfrac{1}{2}\Sigma k_n(x_\nu)^2)| > 3\varepsilon_n s_n^2\}$$
$$\leq Q^n\{\Sigma|k_n(x_\nu)|^3 > 3\varepsilon_n s_n^2\}$$
$$\leq Q^n\{\Sigma(|k_n(x_\nu)|^3 - Q(|k_n|^3)) > \varepsilon_n s_n^2\}$$
$$\leq \varepsilon_n^{-2} s_n^{-4} nQ(|k_n|^6) \leq 2\varepsilon_n^2 s_n^{-2} \downarrow 0 .$$

We may choose $s_n \uparrow \infty$ such that $\varepsilon_n s_n^2 \downarrow 0$.

(iii) We prove that there exists $\delta_n \downarrow 0$ such that

$$g_n := g1_{A_n}$$

fulfills

$$(19.2.22)\qquad Q^n\{|\Sigma k_n(x_\nu) - \tfrac{1}{2}\Sigma k_n(x_\nu)^2 - (\Sigma g_n(x_\nu) - \tfrac{1}{2}n\|g\|^2)| > \delta_n\} = o(n^o).$$

To see this we write

$$\Sigma k_n(x_\nu) - \tfrac{1}{2}\Sigma k_n(x_\nu)^2 = \Sigma g_n(x_\nu) - \tfrac{1}{2}n\|g\|^2$$
$$+ \Sigma(r(x_\nu)1_{B_n}(x_\nu) - Q(r1_{B_n})) + nQ(r1_{B_n})$$
$$- \tfrac{1}{2}\Sigma(k_n(x_\nu)^2 - Q(k_n^2)) + \tfrac{1}{2}n(Q(k_n^2) - \|g\|^2) .$$

We have to show that of the right-hand terms all but the first two can be neglected. By (19.2.15),

$$Q^n\{|\Sigma(r(x_\nu)1_{B_n}(x_\nu) - Q(r1_{B_n}))| > \delta_n\} < \delta_n^{-2} nQ(r^2 1_{B_n}) = o(n^o).$$

By (19.2.14),

$$Q(r1_{B_n}) = -Q(r1_{B_n^c}) = o(n^{-1}) .$$

By (19.2.17),

$$Q^n\{|\Sigma(k_n(x_\nu)^2 - Q(k_n^2))| > \delta_n\}$$

$$\leq \delta_n^{-2} nQ(k_n^4) \leq \delta_n^{-2} n\varepsilon_n^2 Q(k_n^2) = o(n^o)$$

and

$$n(Q(k_n^2) - ||g||^2) = o(n^o) .$$

Hence (19.2.22) holds. The assertion for $\hat{P} = Q$ now follows from (19.2.18), (19.2.21) and (19.2.22).

(iv) In order to prove the assertion for arbitrary $\hat{P}$, we have to replace Q by $\hat{P}$ in the preceding arguments. The corresponding relations will be given numbers with a dash. Assume $s_n \uparrow \infty$, $\rho_n \downarrow 0$, $\varepsilon_n \downarrow 0$ chosen such that

$$\hat{A}_n := \{|\hat{g}| \leq n^{1/2}||\hat{g}||\rho_n\}, \qquad \hat{B}_n := \{|\hat{r}| \leq \varepsilon_n/2\}$$

fulfill the relations corresponding to (19.2.10) - (19.2.15) (with g,r,A_n,B_n replaced by $\hat{g},\hat{r},\hat{A}_n,\hat{B}_n$) uniformly for $||\hat{g}|| \leq n^{-1/2}s_n$. Then

$$Q(\hat{g}1_{A_n^c}) = Q(\hat{g}1_{\hat{A}_n A_n^c}) + Q(\hat{g}1_{\hat{A}_n^c A_n^c}) = o(n^{-1}) ,$$

$$Q(\hat{r}1_{A_n^c}) = Q(\hat{r}1_{\hat{B}_n A_n^c}) + Q(\hat{r}1_{\hat{B}_n^c A_n^c}) = o(n^{-1}) .$$

Hence

(19.2.10') $\quad \hat{P}(A_n^c) = Q((1 + \hat{g} + \hat{r})1_{A_n^c}) = o(n^{-1}) .$

Similarly

(19.2.13') $\quad \hat{P}(B_n^c) = o(n^{-1}) .$

Relation (19.2.18') follows from (19.2.10') and (19.2.13') by the same arguments as for (19.2.18).

Furthermore,

(19.2.17') $$\hat{P}(k_n^2) = Q(k_n^2) + o(n^{-1}) = \|g\| + o(n^{-1}) ,$$

hence

(19.2.20') $$\hat{P}(|k_n|^3) \leq \varepsilon_n \hat{P}(k_n^2) = o(n^{-1}) .$$

Relation (19.2.21') follows from (19.2.19), (19.2.20') and (19.2.17') by the same arguments as for (19.2.21). For (19.2.22') we need $\hat{P}(r^2 1_{B_n}) = o(n^{-1})$ and $\hat{P}(r 1_{B_n}) = o(n^{-1})$. The assertion for arbitrary $\hat{P}$ now follows from (19.2.18'), (19.2.21') and (19.2.22').

Proof of Corollary 19.2.9. We have to standardize g_n so that Theorem 19.1.2 can be applied. With notations as in Theorem 19.2.7 we obtain uniformly for $\|g\|, \|\hat{g}\| \leq n^{-1/2} s_n$

(19.2.23) $$\hat{P}(g_n) = Q((1 + \hat{g} + \hat{r}) g 1_{A_n})$$

$$= Q(\hat{g}g) - Q(\hat{g}g 1_{A_n^c}) + Q(\hat{r}g 1_{A_n}) + o(n^{-1}) = Q(\hat{g}g) + o(n^{-1}),$$

(19.2.24) $$\hat{P}(g_n^2) = \|g\|^2 + o(n^{-1}) .$$

The following corollary expresses a certain contiguity property of P^n and Q^n for p-measures Q,P fulfilling Assumption 19.2.1.

19.2.25. Corollary. *For any sequence* $c_n \uparrow \infty$ *there exist* $s_n \uparrow \infty$ *and* $\delta_n \downarrow 0$ *such that for all* $n \in \mathbb{N}$ *and all p-measures* Q,P *fulfilling Assumption 19.2.1 and the condition* $\|g\| \leq n^{-1/2} s_n$, *and all measurable* $\varphi_n : X^n \to [0,1]$,

$$P^n(\varphi_n) \leq c_n Q^n(\varphi_n) + \delta_n .$$

Proof. Define

$$S_n := \Sigma g_n(x_\nu) - \tfrac{1}{2} n \|g\|^2 ,$$

$$R_n := \Sigma \log (p(x_\nu)/q(x_\nu)) - S_n ,$$

$$E_n := \{|R_n| \leq \delta_n\} .$$

By Theorem 19.2.7, there exist $\delta_n \downarrow 0$ and $s_n \uparrow \infty$ such that uniformly for $||g|| \leq n^{-1/2}s_n$

$$Q^n(E_n^c) = o(n^o), \qquad P^n(E_n^c) = o(n^o) .$$

It follows easily from (19.2.23) and (19.2.24) that for $d_n \uparrow \infty$ we can choose $s_n \uparrow \infty$ such that

$$F_n := \{|S_n| \leq d_n\}$$

fulfills uniformly for $||g|| \leq n^{-1/2}s_n$

$$Q^n(F_n^c) = o(n^o), \qquad P^n(F_n^c) = o(n^o) .$$

Hence uniformly for $||g|| \leq n^{-1/2}s_n$ and $\varphi_n: X^n \to [0,1]$,

$$\begin{aligned} P^n(\varphi_n) &\leq P^n(\varphi_n 1_{E_n F_n}) + P^n(E_n^c) + P^n(F_n^c) \\ &= Q^n(\exp[R_n+S_n]\varphi_n 1_{E_n F_n}) + o(n^o) \\ &\leq \exp[\delta_n+d_n]Q^n(\varphi_n) + o(n^o) . \end{aligned}$$

The following corollary is a variant of Lemma 1 of Hâjek (1970, p. 327).

<u>19.2.26. Corollary</u>. *There exist sequences* $s_n \uparrow \infty$ *and* $\varepsilon_n \downarrow 0$, $\delta_n \downarrow 0$ *(depending only on the null-functions occuring in Assumption 19.2.1) with the following properties:*

For all $n \in \mathbb{N}$ *and all p-measures* Q,P *fulfilling Assumption 19.2.1 and the condition* $||g|| \leq n^{-1/2}s_n$ *, the following relation holds for all measurable* $b_n: X^n \to [-1,1]$*:*

$$|P^n(b_n) - Q^n(b_n \exp[S_n]1_{\{|S_n| \leq s_n^2\}})| \leq \delta_n ,$$

where

$$S_n(\underline{x}) := \sum_{\nu=1}^n g_n(x_\nu) - \tfrac{1}{2}n||g||^2 ,$$

with g_n *defined by (19.2.8).*

<u>Proof</u>. We use the notations of the proof of Corollary 19.2.25. By Čebyšev's inequality,(19.2.23) and (19.2.24), we can choose $s_n \uparrow \infty$ such that

$$G_n := \{|S_n| \le s_n^2\}$$

fulfills uniformly for $\|g\| \le n^{-1/2}s_n$

$$Q^n(G_n^c) = o(n^o), \qquad P^n(G_n^c) = o(n^o) .$$

Furthermore, we can choose $s_n \uparrow \infty$ slowly enough so that uniformly for $\|g\| \le n^{-1/2}s_n$

$$Q^n(\exp[S_n]1_{E_n^c G_n}) \le e^{s_n^2} Q^n(E_n^c) = o(n^o) .$$

Hence we obtain uniformly for $\|g\| \le n^{-1/2}s_n$ and $b_n\colon X^n \to [-1,1]$,

$$|P^n(b_n) - Q^n(\exp[S_n]b_n 1_{G_n})|$$
$$= |P^n(b_n 1_{E_n G_n}) - Q^n(\exp[S_n]b_n 1_{E_n G_n})| + o(n^o)$$
$$= |Q^n((\exp[R_n+S_n] - \exp[S_n])b_n 1_{E_n G_n})| + o(n^o)$$
$$\le (e^{\delta_n} - 1)e^{s_n^2} + o(n^o) .$$

The assertion now follows by choosing $s_n \uparrow \infty$ slowly enough.

REFERENCES

Aitchison, J., and Silvey, S.D. (1958). Maximum-likelihood estimation of parameters subject to restraints. Ann. Math. Statist. 29 813-828.

Amari, S. (1981). Differential geometry of estimation: Higher-order efficiency in curved exponential families. Tech. Rep. METR 81-1. Department of Mathematical Engineering and Instrumentation Physics, Faculty of Engineering, University of Tokyo.

Amari, S. (1982a). Geometrical theory of asymptotic ancillarity and conditional inference. Biometrika 69 1 - 17.

Amari, S. (1982b). Theory of information space: A differential-geometrical foundation of statistics. To appear: Ann. Statist. 10.

Amari, S., and Kumon, M. (1982). Differential geometry of Edgeworth expansions in curved exponential family. To appear: J. Japan Statist. Soc.

Andersen, E.B. (1970). Asymptotic properties of conditional maximum-likelihood estimators. J. Roy. Statist. Soc. Ser. B 32 283 - 301.

Anderson, T.W. (1955). The integral of a symmetric unimodal function over a symmetric convex set and some probability inequalities. Proc. Amer. Math. Soc. 6 170 - 176.

Ash, R.B. (1972). *Real Analysis and Probability*. Academic Press, New York.

Bahadur, R.R. (1964). On Fisher's bound for asymptotic variances. Ann. Math. Statist. 35 1545 - 1552.

Bahadur, R.R., and Savage, L.J. (1956). The nonexistence of certain statistical procedures in nonparametric problems. Ann. Math. Statist. 27 1115 - 1122.

Barlow, R.E., Bartholomew, D.J., Bremner, J.M., and Brunk, H.D. (1972). *Statistical Inference under Order Restrictions*. Wiley, New York.

Barlow, R.E., and Proschan, F. (1966). Tolerance and confidence limits for classes of distributions based on failure rate. Ann. Math. Statist. 37 1593 - 1601.

Begun, J.M., and Wellner, J.A. (1981). A representation theorem for estimates of relative risk. Unpublished manuscript.

Behnen, K. (1972). A characterization of certain rank-order tests with bounds for the asymptotic relative efficiency. Ann. Math. Statist. 43 1839 - 1851.

Beran, R. (1974). Asymptotically efficient adaptive rank estimates in location models. Ann. Statist. 2 63 - 74.

Beran, R. (1977a). Robust location estimates. Ann. Statist. 5 431-444.

Beran, R. (1977b). Minimum Hellinger distance estimates for parametric models. Ann. Statist. 5 445 - 463.

Beran, R. (1978). An efficient and robust adaptive estimator of location. Ann. Statist. 6 292 - 313.

Bhattacharya, R.N., and Rao, R.R. (1976). *Normal Approximation and Asymptotic Expansions*. Wiley, New York.

Bhattacharyya, G.K., and Roussas, G.G. (1969). Estimation of a certain functional of a probability density function. Skand. Aktuarietidskr. 52 201 - 206.

Bhuchongkul, S. (1964). A class of nonparametric tests for independence in bivariate populations. Ann. Math. Statist. 35 138 - 149.

Bickel, P.J. (1981). Quelques aspects de la statistique robuste. In: *Ecole d'Eté de Probabilités de Saint-Flour* IX - 1979 (P.L. Hennequin, ed.), 1 - 72, Lecture Notes in Mathematics 876, Springer-Verlag, Berlin.

Bickel, P.J. (1982). On adaptive estimation. To appear: Ann. Statist. 10.

Bickel, P.J., and Lehmann, E.L. (1975a). Descriptive statistics for nonparametric models. I. Introduction. Ann. Statist. 3 1038-1044.

Bickel, P.J., and Lehmann, E.L. (1975b). Descriptive statistics for nonparametric models. II. Location. Ann. Statist. 3 1045 - 1064.

Bickel, P.J., and Lehmann, E.L. (1976). Descriptive statistics for nonparametric models. III. Dispersion. Ann. Statist. 4 1139-1158.

Bickel, P.J., and Lehmann, E.L. (1979). Descriptive statistics for nonparametric models. IV. Spread. In: *Contributions to Statistics. Jaroslav Hájek Memorial Volume* (J.Jurečková, ed.), 33 - 40, Academia, Prague.

Billingsley, P. (1968). *Convergence of Probability Measures*. Wiley, New York.

Blomqvist, N. (1950). On a measure of dependence between two random variables. Ann. Math. Statist. 21 593 - 600.

Boos, D.D. (1979). A differential for L-statistics. Ann. Statist. 7 955 - 959.

Boos, D.D. (1981). Minimum distance estimators for location and goodness of fit. J. Amer. Statist. Assoc. 76 663 - 670.

Cox, D.R. (1972). Regression models and life-tables. J. Roy. Statist. Soc. Ser. B 34 187 - 220.

Daniels, H.E. (1944). The relation between measures of correlation in the universe of sample permutations. Biometrika 33 129 - 135.

Deheuvels, P. (1979). La fonction de dépendance empirique et ses propriétés, un test non-paramétrique d'indépendance. Acad. Royale Belgique, Bull. de la classe des sciences, 5e Sér. 65 274 - 292.

Dmitriev, Yu.G., and Tarasenko, F.P. (1974). On a class of non-parametric estimates of non-linear functionals of density. Theor. Prob. Appl. 19 390 - 394.

Doksum, K. (1969). Starshaped transformations and the power of rank tests. Ann. Math. Statist. 40 1167 - 1176.

Droste, W. and Wefelmeyer, W. (1982). On Hâjek's convolution theorem. To appear: Statistics and Decisions.

Durbin, J., and Knott, M. (1972). Components of Cramér-von Mises statistics I. J. Roy. Statist. Soc. Ser. B 34 290 - 307.

van Eeden, C. (1970). Efficiency-robust estimation of location. Ann. Math. Statist. 41 172 - 181.

Efron, B. (1975). Defining the curvature of a statistical problem. Ann. Statist. 3 1189 - 1242.

Efron, B. (1977). The efficiency of Cox's likelihood function for censored data. J. Amer. Statist. Assoc. 72 557 - 565.

Fabian, V., and Hannan, J. (1982). On estimation and adaptive estimation for locally asymptotically normal families. Z. Wahrscheinlichkeitstheorie verw. Gebiete 59 459 - 478.

Geman, S. (1981). Sieves for nonparametric estimation of densities and regressions. Reports in Pattern Analysis No. 99. D.A.M., Brown University.

Geman, S., and Hwang, C.R. (1982). Nonparametric maximum likelihood estimation by the method of sieves. To appear: Ann. Statist. 10.

Grenander, U. (1981). *Abstract Inference*. Wiley, New York.

Hâjek, J. (1962). Asymptotically most powerful rank-order tests. Ann. Math. Statist. 33 1124 - 1147.

Hâjek, J. (1970). A characterization of limiting distributions of regular estimates. Z. Wahrscheinlichkeitstheorie verw. Gebiete 14 323 - 330.

Hâjek, J. (1972). Local asymptotic minimax and admissibility in estimation. Proc. Sixth Berkeley Symp. Math. Statist. Probab. 1 175 - 194.

Hâjek, J., and Šidák, Z. (1967). *Theory of Rank Tests*. Academia, Prague.

Hampel, F.R. (1971). A general qualitative definition of robustness. Ann. Math. Statist. 42 1887 - 1896.

Hasminskii, R., and Ibragimov, I.A. (1979). On the nonparametric estimation of functionals. In: *Proceedings of the Second Prague Symposium on Asymptotic Statistics* (P.Mandl and M.Hušková, eds.), 41 - 51, North-Holland, Amsterdam.

Hewitt, E., and Stromberg, K. (1965). *Real and Abstract Analysis*. Springer-Verlag, New York.

Hoadley, B. (1971). Asymptotic properties of maximum likelihood estimators for the independent not identically distributed case. Ann. Math. Statist. 42 1977 - 1991.

Hodges, J.L., and Lehmann, E.L. (1956). The efficiency of some nonparametric competitors of the t-test. Ann. Math. Statist. 27 324 - 335.

Hoeffding, W. (1948). A class of statistics with asymptotically normal distribution. Ann. Math. Statist. 19 293 - 325.

Hogg, R.V. (1974). Adaptive robust procedures: A partial review and some suggestions for future applications and theory. J. Amer. Statist. Assoc. 69 909 - 927.

Huber, P.J. (1964). Robust estimation of a location parameter. Ann. Math. Statist. 35 73 - 101.

Huber, P.J. (1972). Robust statistics: a review. Ann. Math. Statist. 43 1041 - 1067.

Huber, P.J. (1981). *Robust Statistics*. Wiley, New York.

Ibragimov, I.A., and Has'minskii, R.Z. (1981). *Statistical Estimation. Asymptotic Theory*. Springer-Verlag, New York.

Inagaki, N. (1970). On the limiting distribution of a sequence of estimators with uniformity property. Ann. Inst. Statist. Math. 22 1 - 13.

Kale, B.K. (1962). A note on a problem in estimation. Biometrika 49 553 - 557.

Kaufman, S. (1966). Asymptotic efficiency of the maximum likelihood estimator. Ann. Inst. Statist. Math. 18 155 - 178.

Kendall, M.G. (1955). *Rank Correlation Methods*. 2nd ed. Griffin, London.

Kiefer, J., and Wolfowitz, J. (1956). Consistency of the maximum likelihood estimator in the presence of infinitely many incidental parameters. Ann. Math. Statist. 27 887 - 906.

Klaassen, C.A.J. (1979). Nonuniformity of the convergence of location estimators. In: *Proceedings of the Second Prague Symposium on Asymptotic Statistics* (P.Mandl and M.Hušková, eds.), 251 - 258, North-Holland, Amsterdam.

Koshevnik, Yu.A., and Levit, B.Ya. (1976). On a non-parametric analogue of the information matrix. Theor. Probab. Appl. 21 738 - 753.

Kruskal, W.H. (1958). Ordinal measures of association. J. Amer. Statist. Assoc. 53 814 - 861.

Kumon, M., and Amari, S. (1981). Geometrical theory of higher-order asymptotically most powerful two-sided test. Tech. Rep. METR 81-9. Department of Mathematical Engineering and Instrumentation Physics, Faculty of Engineering, University of Tokyo.

Lancaster, H.O. (1969). *The Chi-squared Distribution*. Wiley, New York.

Lawrence, M.J. (1975). Inequalities of s-ordered distributions. Ann. Statist. 3 413 - 428.

LeCam, L. (1953). On some asymptotic properties of maximum likelihood estimates and related Bayes estimates. Univ. California Publ. Statist. 1 277 - 320.

LeCam, L. (1956). On the asymptotic theory of estimation and testing hypotheses. Proc. Third Berkeley Symp. Math. Statist. Probab. 1 129 - 156.

LeCam, L. (1960). Locally asymptotically normal families of distributions. Univ. California Publ. Statist. 3 37 - 98.

LeCam, L. (1966). Likelihood functions for large numbers of independent observations. In: *Research Papers in Statistics. Festschrift for J.Neyman* (F.N.David, ed.), 167 - 187, Wiley, London.

LeCam, L. (1970). On the assumptions used to prove asymptotic normality of maximum likelihood estimates. Ann. Math. Statist. 41 802 - 828.

LeCam, L. (1972). Limits of experiments. Proc. Sixth Berkeley Symp. Math. Statist. Probab. 1 245 - 261.

Lehmann, E.L. (1953). The power of rank tests. Ann. Math. Statist. 24 23 - 43.

Lehmann, E.L. (1959). *Testing Statistical Hypotheses*. Wiley, New York.

Lehmann, E.L. (1966). Some concepts of dependence. Ann. Math. Statist. 37 1137 - 1153.

Lehmann, E.L., and Stein, C. (1948). Most powerful tests of composite hypotheses. I. Normal distributions. Ann. Math. Statist. 19 495 - 516.

Levit, B.Ya. (1974). On optimality of some statistical estimates. In: *Proceedings of the Prague Symposium on Asymptotic Statistics*, Vol. 2 (J.Hájek, ed.), 215 - 238, Charles University, Prague.

Levit, B.Ya. (1975). On the efficiency of a class of non-parametric estimates. Theor. Probab. Appl. 20 723 - 740.

Loève, M. (1977). *Probability Theory I*. 4th ed. Springer-Verlag, New York.

von Mises, R. (1936). Les lois de probabilité pour les fonctions statistiques. Ann. Inst. Henri Poincaré 6 185 - 212.

von Mises, R. (1938). Sur les fonctions statistiques. Conf. Réunion Internat. Math., 1 - 8, Gauthier-Villars, Paris.

von Mises, R. (1947). On the asymptotic distribution of differentiable functions. Ann. Math. Statist. 18 309 - 348.

von Mises, R. (1952). Théorie et application des fonctions statistiques. Rendiconti Mat. Appl. 11 374 - 410.

Moussatat, M. (1976). On the asymptotic theory of statistical experiments and some of its applications. Thesis, University of California, Berkeley.

Neyman, J. (1937). Smooth tests for goodness of fit. Skand. Aktuarietidskr. 20 149 - 199.

Neyman, J. (1941). On a statistical problem arising in routine analyses and in sampling inspections of mass production. Ann. Math. Statist. 12 46 - 76.

Oosterhoff, J., and van Zwet, W.R. (1979). A note on contiguity and Hellinger distance. In: *Contributions to Statistics. Jaroslav Hájek Memorial Volume* (J.Jurečková, ed.), 157 - 166, Academia, Prague.

Parr, W.C., and De Wet, T. (1981). On minimum Cramér-von Mises-norm parameter estimation. Commun. Statist.-Theor. Meth. A 10 1149 - 1166.

Parthasarathy, K.R. (1967). *Probability Measures on Metric Spaces.* Academic Press, New York.

Pearson, K. (1900a). Mathematical contributions to the theory of evolution. VII. On the correlation of characters not quantitatively measurable. Philos. Trans. Roy. Soc. London Ser. A 195 1 - 47.

Pearson, K. (1900b). Mathematical contributions to the theory of evolution. VIII. On the inheritance of characters not capable of exact quantitative measurement. Philos. Trans. Roy. Soc. London Ser. A 195 79 - 150.

Pfanzagl, J. (1964). On the topological structure of some ordered families of distributions. Ann. Math. Statist. 35 1216 - 1228.

Pfanzagl, J. (1974). Nonexistence of tests with deficiency zero. Preprints in Statistics, No. 8. University of Cologne, Cologne.

Pfanzagl, J. (1979). Nonparametric minimum contrast estimators. Selecta Statist. Canadiana 5 105 - 140.

Pfanzagl, J. (1980a). Asymptotic expansions in parametric statistical theory. In: *Developments in Statistics*, Vol. 3 (P.R.Krishnaiah, ed.), 1 - 97, Academic Press, New York.

Pfanzagl, J. (1980b). The errors of risk functions. To appear: Statistics and Decisions.

Pfanzagl, J. (1981). The second order optimality of tests and estimators for minimum contrast functionals. I. Probab. Math. Statist. 2 55 - 70.

Pfanzagl, J., and Wefelmeyer, W. (1978). An asymptotically complete class of tests. Z. Wahrscheinlichkeitstheorie verw. Gebiete 45 49 - 72.

Pitman, E.J.G. (1979). *Some Basic Theory for Statistical Inference.* Chapman and Hall, London.

Rao, B.V., Schuster, E.F., and Littell, R.C. (1975). Estimation of shift and center of symmetry based on Kolmogorov-Smirnov statistics. Ann. Statist. 3 862 - 873.

Rao, C.R. (1973). *Linear Statistical Inference and Its Applications.* 2nd ed. Wiley, New York.

Rasch, G. (1961). On general laws and the meaning of measurement in psychology. Proc. Fourth Berkeley Symp. Math. Statist. Probab. 4 321 - 333.

Reiss, R.-D. (1980). Estimation of quantiles in certain nonparametric models. Ann. Statist. 8 87 - 105.

Rieder, H. (1980). Locally robust correlation coefficients. Commun. Statist.-Theor. Meth. A 9 803 - 819.

Rogge, L. (1970). Konsistenz und asymptotische Normalität von Minimum Kontrast Schätzern bei nicht identisch verteilten Zufallsvariablen. Dissertation, Univ. of Cologne.

Roussas, G.R. (1972). *Contiguity of Probability Measures: Some Applications in Statistics.* Cambridge University Press.

Sacks, J. (1975). An asymptotically efficient sequence of estimators of a location parameter. Ann. Statist. 3 285 - 298.

Schaafsma, W. (1976). Bivariate symmetry and asymmetry. Report TW-170, Mathematical Institute of the University of Groningen.

Schüler, L., and Wolff, H. (1976). Zur Schätzung eines Dichtefunktionals. Metrika 23 149 - 153.

Schweizer, B., and Sklar, A. (1974). Operations on distribution functions not derivable from operations on random variables. Studia Math. 52 43 - 52.

Schweizer, B., and Wolff, E.F. (1981). On nonparametric measures of dependence for random variables. Ann. Statist. 9 879 - 885.

Sen, P.K. (1981). *Sequential Nonparametrics*. Wiley, New York.

Sheppard, W.F. (1899). On the application of the theory of error to cases of normal distribution and normal correlation. Phil. Trans. Roy. Soc. London Ser. A 192 101 - 167.

Sklar, A. (1959). Fonctions de rêpartition à n dimensions et leurs marges. Publ. Inst. Statist. Univ. Paris 8 229 - 231.

Skovgaard, L.T. (1981). A Riemannian geometry of the multivariate normal model. Research Report 81/3, Statistical Research Unit, Danish Medical Research Council, Danish Social Science Research Council.

Snijders, T. (1981). Rank tests for bivariate symmetry. Ann. Statist. 9 1087 - 1095.

Stein, C. (1956). Efficient nonparametric testing and estimation. Proc. Third Berkeley Symp. Math. Statist. Probab. 1 187 - 195.

Stone, C.J. (1975). Adaptive maximum likelihood estimators of a location parameter. Ann. Statist. 3 267 - 284.

Takeuchi, K. (1970). Asymptotically efficient tests for location: nonparametric and asymptotically nonparametric. In: *Nonparametric Techniques in Statistical Inference* (M.L.Puri, ed.), 283 - 296, Cambridge University Press.

Takeuchi, K. (1971). A uniformly asymptotically efficient estimator of a location parameter. J. Amer. Statist. Assoc. 66 292 - 301.

Walter, G.G. (1977). Properties of Hermite series estimation of probability density. Ann. Statist. 5 1258 - 1264.

Walter, G.G., and Blum, J. (1979). Probability density estimation using delta sequences. Ann. Statist. 7 328 - 340.

Wegman, E. (1972a). Nonparametric probability density estimation I. Summary of available methods. Technometrics 14 533 - 546.

Wegman, E. (1972b). Nonparametric probability density estimation II. A comparison of density estimation methods. J. Statist. Comput. Simul. 1 225 - 246.

Weiss, L., and Wolfowitz, J. (1970). Asymptotically efficient nonparametric estimators of location and scale parameters. Z. Wahrscheinlichkeitstheorie verw. Gebiete 16 134 - 150.

Wertz, W. (1978). *Statistical Density Estimation: A Survey*. Vandenhoeck und Ruprecht, Göttingen.

Wolfowitz, J. (1974). Asymptotically efficient non-parametric estimators of location and scale parameters. II. Z. Wahrscheinlichkeitstheorie verw. Gebiete 30 117 - 128.

Yanagimoto, T. (1972). Families of positively dependent random variables. Ann. Inst. Statist. Math. 24 559 - 573.

Yanagimoto, T., and Sibuya, M. (1972a). Stochastically larger component of a random vector. Ann. Inst. Statist. Math. 24 259 - 269.

Yanagimoto, T., and Sibuya, M. (1972b). Test of symmetry of a one-dimensional distribution against positive biasedness. Ann. Inst. Statist. Math. 24 423 - 434.

van Zwet, W.R. (1964). *Convex Transformations of Random Variables*. Mathematical Center, Amsterdam.

NOTATION INDEX

Δ, 91

$\widetilde{f}$, 19

Φ, 19

φ, 19

H, 90

$L(\theta)$, 18

$L_{i,j}(\theta)$, 18

$\mathscr{L}_*(P)$, 23

Λ, 18

$m(P)$, 42

N_α, 19

$N(\mu,\Sigma)$, 19

O, 19

O_P, 19

O_θ, 19

o, 19

o_P, 19

o_θ, 19

$\Psi(P)$, 42

$T(P,\mathfrak{P})$, 23

$T_s(P,\mathfrak{P})$, 24

$T_w(P,\mathfrak{P})$, 24

$T_*(P,\mathfrak{P})$, 24

$T^\perp(P;\mathfrak{P}_o,\mathfrak{P})$, 115

V, 90

$*$, 18

$\circledast$, 18

$\oplus$, 19

$\Rightarrow$, 18

[ ], 19

AUTHOR INDEX

Aitchison, J., 224
Amari, S., 14
Andersen, E.B., 234
Anderson, T.W., 82, 161
Ash, R.B., 50, 291

Bahadur, R.R., 167, 289
Barlow, R.E., 58, 62, 69
Bartholomew, D.J., 58, 62, 69
Begun, J.M., 158, 280
Behnen, K., 274
Beran, R., 114, 157, 158, 190, 245, 246, 277
Bhattacharya, R.N., 289
Bhattacharyya, G.K., 169
Bhuchongkul, S., 49
Bickel, P.J., 7, 15, 59, 87, 157, 159, 171, 203
Billingsley, P., 160
Blomqvist, N., 259
Blum, J., 182
Boos, D.D., 89, 191
Bremner, J.M., 58, 62, 69
Brunk, H.D., 58, 62, 69

Cox, D.R., 267

Daniels, H.E., 262
Deheuvels, P., 259
Dmitriev, Yu.G., 169
Doksum, K., 58
Droste, W., 289
Durbin, J., 120

van Eeden, C., 203, 246, 277
Efron, B., 14, 218

Fabian, V., 15, 167, 171

Geman, S., 6
Grenander, U., 6

Hâjek, J., 9, 49, 157, 203, 264, 292, 298
Hampel, F.R., 17
Hannan, J., 15, 167, 171
Hasminskii, R., 167, 202, 203
Hewitt, E., 35
Hoadley, B., 228
Hodges, J.L., 169
Hoeffding, W., 198, 262
Hogg, R.V., 14
Huber, P.J., 4, 7, 17, 81, 87, 89
Hwang, C.R., 6

Ibragimov, I.A., 167, 202, 203
Inagaki, N., 157

Kale, B.K., 176
Kaufman, S., 157
Kendall, M.G., 260
Kiefer, J., 6
Klaassen, C.A.J., 167
Knott, M., 120
Koshevnik, Yu.A., 10, 73, 76, 81, 150
Kruskal, W.H., 259
Kumon, M., 14

Lancaster, H.O., 91
Lawrence, M.J., 58
LeCam, L., 8, 91, 158, 159, 160, 200, 292
Lehmann, E.L., 7, 59, 63, 87, 159, 161, 169, 259, 262, 267, 291

Levit, B.Ya., 10, 73, 76, 81, 82, 150, 204
Littell, R.C., 245
Loève, M., 179

von Mises, R., 76, 79, 198
Moussatat, M., 158

Neyman, J., 120, 121

Oosterhoff, J., 145

Parr, W.C., 190
Parthasarathy, K.R., 164
Pearson, K., 91
Pfanzagl, J., 1, 13, 58, 82, 132, 153, 154, 214
Pitman, E.J.G., 175
Proschan, F., 58

Rao, B.V., 245
Rao, C.R., 174
Rao, R.R., 289
Rasch, G., 51
Reiss, R.-D., 98, 239
Rieder, H., 264
Rogge, L., 229
Roussas, G.G., 169
Roussas, G.R., 157

Sacks, J., 246
Savage, L.J., 167
Schaafsma, W., 282
Schüler, L., 169
Schuster, E.F., 245
Schweizer, B., 259
Sen, P.K., 254, 255
Sheppard, W.F., 259
Sibuya, M., 61, 282
Šidák, Z., 49, 264
Silvey, S.D., 224
Sklar, A., 259
Skovgaard, L.T., 14
Snijders, T., 282
Stein, C., 67, 76, 171, 246, 275, 277, 291
Stone, C.J., 246
Stromberg, K., 35

Takeuchi, K., 203, 246, 275
Tarasenko, F.P., 169

Walter, G.G., 182
Wefelmeyer, W., 214, 289
Wegman, E., 182
Weiss, L., 277
Wellner, J.A., 158, 280
Wertz, W., 182
De Wet, T., 190
Wolff, E.F., 259
Wolff, H., 169
Wolfowitz, J., 6, 277

Yanagimoto, T., 61, 64, 282

van Zwet, W.R., 58, 145

SUBJECT INDEX*

adaptiveness, 14, 171
approximability by tangent cones, 25
approximation by distance functions, 92
asymptotic efficiency
  of estimators, 178, 196
  of tests, 131
asymptotic envelope power function, 125, 128
asymptotic maximum likelihood estimator, 182
asymptotic median unbiasedness, 154
average failure rate, 62

canonical gradient, 71
concentration of estimators, 151
conditional maximum likelihood estimator, 234
continuity of tangent cones, 27
contrast function, 80
convexity, local asymptotic, 28
copula, 258
co-space, 115, 119
Cramér-von Mises distance, 92, 111
curved families, 41, 221

density estimators, 179
dependence function, 259
derivative of paths, 22
differentiable functional, 65
differentiable path, 22
distribution of losses, 152

efficiency, asymptotic,
  of estimators, 178, 196
  of tests, 131
envelope power function, asymptotic, 125, 128
estimating equation, 204
expected life time, 79

failure rate, 61
  average, 62
  monotone, 61
  proportional, 267, 277
full families, 33

gradient, 65
  canonical, 71
  strong, 66

Hellinger distance, 90

improvement procedure, 200
influence curve, 76
information inequality, 174

Kendall's τ, 259
kernel estimators, 180
Kolmogorov distance, 245
Kolmogorov-Smirnov distance, 92

least favorable direction, 157
life time, expected, 79
likelihood equation, 204
local asymptotic convexity, 28
local asymptotic symmetry, 30

---

* Underlined page numbers refer to definitions.

location functional, 86, 238
loss function, 152

maximum likelihood estimator,
asymptotic, 182
conditional, 234
median, 246
median unbiasedness, asymptotic, 154
minimum contrast functional, 80, 204, 238
minimum distance estimators, 190
von Mises functional, 78, 198, 249
monotone failure rate, 61

Newton-Raphson approximation, 200
null-function, 289

orthogonal series estimators, 181

path, 22
differentiable, 22
strongly differentiable, 23
weakly differentiable, 23
partial likelihood estimator, 218
power function, 122
projection, 100
proportional failure rate, 267, 277

quadrant correlation coefficient, 79, 259
quadrant dependence, 63
quantile, 85, 238

regression dependence, 63
regression residual, 70
robustness, 16

side conditions, 41, 223
slope, 122, 129
Spearman's ρ, 259
stochastic expansion of estimators, 209
strong gradient, 66
strongly differentiable path, 23
sup-distance, 90
symmetry, local asymptotic, 30

tangent cone, 23
transformation models, 267, 283

unrelated functionals, 169
unrelated parameters, 140, 209

variational distance, 90

weakly differentiable path, 23

Lecture Notes in Statistics

Vol. 1: R. A. Fisher: An Appreciation. Edited by S. E. Fienberg and D. V. Hinkley. xi, 208 pages, 1980.

Vol. 2: Mathematical Statistics and Probability Theory. Proceedings 1978. Edited by W. Klonecki, A. Kozek, and J. Rosiński. xxiv, 373 pages, 1980.

Vol. 3: B. D. Spencer, Benefit-Cost Analysis of Data Used to Allocate Funds. viii, 296 pages, 1980.

Vol. 4: E. A. van Doorn, Stochastic Monotonicity and Queueing Applications of Birth-Death Processes. vi, 118 pages, 1981.

Vol. 5: T. Rolski, Stationary Random Processes Associated with Point Processes. vi, 139 pages, 1981.

Vol. 6: S. S. Gupta and D.-Y. Huang, Multiple Statistical Decision Theory: Recent Developments. viii, 104 pages, 1981.

Vol. 7: M. Akahira and K. Takeuchi, Asymptotic Efficiency of Statistical Estimators. viii, 242 pages, 1981.

Vol. 8: The First Pannonian Symposium on Mathematical Statistics. Edited by P. Révész, L. Schmetterer, and V. M. Zolotarev. vi, 308 pages, 1981.

Vol. 9: B. Jørgensen, Statistical Properties of the Generalized Inverse Gaussian Distribution. vi, 188 pages, 1981.

Vol. 10: A. A. McIntosh, Fitting Linear Models: An Application of Conjugate Gradient Algorithms. vi, 200 pages, 1982.

Vol. 11: D. F. Nicholls and B. G. Quinn, Random Coefficient Autoregressive Models: An Introduction. v, 154 pages, 1982.

Vol. 12: M. Jacobson, Statistical Analysis of Counting Processes. vii, 226 pages, 1982.

Vol. 13: J. Pfanzagl (with the assistance of W. Wefelmeyer), Contributions to a General Asymptotic Statistical Theory. vii, 315 pages, 1982.

Vol. 14: GLIM 82: Proceedings of the International Conference on Generalised Linear Models. Edited by R. Gilchrist. v, 188 pages, 1982.